METHODS IN MOL

Series Editor
John M. Walker
School of Life and Medical Sciences
University of Hertfordshire
Hatfield, Hertfordshire, AL10 9AB, UK

For further volumes:
http://www.springer.com/series/7651

Haplotyping

Methods and Protocols

Edited by

Irene Tiemann-Boege

Johannes Kepler University Institute of Biophysics, Linz, Austria

Andrea Betancourt

Vetmeduni Vienna InstitutPopulationsgenetik, Wien, Austria

Publication 1/30/2017

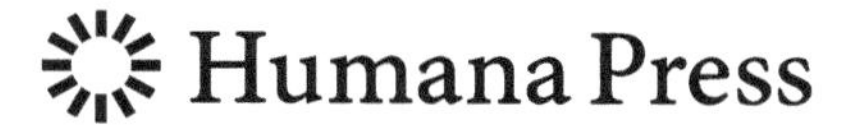

Editors
Irene Tiemann-Boege
Johannes Kepler University Institute of Biophysics
Linz, Austria

Andrea Betancourt
Vetmeduni Vienna InstitutPopulationsgenetik
Wien, Austria

ISSN 1064-3745 ISSN 1940-6029 (electronic)
Methods in Molecular Biology
ISBN 978-1-4939-6748-3 ISBN 978-1-4939-6750-6 (eBook)
DOI 10.1007/978-1-4939-6750-6

Library of Congress Control Number: 2017930813

Printed on acid-free paper

This Humana Press imprint is published by Springer Nature
The registered company is Springer Science+Business Media LLC
The registered company address is: 233 Spring Street, New York, NY 10013, U.S.A.

Preface

In diploid organisms, most loci are present in two copies per genome, with some exceptions, such as most of those on sex chromosomes in the heterogametic sex and those encoded in the mitochondria. For these genomic regions, typing polymorphisms immediately yields knowledge of haplotypes, or the genetic variants that occur on the same chromosome, and which were inherited from the same parent. This haplotype information has valuable applications in many areas of biology. It can be crucial information for implicating a gene in a genetic disease; for example, individuals with two disabling mutations at the locus should be affected only if the mutations are not in the same parental copy of the gene. For population geneticists, phased data allow powerful inferences of population history: haplotype information reveals historical recombination events, which can occur only between chromosomes in the same population at some point in the past. The list goes on: haplotype information can tell us about recombination hotspots, recurrent mutagenesis, targets of recent selection, biased gene conversion, and *cis*-regulation of gene expression, to mention a few areas.

But for most of the genome in diploid or multiploid organisms, haplotype information does not naturally result as a by-product of sequencing, but must be either inferred or obtained via special experimental techniques. Methods for determining haplotypes, also known as "allelic phasing," "haplotype phasing," or just "phasing," have quickly developed in the past decade, in response to technological advances, and are capable of providing haplotype information at a larger scale than previously possible. This book provides detailed protocols for genetic, molecular, cytological, and bioinformatic methods for determining haplotypes and is aimed at scientists and students who wish to have an overview of experimental methods for haplotyping. The presented methods fall into two major groups: short-range haplotyping for given variants or loci of interest and long-range haplotyping for whole chromosomes or genomes. The simplest form of haplotyping is pairwise, or determining which alleles at two sites lie on the same chromosome; though conceptually straightforward, this offers a powerful way to screen thousands of molecules for recombinant types or rare haplotypes.

Some methods in this book take advantage of biological material that is naturally haploid to assist in obtaining haplotype information. Male gametes, such as sperm or pollen, are usually used in these methods, as these are generally available in enough abundance to extract sufficient haploid DNA. Several protocols are offered to infer haplotypes from male gametes from very different organisms. Extracting DNA from highly compacted DNA packaged in resistant and hard shells like pollen is especially challenging; a protocol addressing haplotyping of this difficult material is described in full detail in this book. Instead of using haploid material, one protocol cleverly takes the opposite strategy, inducing aneuploidy, so that the ratio of alleles on each homologue is no longer equal, and quantification of allelic ratios becomes informative about haplotypes for entire chromosomes. While the protocols for these methods are specific to particular organisms, we expect each can be easily adapted to a broad range of experimental systems.

Several specialized methods detail protocols for haplotyping loci that are difficult to resolve, such as those in gene families or in the Major Histocompatibility Complex (MHC). Gene families can be especially challenging, as they are essentially more-than-diploid. In order to phase alleles in these regions, it is necessary to identify both the loci and the haplotype to which they belong. Recently developed methods involving long-read sequencing with PacBio prove useful for this problem. In addition, MHC loci are of special interest due to their role in immunity, disease, and organ transplantation; three chapters concern this complex genomic region, with one explaining its role in disease, another describing targeted sequencing of the region for detailed genotype information which can be used to infer MHC haplotypes, and the last describes a highly specific enrichment method using modified capture oligos for fishing out and then sequencing the MHC region from one of the two parental chromosomes.

A variety of protocols in this book, rather than relying on naturally haploid material, use laboratory techniques to physically separate chromosomes, via microdissection, chromosome sorting, fosmid cloning, or simple dilution and barcoding. These methods are typically followed by short-read sequencing, so the preliminary step of separating the material by chromosomal origin is crucial to obtaining haplotype information. Though conceptually similar, these methods differ widely in equipment requirements and in wet lab techniques. With a fluorescence-activated cell sorter, for example, chromosomes can be quickly sorted and sequenced separately; construction of fosmid libraries, in contrast, requires mainly standard molecular biology laboratory protocols. Other, newly developed techniques for preparing next-generation sequencing libraries result in libraries with different barcodes for each chromosomal haplotype. Sequencing of these libraries, even with short-read methods, results in chromosome-wide or genome-wide haplotypes.

In short, some of the covered methods are based on state-of-the-art experimental technologies, mainly next-generation sequencing data, but others use traditional methods that can be easily adapted in any lab without the need of sophisticated equipment or bioinformatic capabilities. The methods vary in their throughput and resolution, as well as the length of the inferred haplotype, and can be applied for whole genome analysis or for identifying a rare haplotype variant in a population. In cases for which it is not possible to sequence individual samples, this book provides an overview of computational methods for obtaining haplotype information from pools of individuals. Thus, this book provides detailed haplotyping protocols for a broad range of questions, samples, budgets, and resources.

Linz, Austria *Irene Tiemann-Boege*
Wien, Austria *Andrea Betancourt*

Contents

Contributors

CHESTER A. ALPER • *Program in Cellular and Molecular Medicine, Boston Children's Hospital, Boston, MA, USA; Department of Pediatrics, Harvard Medical School, Boston, MA, USA*

SASAN AMINI • *Clear Labs, Menlo Park, CA, USA*

BARBARA ARBEITHUBER • *Institute of Biophysics, Johannes Kepler University, Linz, Austria*

JUDITH BERMAN • *Department of Molecular Microbiology & Biotechnology, Tel Aviv University, Ramat Aviv, Israel*

XI CHEN • *Department of Statistics, Stanford University, Stanford, CA, USA; Bio-X Program, Stanford, CA, USA*

KYUHA CHOI • *Department of Plant Sciences, University of Cambridge, Cambridge, UK*

LENA CHRISTIANSEN • *Advanced Research Group, Illumina, Inc., San Diego, CA, USA*

MARTIN DANZER • *Red Cross Transfusion Service of Upper Austria, Linz, Austria*

RADOJE DRMANAC • *Complete Genomics, Inc., Mountain View, CA, USA; BGI-Shenzhen, Shenzhen, China*

JORGE DUITAMA • *Max Planck Institute for Molecular Genetics, Berlin, Germany; International Center for Tropical Agriculture (CIAT), Cali, Colombia*

CARINA FISCHER • *Red Cross Transfusion Service of Upper Austria, Linz, Austria*

ANJA FORCHE • *Department of Biology, Bowdoin College, Brunswick, ME, USA*

KEVIN L. GUNDERSON • *Advanced Research Group, Illumina, Inc., San Diego, CA, USA*

ANGELIKA HEISSL • *Institute of Biophysics, Johannes Kepler University, Linz, Austria*

IAN R. HENDERSON • *Department of Plant Sciences, University of Cambridge, Cambridge, UK*

MARGRET R. HOEHE • *Max Planck Institute for Molecular Genetics, Berlin, Germany*

THOMAS HUEBSCH • *Max Planck Institute for Molecular Genetics, Berlin, Germany*

HELEN R. IRVING • *Monash Institute of Pharmaceutical Science, Monash University, Parkville, VIC, Australia*

CHARLES E. LARSEN • *Program in Cellular and Molecular Medicine, Boston Children's Hospital, Boston, MA, USA; Department of Medicine, Harvard Medical School, Boston, MA, USA*

WENZHI LI • *Cardiovascular Research Institute and Department of Medicine, Morehouse School of Medicine, Atlanta, GA, USA; Center of Big Data and Bioinformatics, First Affiliated Hospital of Medicine School, Xi'an Jiaotong University, Xi'an, Shaanxi, China*

QUAN LONG • *Department of Biochemistry and Molecular Biology and Medical Genetics, Alberta Children's Hospital Research Institute and O'Brien Institute for Public Health, University of Calgary, Calgary, AB, Canada*

LI MA • *4DGenome Inc, Atlanta, GA, USA; Cardiovascular Research Institute and Department of Medicine, Morehouse School of Medicine, Atlanta, GA, USA*

MARK A. MCELWAIN • *Complete Genomics, Inc., Mountain View, CA, USA*

BIRGIT MENTRUP • *Max Planck Institute for Molecular Genetics, Berlin, Germany;*

JOACHIM MESSING • *Waksman Institute of Microbiology, Rutgers University, Piscataway, NJ, USA*

GABRIELE MICHELITSCH • *Center for Medical Research, Graz, Austria*
LEILA MURESAN • *Cambridge Advanced Imaging Centre, University of Cambridge, Cambridge, UK*
NICHOLAS M. MURPHY • *Monash Institute of Pharmaceutical Science, Monash University, Parkville, VIC, Australia; Preimplantation Genetic Diagnosis, Melbourne IVF, Melbourne, VIC, Australia*
NORBERT NIKLAS • *Red Cross Transfusion Service of Upper Austria, Linz, Austria*
BARBARA PALAORO • *Institute of Biophysics, Johannes Kepler University, Linz, Austria*
ELISABETH PALZENBERGER • *Institute of Biophysics, Johannes Kepler University, Linz, Austria*
BROCK A. PETERS • *Complete Genomics, Inc., Mountain View, CA, USA; BGI-Shenzhen, Shenzhen, China*
COLIN W. POUTON • *Monash Institute of Pharmaceutical Science, Monash University, Parkville, VIC, Australia*
JOHANNES PRÖLL • *Red Cross Transfusion Service of Upper Austria, Linz, Austria*
RONJA REINHARDT • *Institute of Biophysics, Johannes Kepler University, Linz, Austria*
MOSTAFA RONAGHI • *Advanced Research Group, Illumina, Inc., San Diego, CA, USA*
SABRINA SCHULZ • *Max Planck Institute for Molecular Genetics, Berlin, Germany*
HEÏDI SERRA • *Department of Plant Sciences, University of Cambridge, Cambridge, UK*
QING SONG • *4DGenome Inc, Atlanta, GA, USA; Cardiovascular Research Institute and Department of Medicine, Morehouse School of Medicine, Atlanta, GA, USA; Center of Big Data and Bioinformatics, First Affiliated Hospital of Medicine School, Xi'an Jiaotong University, Xi'an, Shaanxi, China; Research Wing D-203, Atlanta, GA, USA*
FRANK J. STEEMERS • *Advanced Research Group, Illumina, Inc., San Diego, CA, USA*
EUN-KYUNG SUK • *Max Planck Institute for Molecular Genetics, Berlin, Germany*
IRENE TIEMANN-BOEGE • *Institute of Biophysics, Johannes Kepler University, Linz, Austria*
WING HUNG WONG • *Department of Statistics, Stanford University, Stanford, CA, USA; Bio-X Program, Stanford, CA, USA; Department of Health Research and Policy, Stanford University, Stanford, CA, USA*
SEN XU • *Department of Biology, University of Texas at Arlington, Arlington, TX, USA*
HONG YANG • *Department of Statistics, Stanford University, Stanford, CA, USA; Bio-X Program, Stanford University, Stanford, CA, USA*
NATALIYA E. YELINA • *Department of Plant Sciences, University of Cambridge, Cambridge, UK*
KIM YOUNG • *Department of Biology, Indiana University, Bloomington, IN, USA*
REBECCA YU ZHANG • *Complete Genomics, Inc., Mountain View, CA, USA*
WEI ZHANG • *Waksman Institute of Microbiology, Rutgers University, Piscataway, NJ, USA*
FAN ZHANG • *Advanced Research Group, Illumina, Inc., San Diego, CA, USA*

Part I

Haplotyping with Long Range PCR

Chapter 1

Haplotyping of Heterozygous SNPs in Genomic DNA Using Long-Range PCR

Barbara Arbeithuber, Angelika Heissl, and Irene Tiemann-Boege

Abstract

To study meiotic recombination products, cis- or trans-association of disease polymorphisms, or allele-specific expression patterns, it is necessary to phase heterozygous polymorphisms separated by several kilobases. Haplotyping using long-range polymerase chain reaction (PCR) is a powerful, cost-effective method to directly obtain the phase of multiple heterozygous sites with standard laboratory equipment in a handful of *loci* for many samples. The method is based on the amplification of large genomic DNA regions (up to ~40 kb) with a reaction mixture that combines a proofreading polymerase with allele-specific primer pairs that preferentially amplify matched templates. The analysis of two heterozygous SNPs requires four reactions, each containing one of the four possible allele-specific primer combinations (two forward and two reverse primers), with the mismatches occurring at the 3′ ends of the primers. The two correct primer combinations will more efficiently elongate the matching alleles than the alternative alleles, and the difference in amplification efficiency can be monitored with real-time PCR.

Key words Genotype, Haplotype, Allele-specific amplification, Long-range PCR, Real-time PCR, Single nucleotide polymorphism, Heterozygous SNP, Proof-reading polymerase, Allelic phase

1 Introduction

In this chapter, we describe an allele-specific long-range polymerase chain reaction (PCR) used for haplotyping (determination of the allelic phase of heterozygous sites on a chromosome) of single nucleotide polymorphisms (SNPs) separated by several kilobases in genomic DNA.

The method is applicable for haplotyping of a handful of SNPs separated by up to ~40 kb. This is done in a very cost-effective way since it can be performed in any lab with access to a real-time PCR compatible thermocycler. In comparison to commonly used methods that rely on the inference or statistical computation of haplotypes from parental or population genotype data (reviewed in [1]), the haplotype can be experimentally determined for an individual with long-range PCR which does not require additional family material. An alternative experimental method for haplotyping of

Irene Tiemann-Boege and Andrea Betancourt (eds.), *Haplotyping: Methods and Protocols*, Methods in Molecular Biology, vol. 1551, DOI 10.1007/978-1-4939-6750-6_1, © Springer Science+Business Media LLC 2017

genomic DNA is based on single molecule dilution (SMD), which results in a physical separation of alleles [2]. With SMD, however, further analysis is needed to infer the amplified allele using, for example, restriction fragment length polymorphism genotyping, allele-specific oligo extension, or sequencing. In comparison, haplotyping with allele-specific long-range PCR can be performed directly on extracted DNA and requires fewer steps.

The low SNP density in the human genome with on average one SNP per ~800 bp [3], makes it necessary to combine allele-specific and long-range PCR for haplotyping of heterozygous SNPs. The principle of haplotyping with long-range PCR relies on the ability to (1) amplify long PCR products (>3 kb), and (2) make this amplification specific for the alleles of interest at heterozygous sites, i.e., allele-specific amplification [2, 4–9]. Long-range PCR generally refers to the amplification of DNA fragments at a length that cannot be achieved with standard DNA polymerases and amplification conditions. By using special amplification parameters PCR product lengths of up to ~40 kb can be achieved for simple, high-quality templates depending on the polymerase used [10, 11].

While the amplification of some genomic DNA fragments (up to about 30 kb) works sufficiently well with a single polymerase type (such as *Taq* or *Pfu*), specially modified for the amplification of long templates (e.g. [12, 13]), the mixture of a proofreading with a non-proofreading polymerase greatly increases amplification efficiency for extremely long templates, up to ~40 kb [10, 11]. Many commercially available kits optimized for long-range PCR consist of such a mixture (e.g., Expand Long Range Enzyme Mix (Roche) or LongAmp Taq DNA Polymerase (NEB)). This improved amplification efficiency is achieved by allowing the correction of errors, i.e., mismatches, introduced by the non-proofreading polymerase. The longer the template, the more likely these errors become. As mismatches are inefficiently elongated by the non-proofreading polymerase—the polymerase falls off producing truncated products that do not contribute to exponential amplification—their removal increases product yield. Proofreading polymerases remove wrongly inserted bases with their 3′ → 5′ exonuclease activity, allowing templates to be further elongated and producing up to ~40 kb amplicons [10, 11]. In addition to this special mix of polymerases, long denaturation, and elongation times, and protection of the template from DNA damage, enhance the amplification of long DNA template molecules [10].

Haplotypes of genomic DNA are inferred by the selective amplification of the two homologs using allele-specific PCR, in which primers are designed such that the base at the 3′ end of the primer anneals to the SNP of interest [14]. With this design, only the perfectly matching primer will be efficiently extended, resulting in more amplification product during the exponential phase (Fig. 1)

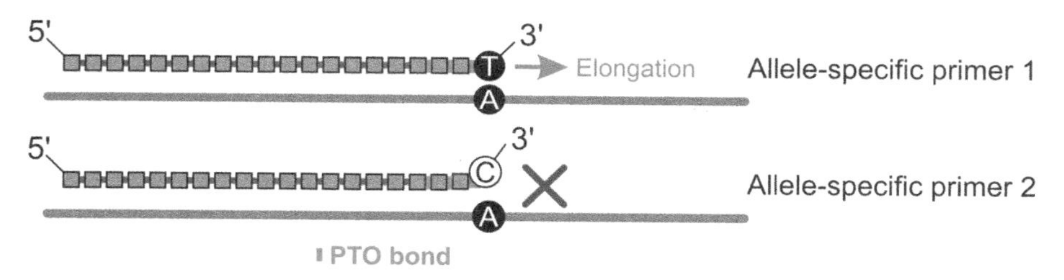

Fig. 1 Principle of allele-specific PCR. Allele-specific primers are designed differing only in the nucleotide at the 3′ end (T/C used here as an example), which anneals to the SNP (A used here as an example). While the perfect match (A-T: allele-specific primer 1) can be efficiently elongated, amplification of the mismatch (A-C: allele-specific primer 2) is very inefficient. Substitution of the last three phosphodiester bonds at the 3′ end by phosphorothioate (PTO) bonds (*red rectangles*) increases specificity of the allele-specific amplification

[15]. The specificity of the allele-specific primer can be further increased by exchanging the last three phosphodiester bonds at the 3′ end with phosphorothioate (PTO) bonds, which protects the allele-specific site from removal by the exonuclease activity of proof-reading polymerases [16, 17]. Moreover, PTO bonds increase the rigidity of the DNA backbone, thus enhancing the conformational DNA distortion of a mismatch [18].

To obtain the phase of two heterozygous SNPs, four PCR reactions need to be set up (one reaction per forward and reverse primer combination). Amplification efficiency is measured using real-time PCR [19], a powerful method first reported in 1992. While in conventional PCR [20], the PCR products are analyzed after the amplification is finished (end-point method) (e.g., via gel electrophoresis [21, 22]), in real-time PCR amplification can be monitored at each cycle. By adding a fluorescent intercalating dye (such as SYBR Green I or EvaGreen [23–25]), or fluorescently labeled probes, the increase in signal, which represents the increase in the amount of amplified DNA, can be measured during the reaction. This method has the advantage that different amounts of starting molecules can be quantified, and different amplification efficiencies can be monitored. A general overview of the method for haplotyping of two SNPs is shown in Fig. 2.

An example for an application of this method is the analysis of single meiotic recombination products, in which DNA is exchanged between paternal and maternal homologs leading to a new combination of alleles at heterozygous sites [16, 17]. Here, single meiotic recombination products (crossovers) are collected by the same principle of haplotyping with long-range PCR, but using two pairs of flanking SNPs representing the crossover haplotype, which are selectively amplified with allele-specific PCR. The phase of these four heterozygous SNPs flanking a recombination hotspot (region with high recombination frequency compared to the genome

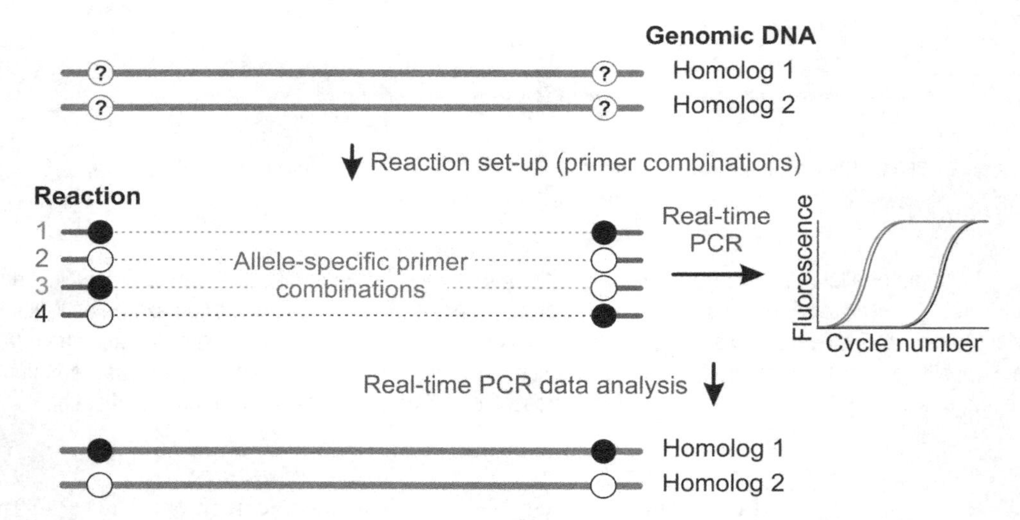

Fig. 2 General scheme for haplotyping with long-range PCR. To obtain the haplotype of two heterozygous SNPs (shown with *circles*) four reactions need to be set up, one for each possible allele (allele-specific primer) combination. Reactions in which the allele-specific primer combinations represent the haplotypes of the genomic DNA (one for each homolog) amplify the template with higher efficiency and can be identified with real-time PCR (*green and blue curves*) by an earlier amplification curve

average) needs first to be determined to correctly choose the allele-specific primers of the recombinant molecule. Therefore, in this chapter, we describe a haplotyping application with long-range PCR for two pairs of SNPs separated by about 3–4 kb (as required in the analysis of meiotic recombination products). Generally, the number of analyzed SNPs can also be decreased or increased, depending on the application. Once allele-specific long-range PCR was performed for a pair of distant SNPs, the haplotype of all internal heterozygous sites can either be analyzed by sequencing, or by genotyping the two PCR products that showed the most efficient amplification. Alternatively, internal haplotypes can be determined by performing additional allele-specific PCR reactions of internal polymorphisms.

2 Materials

2.1 Reagents

1. Polymerase suited for long-range PCR (see customer information): Expand Long Range Enzyme Mix (Roche) and corresponding buffer (*see* **Note 1**).
2. dNTPs (10 mM each).
3. PCR-grade water (*see* **Note 2**).

4. Allele-specific primers (diluted to 5 μM in PCR-grade water) (*see* **Note 3**).
5. SYBR Green I (10× in DMSO) (*see* **Note 4**).
6. Purified genomic DNA (*see* **Note 5**).

2.2 Equipment

1. PCR hood with filtered air-flow and UV light (*see* **Note 6**).
2. Pipettes.
3. Multichannel pipettes (for pipetting of 5 μl volumes).
4. DNase/RNase-free aerosol barrier pipette tips.
5. Microcentrifuge tubes.
6. 200 μl PCR tubes.
7. Vortex mixer.
8. Table centrifuge to spin down the reagents and master mixes.
9. Real-time PCR thermocycler (for usage with 384-well plates) (*see* **Note 7**).
10. Real-time PCR compatible 384-well plates (*see* **Note 8**).
11. Optically clear PCR-plate sealing foils (compatible for real-time PCR) (*see* **Note 9**).
12. Microcentrifuge compatible to spin down plates (*see* **Note 10**).

3 Methods

As an example, haplotyping with long-range PCR is described here for four heterozygous SNPs, two on each side of a recombination hotspot (*see* **Note 11**). An overview of the different steps involved is shown in Fig. 3. However, the method can be adapted to different applications and numbers of SNPs, depending on the requirements.

3.1 Primer Design and Optimization of PCR Cycling Conditions

Design allele-specific primers for each SNP that will be analyzed (four heterozygous SNPs, two of them upstream, and two of them downstream of the recombination hotspot) (*see* **Note 12**):

- The directionality of the primers (forward or reverse) depends on the location of the SNP relative to the recombination hotspot, with all primers to be extended toward the hotspot center (Fig. 4).
- The melting temperature (T_m) of each primer needs to be similar (*see* **Note 13**).
- For each SNP, two primers need to be designed overlapping the SNP (one primer for each allele). Only the last base at the 3′ end (which needs to hybridize to the SNP (Fig. 1) should differ in the two allele-specific primers.

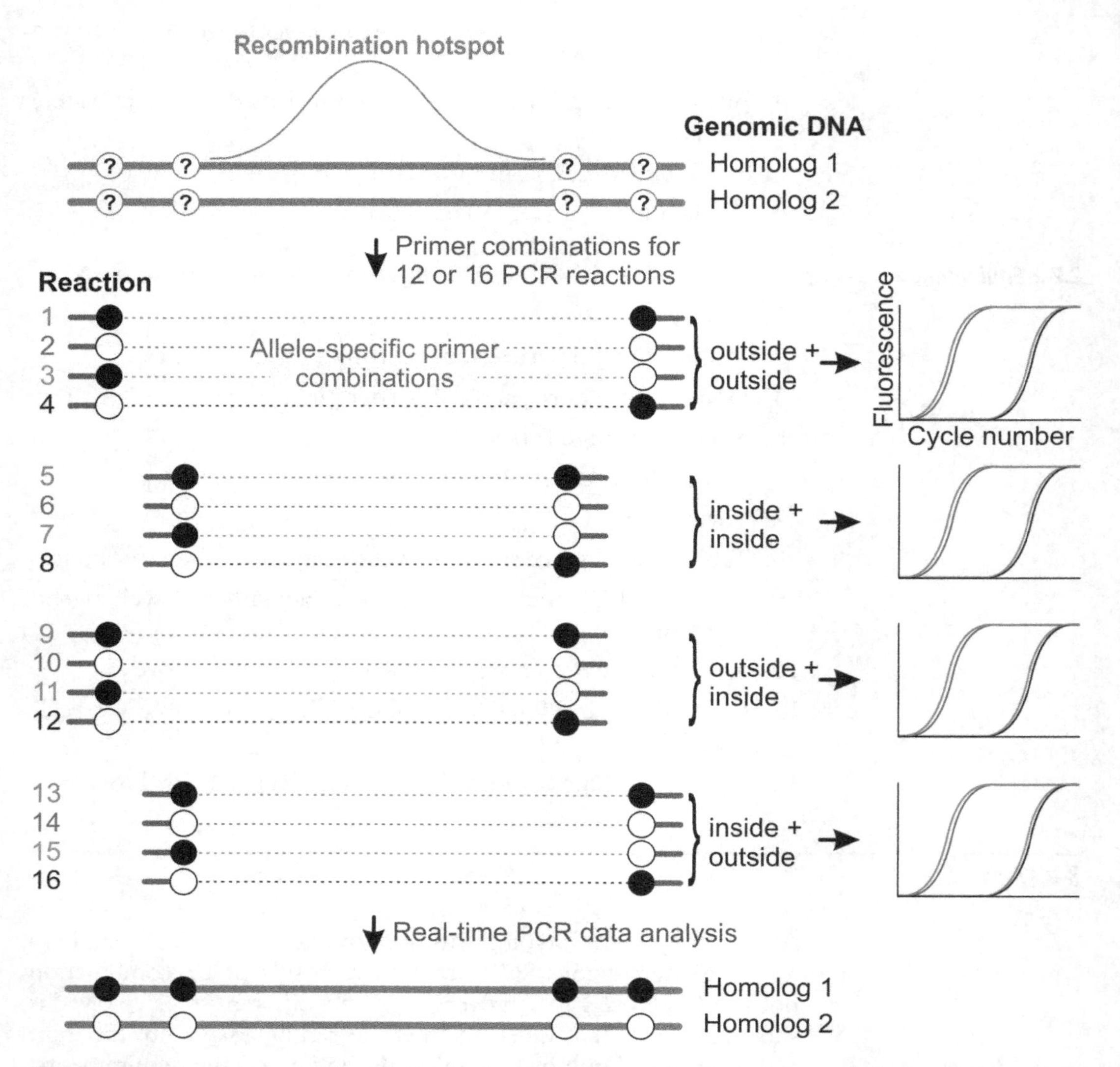

Fig. 3 Haplotyping scheme for four SNPs. For the haplotyping of four heterozygous SNPs, a minimum of 12 long-range PCR reactions (or 16 for a more accurate phasing) need to be set up, each of them containing a different primer combination of the outside (forward + reverse) and inside (forward + reverse) primers. When amplifying genomic DNA (heterozygous for the four selected SNPs), primer combinations matching the haplotype amplify with a higher efficiency (amplification curves that start to rise at lower cycle numbers, and therefore have a lower Cq-value), shown here in green and blue. From the obtained haplotypes of the different allele-specific primer combinations used in the long-range PCR (representing the different SNP-pairs), the phase of the four SNPs can be inferred

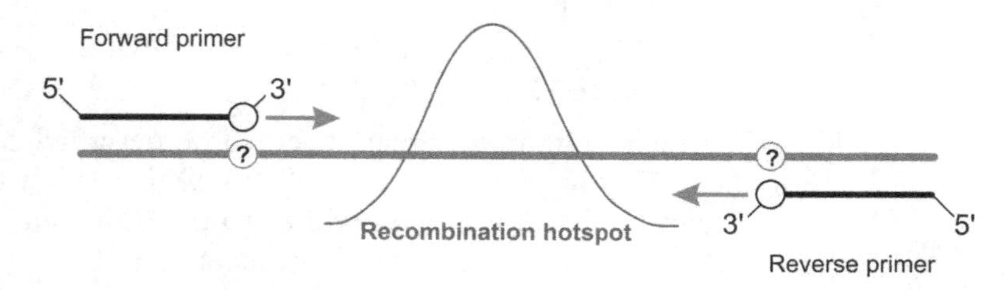

Fig. 4 Amplification directionality for haplotyping with long-range PCR. All primers need to be designed that they amplify toward the middle of the region (recombination hotspot), with the 3′-end of the primers annealing to the SNP

- The last three phosphodiester bonds at the 3′ end should be substituted by phosphorothioate bonds (PTOs), which increases allele-specificity during amplification (*see* **Note 14**).
- Primers should be designed according to standard guidelines (*see* **Note 15**) that include a length between 18 and 25 bases, a GC-content between 40% and 60%, a T_m >45 °C. In addition, primers should have no mononucleotide runs with more than four repeats, no regions within a primer that will hybridize to each other, or complementary sequences between the different forward and reverse primers, which could result in the formation of primer-dimers, and low probability for the formation of secondary structures. The different parameters mentioned here can be checked with online tools provided for free by different companies such as IDT or Eurofins Genomics (*see* **Note 16**).
- Amplification conditions need to be optimized prior to sample haplotyping. Different aspects need to be considered for the optimization: (1) The presence of the correct amplification product needs to be checked (requires gel electrophoresis additionally to the inspection of the melting curves (*see* **Note 17**); (2) The PCR products should be free of primer dimers, which could otherwise bias the results; (3) Since allele-specificity is not 100% efficient, it is important to find conditions that provide a good separation of real-time PCR amplification curves of reactions that truly represent the haplotype of the sample (with a lower Cq-value) from curves due to unspecific amplification (*see* **Note 18**).

3.2 Real-Time PCR—Haplotyping

Sample dilutions and master mixes can be prepared at room temperature (*see* **Note 19**). Set up the haplotyping reaction in a clean PCR hood (*see* **Note 6**). Vortex and spin down all reagents before usage, except for the polymerase and genomic DNA, which should be mixed by inverting or flicking of the tube (*see* **Note 5**).

1. Prepare sample dilutions: Dilute each genomic DNA sample to a final concentration of 10 ng/μl (dilute in PCR-grade water, 80 μl of each sample are needed for 16 reactions, to account for pipetting errors, prepare at least 90 μl). Also prepare a tube with PCR-grade water for the pipetting of the no-template control (NTC). If more than two samples need to be haplotyped, prepare the sample dilutions in 200 μl PCR tubes. Mix the sample dilutions by inverting or flicking of the tubes and spin the fluids down (*see* **Note 20**).
2. Prepare a master mix (total reactions = 16×[number of samples + 1 NTC] × 1.2) that contains all components listed in Table 1, except for those in parentheses since primers and genomic DNA are added at a later time point. Listed reagent amounts are for one reaction): The volume (μl) of each component multiplied by the number of total reactions, except for the items in parenthesis (*see* **Note 21**).

Table 1
Final master mix composition for haplotyping PCR. The master mix composition is shown for haplotyping reactions when using the Expand Long Range Enzyme Mix. Volumes are calculated for 10 μl final reactions. Components that are not part of the initial master mix are shown in parentheses

Reagent	Initial concentration	Final concentration	Volume per final reaction (10 μl) [μl]
Expand Long Range Enzyme Mix	5 U	0.35 U	0.07
Expand Long Range Buffer with $MgCl_2$	5×	1×	2.00
dNTPs	10 mM each	0.2 mM each	0.20
(Forward allele-specific primer)	5 μM	0.5 μM	(1.00)
(Reverse allele-specific primer)	5 μM	0.5 μM	(1.00)
SYBR Green I	10×	0.1×	0.10
PCR-grade water	-	-	0.63
(Genomic DNA)	10 ng/μl	5 ng/μl	(5.00)

3. Shortly vortex the master mix and spin the fluid down.
4. Pipette 16 aliquots into 200 μl PCR tubes (each aliquot: 3 μl × [number of samples + 1 NTC] × 1.1), and add the different allele-specific primer combinations (four different SNP-pairs—four reaction per SNP-pair) to the aliquots: 1 μl forward allele-specific primer × [number of samples + 1 NTC] × 1.1 + 1 μl reverse allele-specific primer × [number of samples + 1 NTC] × 1.1, as shown in Figs. 3 and 4.

   For example, when nine samples need to be haplotyped, make 16 aliquots of 33 μl each (3 μl × 10 [samples + NTC] × 1.1), and add 11 μl of each primer (1 μl × [samples + NTC] × 1.1) per aliquot (16 different primer combinations) (*see* **Note 22).**
5. Shortly vortex the master mixes and spin the reagents down.
6. After primer addition, aliquot 5 μl of the master mixes into a 384-well PCR plate, as shown in Fig. 5, with one aliquot of each master mix (1–16) per row, and one row per sample. Additionally, add one row of each master mix for NTCs (*see* **Note 23**).
7. Add 5 μl of the diluted genomic DNA to each master mix in a row. Use a separate row for each sample (*see* Fig. 5). Add 5 μl of PCR-grade water to each master mix in the row reserved for the NTC (*see* **Note 24**).
8. Properly seal the PCR plate with an optically clear PCR-plate sealing foil (*see* **Note 25**).
9. Mix the plate contents by turning the plate up-side down and tapping for several times, spin the reaction down by centrifugation (2000 × *g* for 2 min).

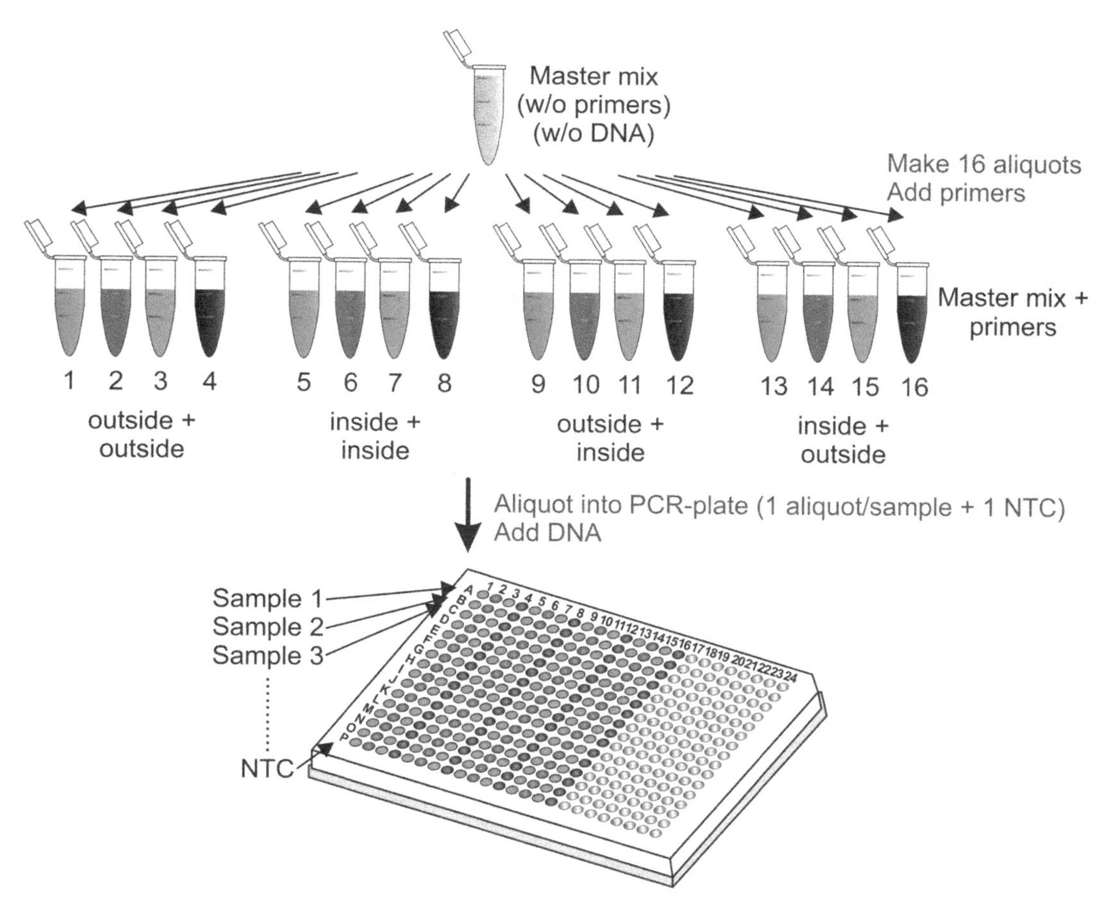

Fig. 5 Pipetting scheme for haplotyping PCR. First, a master mix without primers and genomic DNA is prepared. The master mix is then aliquoted into 16 reactions and the different primer combinations are added. In all reactions, a forward primer is combined with a reverse primer (e.g., outside (forward) + outside (reverse)). The outside primers are the allele-specific primers with the greater distance to the hotspot center, the inside primers represent the primers closer to the hotspot (one upstream, and one downstream). After primer addition, the new master mixes are then aliquoted into a PCR plate, one aliquot per sample plus one NTC. The haplotyping reaction setup is shown here for 15 different samples (row A-O). For each primer combination a no-template control (NTC) is also included. In the plate, the samples are then added to the master mixes containing the different primer combinations (columns 1–16)

10. Transfer the sealed plate into a suitable real-time PCR machine and run the PCR reactions with a suitable program, using denaturation, annealing/extension times, and temperatures that have been optimized for the used allele-specific primers. A recommended program when using the Expand Long Range Enzyme Mix with the primers of **Note 12** is shown in Table 2 (*see* **Note 26**).

Table 2
Cycling program for long-range PCR with Expand Long Range Enzyme Mix. Temperatures and durations are shown for the different steps during amplification. The annealing temperature depends on the T_m of the used allele-specific primers (*see* examples in Note 12), and represents the temperature that showed the best performance during optimization. The extension time depends on the length of the target region. For the Expand Long Range Enzyme Mix, it is recommended to use 1 min/kb

Step	Temperature	Time	Cycles
1. Initial denaturation	92 °C	2 min	
2. Denaturation	92 °C	10 s	
3. Annealing	T_m	15 s	
4. Extension	68 °C	1 min/kb	go to step 2; repeat 55×
5. Final extension	68 °C	5 min	
6. Melting curve	65–95 °C	5 s	0.5 °C increment

3.3 Data Analysis

Examine the Cq-values generated from each real-time PCR reaction. Evaluate the reactions with the two lowest Cq-values of each SNP-pair and note the haplotypes. Combine the haplotypes of the different SNP-pairs to a total haplotype (*see* Figs. 3 and 6, and Table 3). The NTC reactions should not yield a Cq-value (or only a very high Cq-value close to 50). Also examine the melting curves of all reactions to ensure the presence of the correct PCR product, without contamination from unspecific amplification or primer dimers (*see* **Note 27**).

If the phase of the heterozygous polymorphisms in the inside region is required, different strategies can be followed: sequencing of haplotyping PCR products; design of additional long-range allele-specific PCR; or genotyping the long-range haplotyping reactions. The decision may depend on the number of samples to haplotype and on the number of SNPs.

4 Notes

1. Any polymerase system suitable for the amplification of long regions can be used for haplotyping with long-range PCR. Different polymerases perform differently on distinct samples. For our applications, the Expand Long Range Enzyme Mix [26] showed the best performance for human genomic DNA extracted from blood or sperm; however, the performance is also strongly dependent on the amplicon size. For smaller regions (up to about 5 kb), haplotyping might also work with

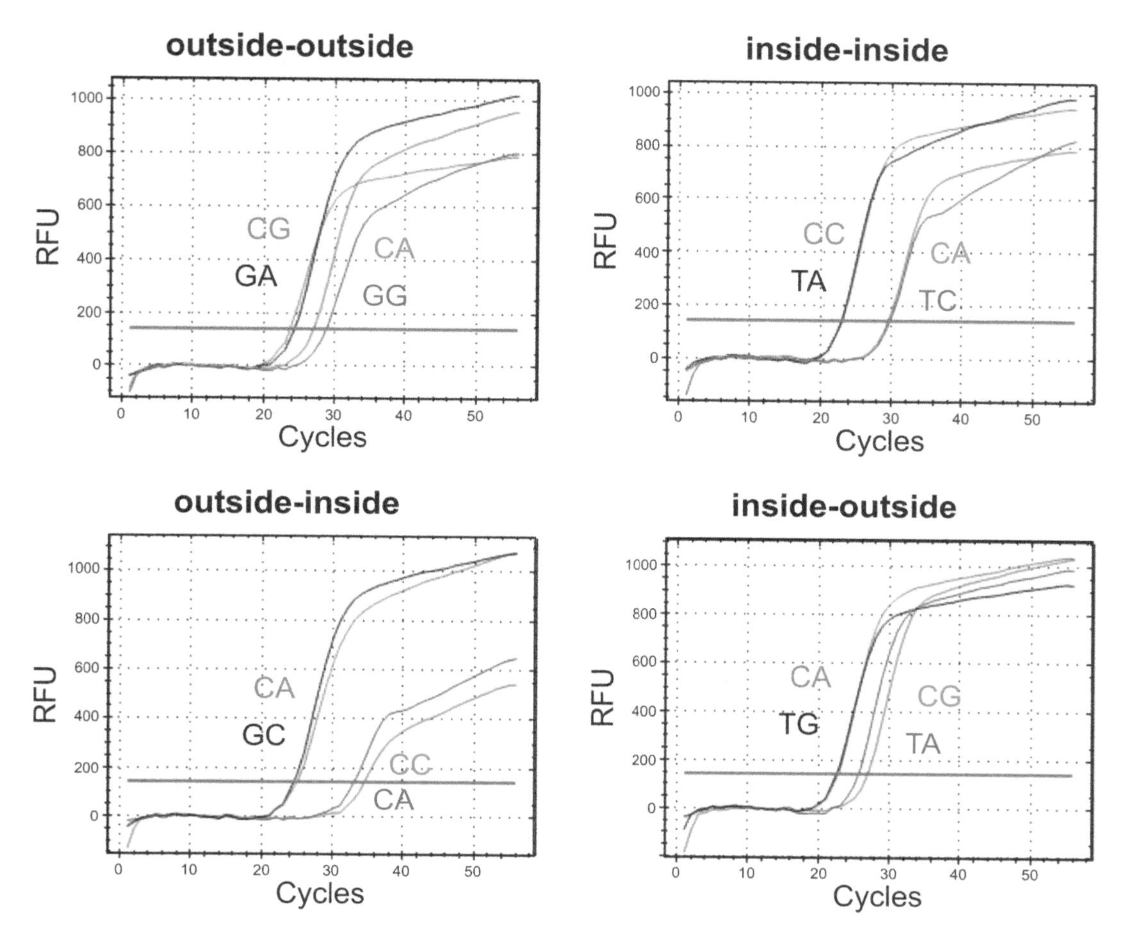

Fig. 6 Amplification curves for four haplotyped SNPs. Haplotyping was performed for the four SNPs described in **Notes 11** and **12**, using the cycling conditions from Table 2. The cycle number is shown on the x-axes, relative fluorescent unit (RFU) is shown on the y-axis. The corresponding Cq-values are listed in Table 3. Amplification curves for the different primer combinations (outside and inside primers) are shown separately, with the different colors representing the different allele combinations

simple proofreading polymerases such as Phusion Hot Start II DNA Polymerase or (with a weaker proofreading activity) Phire Hot Start II DNA Polymerase (both Thermo Scientific). Further long-range PCR polymerases are evaluated, e.g., in Jia et al. [12].

2. The usage of PCR-grade water, which is free of any nucleases, is strongly recommended. Residual traces of DNase can otherwise impair the performance of the amplification. In our experiments we use Sigma water (Sigma-Aldrich).
3. The design of the allele-specific primers is described in detail in the Methods section. We ordered our primers from Eurofins

Table 3
Example of haplotyping results for four SNPs. Results are shown for the haplotyping of four SNPs flanking hotspot II. The two reactions of a SNP-pair with the lowest Cq-values represent the haplotype of a sample. All haplotypes of the SNP-pairs can then be combined to a total haplotype

Allele-specific primer combination	Well	Cq	Haplotype
outside + outside	A01	27.05	
	A02	**23.71**	CG
	A03	**24.39**	GA
	A04	28.81	
inside + inside	A05	29.50	
	A06	**23.08**	CC
	A07	**23.03**	TA
	A08	29.75	
outside + inside	A09	**25.05**	CA
	A10	34.54	
	A11	33.06	
	A12	**24.56**	GC
inside + outside	A13	**22.75**	CA
	A14	26.91	
	A15	25.62	
	A16	**22.56**	TG
Total haplotype		CTAG/GCCA	

Genomics as 0.01 μmol DNA oligos with three PTO bonds connecting the last bases 3′, purified by HPSF; however, primers from any other company that offers the PTO modification can be used. For the purification option, desalted primers are fine since primer synthesis starts 3′ to 5′. Thus, the allele-specific base is always the first nucleotide in the synthesis.

4. For our assays, 10× SYBR Green I was prepared by diluting in DMSO from a 10,000× stock solution (Invitrogen). Alternatively to SYBR Green I, equivalent intercalating fluorescent dyes such as EvaGreen (Jena Bioscience) can be used. Always protect the dye from light. Store at −20 °C, several repeated freeze-thaw cycles did not influence performance.
5. The use of high quality genomic DNA is of great importance. Fragmentation greatly reduces the amplification efficiency. Avoid vortexing and repeated freezing and thawing of the DNA. Freshly extracted DNA works the best. Residual reagents from DNA extraction can also inhibit amplification. In our experiments, we used the PAXgene Blood DNA Kit (Qiagen) for the DNA extraction from frozen blood, and the

Gentra Puregene Cell Kit (Qiagen) for DNA extraction from sperm (both methods are described in [17, 27]).

6. The setup of the PCR reactions, as well as the extraction of genomic DNA, is suggested to be performed in PCR hoods with laminar flow, to avoid contamination from PCR products or plasmids. Never open tubes containing the long-range PCR products of your haplotyping experiment in the same room/PCR hood used also to set up the reactions (pre-PCR) or to extract the DNA to avoid carry-over contamination.
7. The method is also suitable for 96-well real-time PCR thermocyclers; however, here the reaction volume should be scaled up to 20 μl ensuring equal accuracy. If haplotyping is required for a higher number of samples, split the samples between different plates. Never split the reactions of one sample (especially the four reactions of one primer pair) between plates since this can introduce errors due to differences in amplification efficiencies between experiments.
8. PCR-plates need to be chosen according to the requirements of the used thermocycler. For 96-well real-time PCR thermocyclers, suitable 96-well plates need to be used. The use of plates with white wells is suggested since sensitivity is increased due to an increase in reflection of fluorescence. When an assay was optimized for one type of plates, switching to a different brand/type might change the required cycling conditions and make a new optimization necessary. We used Hard-Shell Thin-Wall 384-Well Skirted PCR Plates (BioRad) or FrameStar 384 plates (VWR) for our assays, which showed equal behavior during amplification.
9. As for PCR-plates, different sealing foils than used during initial assay optimization might require a new optimization. We used Microseal "B" Adhesive Seals (BioRad).
10. For 96-well plates, a normal table centrifuge compatible with spin down plates is sufficient. When using 384-well plates, it is recommended to spin the plates with $2000 \times g$ for 2 min (necessary to ensure that the PCR reaction is at the bottom of the plates, and air bubbles are removed, which could disturb the PCR/fluorescence signal); therefore, a suitable centrifuge for these conditions is required.
11. As an example, we show the haplotyping setup and results of four SNPs located on chromosome 16 (rs7201177, rs12149730, rs1861187, and rs4786855), flanking hotspot II, as published in [17].
12. The sequences of the allele-specific primers designed for the SNPs rs7201177, rs12149730, rs1861187, and rs4786855 are shown in Table 4. Further details on primer design are described in **Notes 13** and **14**.

Table 4
Allele-specific primer sequences used for haplotyping of four SNPs flanking hotspot II. For each SNP, two primers were designed that only differ in the last nucleotide at the 3′ end (shown in parenthesis). For the SNPs upstream of the hotspot, primers were designed in the forward direction, for the SNPs downstream of the hotspot, in the reverse direction. Phosphodiester bonds substituted by PTO bonds are shown as asterisks, nucleotides that were added to the primers to increase the amplification efficiency are shown in lower case. An annealing temperature of 57 °C was used for haplotyping with this set of primers

Primer name	SNP	Primer sequence	Length [bases]	T_m [°C]	%GC
Forward_out	rs7201177	taGGACGTCTCTCTGC *T*T*(C/G)	19	56.7	52.6
Reverse_out	rs12149730	GTAAGTGCTATGTTC AGAACA*G*A*(T/C)	24	57.6/59.3	37.5/41.7
Forward_in	rs1861187	gcGATTGAAATAATCA GGTTT*C*A*(C/T)	24	57.6/59.3	37.5/33.3
Reverse_in	rs4786855	GAAGTAGCAATGAGA GAGAGAAG*A*A*(T/G)	26	60.1/61.6	38.5/42.3

13. It is important that all primers have similar melting temperatures (T_m) such should not differ by more than 5 °C, such that every forward primer can be combined with every reverse primer for amplification. Additionally, all PCR reactions of a sample need to be amplified on the same plate.
14. The substitution of the last three phosphodiester bonds at the 3′ end by phosphorothioate (PTO) bonds has two major advantages: (1) 3′-ends get protected from digestion by the 3′ → 5′ exonuclease activity of proofreading polymerases, which would otherwise remove allele-specificity; (2) PTOs make the primer more rigid, resulting in less efficient amplification of mismatches, and therefore increase allele-specificity [18]. With the introduction of PTOs, specificity could be increased in several applications of allele-specific PCR [16, 17]. The introduction of a mismatch at the nucleotide two bases before the 3′-end can also improve the allele-specificity.
15. Due to the restriction to specific sites in the genomic DNA containing the SNPs that need to be analyzed at the 3′-end of the primers, and the requirement of more than one compatible primer pair, it is not always possible to fulfill all primer design guidelines. If not otherwise possible, primers that slightly differ from the recommended values can be tested (e.g., primers with a length of 26 bases, or a GC-content below 40% were used for the haplotyping of hotspot II (*see* Table 4)). Additionally, for regions with unusually high or low GC-content, the addition of a few bases at the 5′ end can

improve the performance. Furthermore, this can increase primer hybridization stability, if e.g. the original nucleotides result in primer dimerization at the 5′ end. As shown in Table 4, two nucleotides (gc) were added to the 5′ end of the primer "Forward_in," which is located in a very AT-rich region, to increase the GC-content of the primer.

16. We can recommend the IDT OligoAnalyzer 3.1 (https://eu.idtdna.com/calc/analyzer) [28] as a tool to check the different primer design parameters (self-complementarity, primer dimerization, loop formation, T_m estimation, etc.). Moreover, the primers need to be checked if they amplify additional regions in the genome. This can be done for example with Primer-BLAST (http://www.ncbi.nlm.nih.gov/tools/primer-blast) [29]. Primers that amplify additional regions with a similar length or shorter regions need to be redesigned; otherwise, short amplicons will preferentially get amplified and interfere with the production of the longer-sized expected PCR products. Unspecific amplification of regions that are longer than the region of interest can be avoided by using shorter elongation times than required for these regions during the PCR. As already described in **Note 3**, primers can be ordered from any company that provides synthesis with PTO modifications.

17. Always include the measurement of a melting curve as the last step of the cycling program. To obtain a melting (dissociation) curve, the temperature is slowly raised (e.g., from 65 °C to 95 °C), leading to denaturation of the amplified double-stranded DNA depending on the T_m of the fragment, and fluorescence is measured (e.g., every 0.5 °C). When the temperature reaches the T_m of the fragment, 50% of the DNA is denatured, and with further increase, the total DNA denatures, leading to a decrease in fluorescence due to the release of the intercalating dye. When plotting fluorescence signal (RFU) against the temperature (T), the point of inflection represents the T_m of the fragment. By plotting the negative first derivative of the melting-curve ($-\Delta RFU/\Delta T$ against T) the T_m can be referred easier as the maximum of the obtained peak [24]. Different peaks represent different fragments, which allows the detection of primer dimers and nonspecific products in addition to the desired PCR product.

18. Optimization should always be performed with the same real-time thermocycler as intended for the final haplotyping experiments. The usage of a different thermocycler can require new optimization. If available, use heterozygous samples with known haplotypes, and/or samples homozygous for the different haplotypes as controls during optimization. If controls are not available, haplotyping has to be performed in 16 reactions, and only for samples in which the haplotype is in accordance with all primer combinations should be considered reliable.

For the optimization, set the reactions up as described for the haplotyping of the samples (*see* Table 1 and Fig. 5), except that it should be performed with only a few samples representing different known haplotypes.

In the first step of the optimization, amplification of the correct fragments without additional products (e.g., unspecific amplification or primer dimers) needs to be checked by gel electrophoresis (for long-range PCR, we use 0.5 % agarose gels), in addition to a melting curve analysis (*see* **Note 17**—melting curve analysis). The cycling program should rely on the manufacturer's instructions; different polymerases require different cycling conditions. A recommended cycling program for the Expand Long Range Enzyme Mix is shown in Table 2.

In the second optimization step, the annealing temperatures are tested to provide the largest difference in the amplification efficiencies between the matched and unmatched primer pairs resulting in a better specificity. Try different annealing temperatures during the cycling program: use the calculated T_m (of the primer with the lowest $T_{m,}$ values can be rounded); T_m + 3 °C; T_m + 6 °C. For further PCR runs, choose a temperature that provides the amplification of the correct fragment (without interference of additional unspecific fragments or primer dimer formation) and also a good separation of curves from reactions with primer combinations representing the haplotype, from curves due to non-allele-specific amplification. A difference in the Cq-values of at least 4 cycles is recommended. The separation of allele-specific and non-allele-specific amplification can be increased by the addition of DMSO (it decreases the melting temperature, which can be of interest in GC-rich regions), or additional $MgCl_2$ (an increase in the $MgCl_2$ concertation can increase primer annealing, concentrations up to ~4 mM can be used) (reviewed in [30]). A two-step cycling program using initially a lower T_m for 5–10 cycles and the switching to a higher T_m for the remaining cycles can also help to increase the specific product yields.

19. If a non-hot start DNA polymerase is used for amplification, the master mixes and plates need to be set up on ice. For the Expand Long Range Enzyme Mix, which is a hot start DNA polymerase, all reactions can be set up at room temperature. Only the polymerase itself should be kept at −20 °C until shortly before usage, and frozen again immediately after usage to avoid a reduction in performance.

20. For our tested assays, a DNA concentration of 10 ng/μl provided a good performance. However, if the amplification efficiency of an assay is rather low (bad sample quality, bad performance of the primers) the concentration can be increased, or decreased if sufficient efficiency is reached with lower sample amounts. Diluted samples can be stored at 4 °C

for several days, or at −20 °C for longer storage, if prepared in excess and needed for further experiments. It is suggested to additionally include a sample with a known haplotype (as used for the optimization) in every run as a control of the assay.

21. If a different polymerase as the Expand Long Range Enzyme Mix is used, the master mix (required amount of enzyme, dNTPs, primers, and additional supplements) needs to be adapted according to manufacturer's instructions. Depending on the amount of samples that will be haplotyped, the master mix needs to be prepared in a suitably sized tube. It is necessary to prepare 20% excess of master mix (factor 1.2) to account for pipetting errors. Protect SYBR Green I from excessive exposure with light, which can otherwise reduce the intensity of the fluorescent dye. Note that the SYBR Green I used in this protocol is dissolved in DMSO, when the dye is substituted by an intercalating fluorescent dye that is dissolved in water (e.g., EvaGreen), the respective amount of DMSO needs to be added to ensure equal performance since DMSO has an effect on the DNA melting temperature. The haplotyping assay can also be optimized with a water-dissolved fluorescent dye without the addition of any DMSO.

22. In general, 12 reactions would be sufficient to infer the haplotype of 4 SNPs. However, it is recommended to run 16 reactions (with all possible forward + reverse primer combinations) to have additional information of the haplotype. If only 12 reactions are used, the number 16 has to be substituted by the number 12 in the formula for master mix preparation in Subheading 3.2, **step 2**. It is necessary to prepare 10% excess of the master mix aliquots (factor 1.1) to account for pipetting errors. Primer combinations can be premixed prior to addition (combine equal volumes of 5 μM forward and reverse primer, and add 2 μl of the primer mix per reaction) to reduce the number of pipetting steps in the reaction setup.

23. The pipetting scheme described here, and also shown in Fig. 5, is an example for 384-well plates, but can be adapted according to personal preferences and also to 96-well plates. The empty wells in the example plate can be used for haplotyping of additional samples. We preferred using the described scheme since it can be easily adapted to the usage of multichannel pipettes, which is recommended for haplotyping of more than two samples. This reduces the time required for pipetting, and therefore also evaporation of the reaction mixes. When working with 384-well plates, you have to consider that standard 8- or 12-channel pipettes only reach every second well. To aliquot the master mixes after primer addition (which are prepared in 200 μl tubes), in the first step, we aliquot the reactions 1, 3, 5, 7, 9, 11, 13, and 15 into the corresponding columns

of the plate, and in the next step, the remaining reactions. Different samples can be amplified on different plates; however, never split the reactions for one sample between plates.

24. The same pipette tip can be used for the addition of the same sample to all master mixes in a row containing different primer combinations. The amount of master mix that will be transferred from one well to the next is negligible and does not influence the results. Similar as described for the master mixes, samples (and NTC) should be aliquoted with a 8- or 12-channel multichannel pipette (in the case of a 12-channel pipette, use only eight tips). In the first step, add samples 1, 3, 5,... (rows A, C, E,...), in the second step add samples 2, 4, 6,... (rows B, D, F,...).

25. Make sure that the plate is properly sealed, especially at the outer rows and columns. The usage of a PCR seal hand applicator or a similar tool is recommended, however, if not available, proper sealing can also be achieved with a pen. Improper sealing can lead to evaporation of the reagent and distortion of the real-time PCR signal.

26. The cycling program depends on the polymerase used in the reaction, and needs to be adapted for other polymerases according to manufacturer's instructions. For the annealing step use the temperature that showed the best performance during the optimization. For haplotyping the four SNPs flanking hotspot II, a T_m of 57 °C (the lowest T_m of all allele-specific primers, as shown in Table 4) was used. The extension time needs to be adapted according to the length of the longest amplified fragment. For hotspot II, 270 s were used.

27. Use the analysis software provided with the real-time PCR machine to obtain the Cq-values and melting curves. Two different methods exist to obtain a Cq-value: linear regression or setting manually the threshold; however, both methods should yield similar results. In our experiments, the data analysis was performed with the CFX Manager Software, provided by the CFX Real-Time PCR System (both BioRad), and the manual threshold setting to obtain the Cq-values. A lower Cq-value represents a better amplification efficiency, and therefore the presence of the corresponding haplotype of the reaction in the sample. When the melting curve of a reaction shows the amplification of a different fragment (the peak is shifted to the left or right), or additional peaks are present, discard the reaction from analysis. If the haplotype can be inferred anyway (e.g., if only primers that do not represent the haplotype of the sample lead to this amplification, or the haplotype can be inferred from the other reactions), haplotyping of the sample does not need to be repeated. Otherwise repeat the haplotyping of the sample.

If, for example, the two reactions of a SNP-pair with both SNPs being heterozygous for A/G show preferential amplification of the reactions representing AG and GG, then the second SNP (SNP of the reverse primer) is most likely homozygous, and there was most likely a mistake in genotyping (however, an error in haplotyping cannot be ruled out; therefore, the reaction should be repeated for verification).

Acknowledgements

This work was supported by the "Austrian Science Fund" (FWF) P25525-B13 and P23811-B12 to I.T-B., and DOC Fellowships of the Austrian Academy of Sciences to B.A. and A.H.

References

1. Browning SR, Browning BL (2011) Haplotype phasing: existing methods and new developments. Nat Rev Genet 12(10):703–714. doi:10.1038/nrg3054
2. Ruano G, Kidd KK, Stephens JC (1990) Haplotype of multiple polymorphisms resolved by enzymatic amplification of single DNA molecules. Proc Natl Acad Sci U S A 87(16):6296–6300
3. Zhao Z, Fu YX, Hewett-Emmett D, Boerwinkle E (2003) Investigating single nucleotide polymorphism (SNP) density in the human genome and its implications for molecular evolution. Gene 312:207–213
4. Newton CR, Graham A, Heptinstall LE, Powell SJ, Summers C, Kalsheker N, Smith JC, Markham AF (1989) Analysis of any point mutation in DNA. The amplification refractory mutation system (ARMS). Nucleic Acids Res 17(7):2503–2516
5. Nichols WC, Liepnieks JJ, McKusick VA, Benson MD (1989) Direct sequencing of the gene for Maryland/German familial amyloidotic polyneuropathy type II and genotyping by allele-specific enzymatic amplification. Genomics 5(3):535–540
6. Okayama H, Curiel DT, Brantly ML, Holmes MD, Crystal RG (1989) Rapid, nonradioactive detection of mutations in the human genome by allele-specific amplification. J Lab Clin Med 114(2):105–113
7. Wu DY, Ugozzoli L, Pal BK, Wallace RB (1989) Allele-specific enzymatic amplification of beta-globin genomic DNA for diagnosis of sickle cell anemia. Proc Natl Acad Sci U S A 86(8):2757–2760
8. Ehlen T, Dubeau L (1989) Detection of ras point mutations by polymerase chain reaction using mutation-specific, inosine-containing oligonucleotide primers. Biochem Biophys Res Commun 160(2):441–447
9. Ruano G, Kidd KK (1989) Direct haplotyping of chromosomal segments from multiple heterozygotes via allele-specific PCR amplification. Nucleic Acids Res 17(20):8392
10. Cheng S, Fockler C, Barnes WM, Higuchi R (1994) Effective amplification of long targets from cloned inserts and human genomic DNA. Proc Natl Acad Sci U S A 91(12):5695–5699
11. Barnes WM (1994) PCR amplification of up to 35-kb DNA with high fidelity and high yield from lambda bacteriophage templates. Proc Natl Acad Sci U S A 91(6):2216–2220
12. Jia H, Guo Y, Zhao W, Wang K (2014) Long-range PCR in next-generation sequencing: comparison of six enzymes and evaluation on the MiSeq sequencer. Sci Rep 4:5737. doi:10.1038/srep05737
13. Venturini G, Rose AM, Shah AZ, Bhattacharya SS, Rivolta C (2012) CNOT3 is a modifier of PRPF31 mutations in retinitis pigmentosa with incomplete penetrance. PLoS Genet 8(11), e1003040. doi:10.1371/journal.pgen.1003040
14. Michalatos-Beloin S, Tishkoff SA, Bentley KL, Kidd KK, Ruano G (1996) Molecular haplotyping of genetic markers 10 kb apart by allele-specific long-range PCR. Nucleic Acids Res 24(23):4841–4843
15. Gaudet M, Fara AG, Beritognolo I, Sabatti M (2009) Allele-specific PCR in SNP genotyping. Methods Mol Biol 578:415–424. doi:10.1007/978-1-60327-411-1_26

16. Tiemann-Boege I, Calabrese P, Cochran DM, Sokol R, Arnheim N (2006) High-resolution recombination patterns in a region of human chromosome 21 measured by sperm typing. PLoS Genet 2(5):e70. doi:10.1371/journal.pgen.0020070
17. Arbeithuber B, Betancourt AJ, Ebner T, Tiemann-Boege I (2015) Crossovers are associated with mutation and biased gene conversion at recombination hotspots. Proc Natl Acad Sci U S A 112(7):2109–2114. doi:10.1073/pnas.1416622112
18. de Noronha CM, Mullins JI (1992) Amplimers with 3'-terminal phosphorothioate linkages resist degradation by vent polymerase and reduce Taq polymerase mispriming. PCR Methods Appl 2(2):131–136
19. Higuchi R, Dollinger G, Walsh PS, Griffith R (1992) Simultaneous amplification and detection of specific DNA sequences. Biotechnology (N Y) 10(4):413–417
20. Saiki RK, Scharf S, Faloona F, Mullis KB, Horn GT, Erlich HA, Arnheim N (1985) Enzymatic amplification of beta-globin genomic sequences and restriction site analysis for diagnosis of sickle cell anemia. Science 230(4732):1350–1354
21. Aaij C, Borst P (1972) The gel electrophoresis of DNA. Biochim Biophys Acta 269(2):192–200
22. Gregson S (1972) Polyacrylamide gel electrophoresis of DNA. Anal Biochem 48(2):613–616
23. Ponchel F, Toomes C, Bransfield K, Leong FT, Douglas SH, Field SL, Bell SM, Combaret V, Puisieux A, Mighell AJ, Robinson PA, Inglehearn CF, Isaacs JD, Markham AF (2003) Real-time PCR based on SYBR-Green I fluorescence: an alternative to the TaqMan assay for a relative quantification of gene rearrangements, gene amplifications and micro gene deletions. BMC Biotechnol 3:18. doi:10.1186/1472-6750-3-18
24. Ririe KM, Rasmussen RP, Wittwer CT (1997) Product differentiation by analysis of DNA melting curves during the polymerase chain reaction. Anal Biochem 245(2):154–160. doi:10.1006/abio.1996.9916
25. Wang W, Chen K, Xu C (2006) DNA quantification using EvaGreen and a real-time PCR instrument. Anal Biochem 356(2):303–305. doi:10.1016/j.ab.2006.05.027
26. Fan W, Waymire KG, Narula N, Li P, Rocher C, Coskun PE, Vannan MA, Narula J, Macgregor GR, Wallace DC (2008) A mouse model of mitochondrial disease reveals germline selection against severe mtDNA mutations. Science 319(5865):958–962. doi:10.1126/science.1147786
27. Meyer WK, Arbeithuber B, Ober C, Ebner T, Tiemann-Boege I, Hudson RR, Przeworski M (2012) Evaluating the evidence for transmission distortion in human pedigrees. Genetics 191(1):215–232. doi:10.1534/genetics.112.139576
28. Owczarzy R, Tataurov AV, Wu Y, Manthey JA, McQuisten KA, Almabrazi HG, Pedersen KF, Lin Y, Garretson J, McEntaggart NO, Sailor CA, Dawson RB, Peek AS (2008) IDT SciTools: a suite for analysis and design of nucleic acid oligomers. Nucleic Acids Res 36(Web Server issue):W163–W169. doi:10.1093/nar/gkn198
29. Ye J, Coulouris G, Zaretskaya I, Cutcutache I, Rozen S, Madden TL (2012) Primer-BLAST: a tool to design target-specific primers for polymerase chain reaction. BMC Bioinformatics 13:134. doi:10.1186/1471-2105-13-134
30. Kramer MF, Coen DM (2001) Enzymatic amplification of DNA by PCR: standard procedures and optimization. Curr Protoc Immunol Chapter 10:Unit 10.20. doi:10.1002/0471142735.im1020s24

Chapter 2

Quantification and Sequencing of Crossover Recombinant Molecules from *Arabidopsis* Pollen DNA

Kyuha Choi, Nataliya E. Yelina, Heïdi Serra, and Ian R. Henderson

Abstract

During meiosis, homologous chromosomes undergo recombination, which can result in formation of reciprocal crossover molecules. Crossover frequency is highly variable across the genome, typically occurring in narrow hotspots, which has a significant effect on patterns of genetic diversity. Here we describe methods to measure crossover frequency in plants at the hotspot scale (bp–kb), using allele-specific PCR amplification from genomic DNA extracted from the pollen of F_1 heterozygous plants. We describe (1) titration methods that allow amplification, quantification and sequencing of single crossover molecules, (2) quantitative PCR methods to more rapidly measure crossover frequency, and (3) application of high-throughput sequencing for study of crossover distributions within hotspots. We provide detailed descriptions of key steps including pollen DNA extraction, prior identification of hotspot locations, allele-specific oligonucleotide design, and sequence analysis approaches. Together, these methods allow the rate and recombination topology of plant hotspots to be robustly measured and compared between varied genetic backgrounds and environmental conditions.

Key words Meiosis, Recombination, Crossover, Allele-specific PCR, Titration, Pollen DNA

1 Introduction

Meiotic recombination is conserved throughout the majority of eukaryotes, and has importance for human fertility, agricultural breeding, and understanding patterns of genetic diversity [1–4]. During meiosis, programmed DNA double strand breaks are induced by SPO11, coincident with pairing of homologous chromosomes [3, 5, 6]. The broken DNA ends are then repaired using the homologous chromosome as a template, which can result in reciprocal exchange, or "crossover" [3, 5–7]. Crossover numbers and distributions are under tight genetic control, usually consisting of one or a small number occurring per chromosome arm, regardless of the physical size of the chromosome [7]. Meiotic recombination rate is also highly heterogeneous along chromosomes, with many species having narrow "hotspots", typically kilobases (kb) in width, with high recombination rates relative to the genome

Irene Tiemann-Boege and Andrea Betancourt (eds.), *Haplotyping: Methods and Protocols*, Methods in Molecular Biology, vol. 1551,
DOI 10.1007/978-1-4939-6750-6_2, © Springer Science+Business Media LLC 2017

average [2, 8–10]. Mapping >100 s of crossovers at hotspot scale (kb) can be challenging, due to the low number of total crossovers per meiosis. For example, measurement of crossover hotspots in *Arabidopsis* and maize has shown hotspot genetic map lengths that are close to 0.1 cM (or 0.1 % recombination) [9, 11–13]. Therefore, to see >100 crossovers at such a hotspot requires screening of at least 100,000 meioses. For these reasons, specific methods are required for quantification and sequencing of crossover molecules at the physical scale of hotspots.

One approach to study crossover hotspots is to screen very high numbers of meioses by isolating gamete DNA from individuals that are heterozygous in the hotspot region of interest (Fig. 1). In humans and mice this has primarily been achieved using sperm [14–20], but eggs have also been analyzed in mice [21, 22]. In *Arabidopsis*, an equivalent approach uses genomic DNA isolated from pollen (multicellular haploid gametophytes) [11–13, 23, 24]. These methods rely on recombinant crossover and non-recombinant parental molecules being distinguishable by allele-specific PCR amplification (Fig. 1). Titration can be used to estimate the relative concentrations of parental to recombinant molecules, in order to measure crossover frequency (Fig. 1) [18, 23]. Amplified single crossover molecules can then be internally Sanger sequenced, or otherwise genotyped, to identify recombination sites to the resolution of individual polymorphisms (Fig. 1) [18, 23]. Varying allele-specific oligonucleotide (ASO) and non-allele-specific universal oligonucleotide (UO) configurations can also be used to detect non-crossovers and gene conversion events [18, 23, 25]. Together, these methods allow high-resolution studies of recombination frequency and topology within individual crossover hotspots.

Here, we report detailed methodologies for pollen typing that allow the study of crossover hotspots in plant genomes. First, we describe a simplified and robust allele-specific PCR amplification method that is highly sensitive. We provide information on critical steps including pollen genomic DNA extraction, choice of target hotspot locations and allele-specific oligonucleotide (ASO) design. Next, we describe a pollen typing qPCR technique that allows rapid estimation of hotspot crossover frequency, without the need

Fig. 1 (continued) amplification (Subheading 3.6, **steps 10–14**). (**e**) Crossover and parental molecules are quantified using titration (Subheading 3.6). (**f**) Single crossover molecule amplification products are Sanger sequenced to identify internal recombination sites (Subheading 3.7). (**g**) Quantitative PCR analysis of crossover frequency, following the first round of allele-specific PCR amplification (Subheading 3.8). These data are for the *3a* hotspot and are shown in Table 4 [12]. (**h**) Mass amplification and high-throughput sequencing of crossover molecules (Subheading 3.9). (**i**) Identification of crossovers using paired-end sequencing reads (Subheadings 3.9 and 3.10)

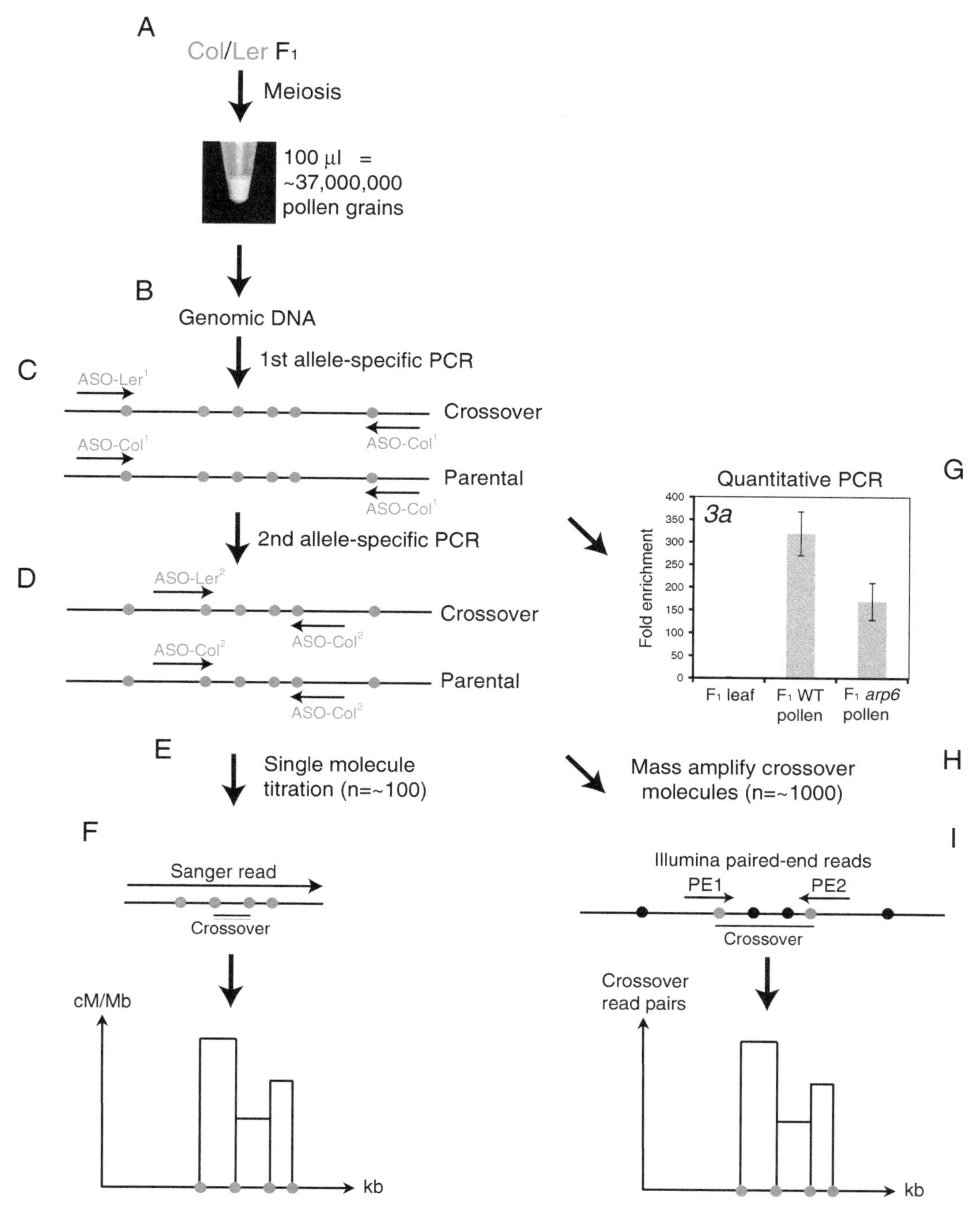

Fig. 1 Diagram illustrating pollen typing methods for crossover hotspot analysis. (**a**) Purification of pollen grains (Subheading 3.1). 100 µl of pollen suspension is shown, which is equivalent to approximately 37,000,000 pollen grains. (**b**) Genomic DNA extraction from pollen (Subheading 3.2). (**c**) 1st round of allele-specific PCR using oligonucleotides specific for Ler (*red*) or Col (*blue*) templates. ASO configurations for crossover or parental molecule amplifications are indicated. Polymorphic sites are represented by *colored circles*. (Subheading 3.6, **steps 1–9**). (**d**) 2nd allele-specific PCR as for (C), but using a nested set of ASOs for

for extensive titration experiments (Fig. 1). Finally, we describe a new method to convert large numbers of amplified crossover molecules into high-throughput sequencing libraries, enabling deep sampling of recombination events within hotspots (Fig. 1).

2 Material

2.1 Pollen Purification and DNA Extraction

1. Blender (Elecheck).
2. 80 μm nylon mesh (Normesh).
3. 1 mm glass beads (Sigma-Aldrich Z273619).
4. 3 mm glass beads (Sigma-Aldrich Z143928).
5. 10% sucrose.
6. Lysis buffer: 100 mM NaCl, 50 mM Tris–HCl (pH 8), 1 mM EDTA, 1% SDS. Add dithiothreitol (DTT) to 1 mM immediately prior to use.
7. Proteinase K (20 mg/ml) (Melford MB2005).
8. Liquid phenol, saturated with 1 M Tris–HCl (pH 8).
9. Chloroform–isoamyl alcohol (24:1 volume:volume).
10. Isopropanol.
11. 70% ethanol.
12. TE buffer: 10 mM Tris–HCl (pH 8), 1 mM EDTA.

2.2 DNA Purification

1. DNase-free RNase A (10 mg/ml) (Qiagen 19101).
2. 3 M sodium acetate (pH 5.2).
3. 100% ethanol.
4. 70% ethanol.
5. TE buffer: 10 mM Tris–HCl (pH 8), 1 mM EDTA.
6. Qubit dsDNA BR Assay Kit (Invitrogen, Q32850).
7. Qubit fluorometer (Invitrogen).

2.3 Allele-Specific PCR

1. TaKaRa Ex Taq DNA Polymerase (TaKaRa, RR001C).
2. 10× Ex Taq buffer (TaKaRa, RR001C).
3. dNTP, 2.5 mM (TaKaRa, RR001C).
4. Nuclease-free W=water (Ambion, AM9937).
5. 96-well PCR plates (STARLAB, E1403-100).
6. 96-well PCR plate seals (Thermo Scientific, AB0558).
7. 96-well plate centrifuge (Labnet).
8. 96-well PCR thermal cycler (Bio-Rad Tetrad2).
9. Agarose (Melford, MB1200).

10. Ethidium bromide.
11. HyperLadder 1 kb (Bioline, H1-415106).
12. 5× DNA loading buffer (Bioline, BIO37045).

2.4 Mapping Crossover Sites by Sanger Sequencing

1. Exonuclease (EXOI, New England Biolabs, M0293L 20 U/μl).
2. Shrimp alkaline phosphatase (SAP, Takara, 2660A 1U/μl).
3. BigDye Terminator v3.1 Cycle Sequencing Kit (Applied Biosystems).
4. 5× BigDye sequencing buffer (Applied Biosystems).

2.5 Pollen Typing qPCR

1. 10× SYBR green (Sigma-Aldrich, S9430).
2. 10× DNA Polymerase PCR Buffer (200 mM Tris–HCl (pH 8.4), 500 mM KCl, 15 mM $MgCl_2$).
3. GoTaq DNA Polymerase (Promega, M3001).
4. Real time PCR thermal cycler (Bio-Rad CFX96).

2.6 Pollen Typing Sequencing Library Construction

1. QIAquick Gel Extraction Kit (Qiagen, 28704).
2. Bioruptor standard (Diagenode).
3. TruSeq DNA Sample prep kit v2 (Illumina).

3 Methods

3.1 DNA Extraction

1. Collect whole *Arabidopsis thaliana* inflorescences and immerse them in ice-cold 10% sucrose. For example, 20 ml 10% sucrose is typically used for one densely packed 50 ml Falcon tube of collected inflorescences. The submerged material can be frozen at −20 °C, or used directly in **step 2**.
2. Disrupt inflorescences using a blender with 3×2 s pulses at maximum speed. Pollen grain walls will remain intact after this treatment.
3. Filter the tissue homogenate through 80 μm nylon mesh into 50 ml Falcon tubes. If multiple tubes are collected, the pollen can be combined following **step 5**.
4. Centrifuge the filtrate at 350×*g* for 10 min at 4 °C.
5. Discard the supernatant and resuspend the pellet, which contains pollen grains and tissues fragments, in 30 ml of ice-cold 10% sucrose.
6. Centrifuge the resuspended pellet at 100×*g* for 10 min at 4 °C.
7. Discard the supernatant and resuspend the pellet in 30 ml of ice-cold 10% sucrose.
8. Centrifuge the resuspended pellet at 100×*g* for 3 min at 4 °C.

9. Repeat **steps 7** and **8**. This process should be repeated until sufficient pollen for downstream experiments has been collected.
10. By this step, a bright yellow pellet enriched for pollen should be obvious (Fig. 1a). Remove the supernatant and resuspend the remaining pollen pellet/suspension in 100 μl aliquots into 1.5 ml eppendorf tubes. From cell counting experiments we estimate that 100 μl contains approximately 37,000,000 pollen grains (Fig. 1a). Pollen aliquots can be stored at −20 °C, or used directly in **step 11**.
11. Resuspend the pollen grain pellet in 4 vol of Lysis Buffer, containing freshly added dithiothreitol (DTT).
12. Add proteinase K to a final concentration of 20 μg/ml.
13. Incubate overnight at 65 °C. Following overnight digestion, the color of the pollen suspension will change from yellow to brown.
14. Add 200 μl of 1 mm glass beads and 4 × 3 mm glass beads.
15. Vortex at full speed for 6 min (*see* **Note 1**).
16. Centrifuge at 17000 × *g* for 5 min.
17. Transfer the supernatant (400 μl) to a new tube containing 100 μl of Lysis Buffer (=A tube).
18. Add 500 μl of fresh Lysis Buffer without SDS to the bead pellet (=B tube).
19. Vortex the B tube at full speed for 3 min.
20. Add 1 volume (500 μl) of phenol saturated with 1 M Tris–HCl (pH 8) to each of the A and B tubes, followed by mixing by hand inversion.
21. Gently mix for 30 min on a rocking wheel.
22. Centrifuge at 15,000 × *g* for 10 min.
23. Carefully transfer the supernatant to a new Eppendorf tube.
24. Add an equal volume of chloroform–isoamyl alcohol (24:1 volume:volume). Homogenize by gentle hand shaking.
25. Centrifuge at 15,000 × *g* for 10 min.
26. Carefully transfer the supernatant to a new Eppendorf tube.
27. Add 0.7 vol of isopropanol. Incubate for 10 min at room temperature.
28. Centrifuge at 15,000 × *g* for 10 min at 4 °C. Discard the supernatant.
29. Wash the pellet with 1 ml of 70% ethanol.
30. Centrifuge at 15,000 × *g* for 5 min at 4 °C. Discard the supernatant.

31. Repeat **steps 29** and **30**.
32. Allow the pellet to air-dry at room temperature for 5 min. Do not overdry the DNA pellet, which will make it difficult to redissolve.
33. Carefully resuspend each pellet in 100 μl of TE buffer and combine the DNA from the A and B tubes (*see* **Note 2**).

3.2 DNA Purification

1. Add RNaseA to a final concentration of 100 μg/ml.
2. Incubate for 30 min at room temperature.
3. Add 0.1 volume of 3 M sodium acetate. Homogenize by gentle hand shaking.
4. Add 2.5 volumes of 100% ethanol. Homogenize by gentle hand shaking.
5. Incubate on ice for at least 30 min.
6. Centrifuge at 15,000 × *g* for 20 min at 4 °C.
7. Wash the pellet with 1 ml of 70% ethanol.
8. Centrifuge at 15,000 × *g* for 5 min at 4 °C. Discard the supernatant.
9. Repeat **steps** 7 and **8**.
10. Allow the pellet to air-dry at room temperature for 5 min. Do not overdry the DNA pellet, which will make it difficult to redissolve.
11. Carefully resuspend the pellet in 200 μl of TE buffer.
12. Repeat **steps 3–10**.
13. Carefully resuspend the pellet in 100 μl of TE buffer (*see* **Note 2**).
14. Quantify DNA concentrations using a fluorometer. DNA quality should also be assessed using 1% agarose gel electrophoresis and ethidium bromide staining, which is added to a final concentration of 0.5 μg/ml (Fig. 2).
15. DNA is then either used directly for pollen typing (Subheadings 3.6–3.9) or frozen in 5–20 μl aliquots.

3.3 Identification of Candidate Crossover Hotspots for Pollen Typing

The design and validation of hotspot allele-specific oligonucleotides (ASOs) represents a considerable amount of work, and so it is important to carefully select hotspots before the start of the experiment. A variety of methods, data and technical considerations are important in this selection procedure. First, an attempt should be made to identify regions that show evidence for high crossover levels using more than one kind of evidence. Hotspots can be identified using population genetics approaches that analyze patterns of linkage disequilibrium, to provide evidence of historical crossovers [19, 26, 27]. This kind of evidence relies on the

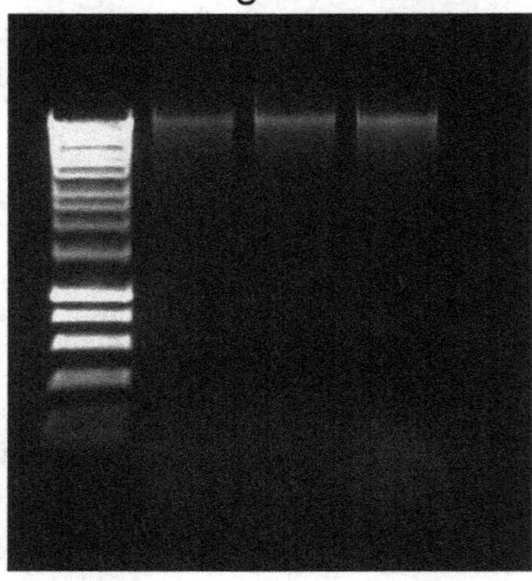

Fig. 2 Gel electrophoresis analysis of pollen genomic DNA. *Arabidopsis* pollen genomic DNA separated by electrophoresis using a 1 % agarose gel containing ethidium bromide. The *three lanes* show 1 μl of DNA, with the concentrations of 34.4, 42.1 and 53.2 ng/μl

availability of high quality natural polymorphism datasets, such as the 1001 *Arabidopsis* Genomes Project [28]. Hotspots can be inferred from these datasets using a variety of statistical methods, which estimate the population-scaled recombination rate *4Ner*, where *Ne* is the effective population size and *r* is the per generation crossover rate [26, 27, 29, 30]. While hotspots estimated via these approaches have been successfully predicted and validated using pollen typing [11, 12], they have important caveats. These approaches use recombination estimated over many haplotype combinations from both male and female meiosis. Therefore, predicted hotspots may not occur in a specific cross (e.g., Col-0 × Ler-0), for example if structural heterozygosity exists at the hotspot, or if hotspot activity is specific to female meiosis.

Another means of identifying high recombination regions is by using existing large datasets of experimentally mapped crossovers [31–35]. These datasets are typically not deep enough to allow prediction of fine-scale (kb) hotspot locations. However, the likely hotspot region can be identified using additional information. Plant hotspots are known to occur primarily at gene promoters and terminators [9, 12, 36–39], so gene annotation can be used to predict likely hotspot locations within high recombination regions. Additionally, low nucleosome density in promoters, and high $H3K4^{me3}$ and H2A.Z within genes have been associated with plant crossover hotspots, which can be identified via analysis of available epigenomic datasets [12, 40–43]. Finally, repetitive sequences in plant genomes, such as densely DNA methylated transposons, are generally silent for recombination [11, 24, 44]. Hence, heterochromatin and repetitive regions should be avoided for pollen typing analysis, unless strong prior evidence for crossovers exists and structural polymorphism is well understood.

Once candidate hotspot regions have been identified, it is critically important to have accurate knowledge of polymorphisms in this region for the two accessions to be crossed (e.g., Col-0 and Ler-0). It is recommended that target hotspot regions are PCR amplified and Sanger sequenced from the laboratory parent lines to be used for crossing, in order to identify all insertions/deletions (indels) and single nucleotide polymorphisms (SNPs), which are required for allele-specific PCR amplification. It may also be informative to perform Southern blotting and hybridization in order to understand structural variation, which can inhibit recombination [15, 16, 37].

3.4 Allele-Specific Oligonucleotide (ASO) Design

Two nested pairs of ASOs are required to PCR-amplify, quantify, and sequence crossover molecules [18, 23] (Fig. 1). This key step must be carefully performed and the specificity and efficiency of ASOs optimized. To test ASO specificity and efficiency, the primer pairs are combined with non-allele-specific universal oligonucleotides (UOs) and used to amplify from parental DNA templates separately (e.g., Col-0 and Ler-0). ASOs are considered unsuccessful if multiple amplification bands are observed, or poor template specificity is seen using the parent accession template DNA. Examples of successful ASOs used for pollen typing in *Arabidopsis* are shown in Table 1. Figure 3 shows the results of test amplifications with successful (e.g., ASO1, 2, and 6) and unsuccessful (e.g., ASO3, 4, 5, 7–11) ASOs. The successful ASOs 1, 2 and 6 correspond to KC417, KC418, and KC465 (Table 1). Finally, pairs of optimized ASOs must then amplify crossovers specifically from F_1 pollen DNA and not F_1 leaf DNA (Fig. 4).

3.4.1 Length of Amplicons

ASOs should be designed to anneal to sequences flanking candidate hotspot regions. Our protocol allows amplicons between 6 and 10 kb in length to be routinely analyzed from pollen genomic DNA. The analysis described in Subheading 3.3 should be used to predict suitable locations for pollen typing.

3.4.2 Allele-Specific Oligonucleotide Selection

ASOs must strongly discriminate between the polymorphic parental templates during PCR amplification (Fig. 3). ASOs should be designed such that their 3′-ends anneal to polymorphic sites, where mismatches will have the greatest effect on the extension of the probe (Table 1). ASOs to be used as pairs in crossover analysis need to amplify equally efficiently from their specific template, at a given melting temperature (T_m) (Fig. 3). The following features should be considered during ASO design.

Table 1
Sequences of pollen typing allele-specific and universal oligonucleotides

Hotspot	ASO/UO	Name	Sequence (5′-3′)	Length	GC %	Mismatch	Poly-dC
3a	ASO	6339—ColF 1st	GAGAAACCAACCACTTCT	18	44	–	–
3a	ASO	6401—LerR 1st	CAAGGATTCTCTATATCTAC	20	35	–	–
3a	ASO	6401—ColR 1st	GGGGATTCTCTATATCTAGGA	21	43	–	–
3a	ASO	6341—ColF 2nd	CCCCCTTTCAAATTGATACAACAA	24	38	–	Yes
3a	ASO	6399—LerR 2nd	CCCAAGTTTTCTTCTCAAGCCA	22	45	Yes	Yes
3a	ASO	6399 – ColR 2nd	CCCAAGTTTTCTTCTCAAGCCT	22	45	Yes	Yes
3a	UO	3a-UF	GACGCTAGGCGCTGGTAAG	19	63	–	–
3a	UO	3a-UR	CTCGACCGGGGTACACCATC	20	65	–	–
3b	ASO	KC156—LerF 1st	AGAAAATAGATAACTCTTCG	20	30	–	–
3b	ASO	KC167—ColF 1st	CTTCTTGATTCACACCTTA	19	37	–	–
3b	ASO	KC166—ColR 1st	ATCGAATTCGCGACTAG	17	47	Yes	–
3b	ASO	KC160—LerF 2nd	CCCACTAAGACACAAATAC	19	42	–	Yes
3b	ASO	KC168—ColF 2nd	AATGTAATGCTCTGCTCC	18	44	–	–
3b	ASO	KC152—ColR 2nd	GCTTTGAAAATTCTTGTC	18	33	Yes	–
3b	UO	3b-UF	GCGGTACACCTCATGTCTAC	20	55	–	–
3b	UO	3b-UR	GGATTCCGCTGCTTCAAGTC	20	55	–	–
RAC1	ASO	KC459—LerF 1st	CTGACTTGAGTGATCGCAA	19	47	–	–
RAC1	ASO	KC493—ColF 1st	AAAACGTGCAACCTAAGAAC	20	40	–	–

RAC1	ASO	KC418—ColR 1st	ATTTCACCCGATGTAGTCC	19	47	–	–
RAC1	ASO	KC465—LerF 2nd	GTGGCCGCAAGCAAAAATAT	20	45	–	–
RAC1	ASO	KC495—ColF 2nd	AACAGATTGGTCTCATTG	18	39	–	–
RAC1	ASO	KC417—ColR 2nd	TAGTTTTTCTGACCCCAC	18	44	–	–
RAC1	UO	RAC1-UF	CTACAAAGCCATATAGCCAGAGCC	24	50	–	–
RAC1	UO	RAC1-UR	GATTCACTAAAGGAGAGCCAAAGTT	25	40	–	–

These ASOs have been successfully used to analyze crossover hotspots by pollen typing in *Arabidopsis* [11, 12, 24]. The relevant Col-0/Ler-0 polymorphic positions are highlighted in ASOs in *red*. *Blue* indicates universal mismatches added to improve amplification specificity. *Green* cytosines (Poly-dC) were added to the 5′ end of primers to increase amplification efficiency

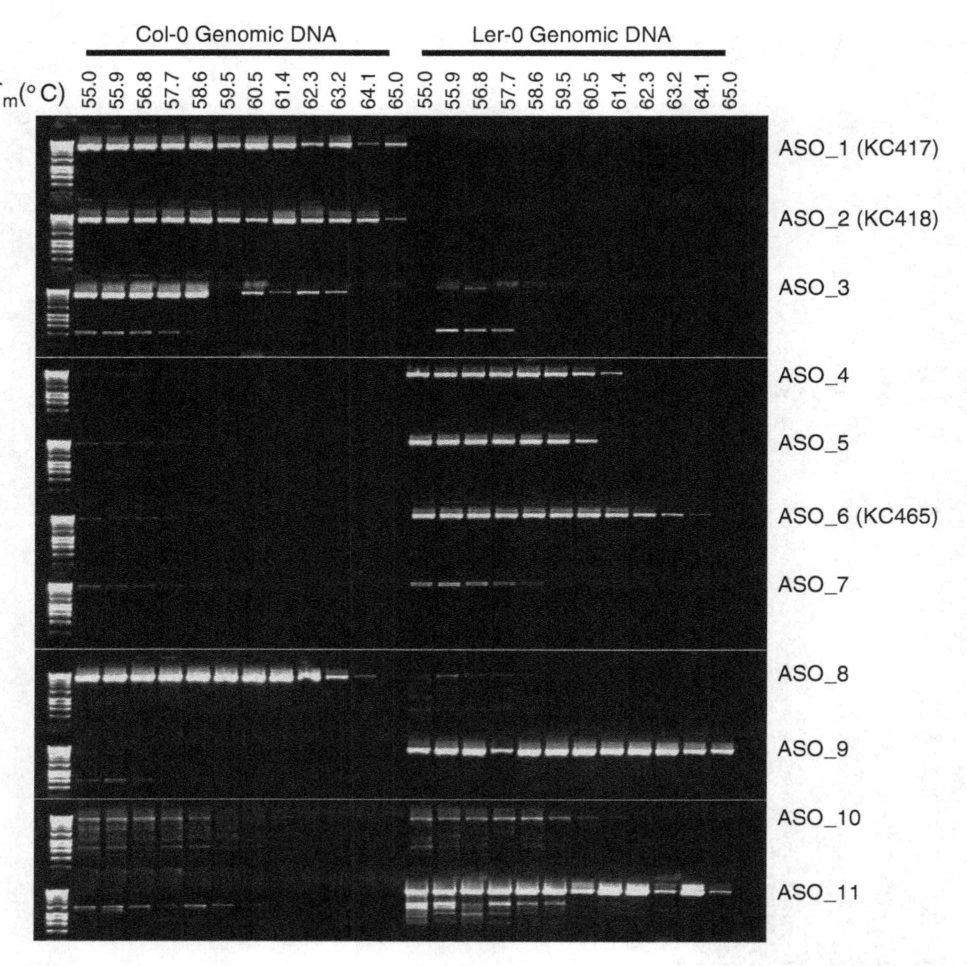

Fig. 3 Optimizing allele-specific PCR amplification. PCR amplification products from ASO optimization experiments were separated by electrophoresis using a 1 % agarose gel containing ethidium bromide. Each ASO was used with a second universal oligonucleotide (UO) at a range of melting temperatures, with either Col-0 or Ler-0 genomic DNA templates. ASO_1 and ASO_2 are successful, showing high template specificity (Table 1). ASO_6 is also successful, though only when 63 °C is used as an annealing temperature (Tables 1 and 2). ASO_3, 4, 5, 8, 9, 10 and 11 are not suitable for pollen typing, due to poor template specificity or amplification of non-specific products

1. Design ASOs to anneal to polymorphic indel sequences between 5 and 1,000 base pairs in length, located in non-repetitive sequence.
2. Alternatively, if no suitable indels are available, identify a site with multiple adjacent SNPs.
3. Design ASOs with 30–50% GC content, a length between 16 and 21 nucleotides (nt) and an estimated T_m between 57 and 62 °C. These parameters can be calculated using websites such as http://www.basic.northwestern.edu/biotools/

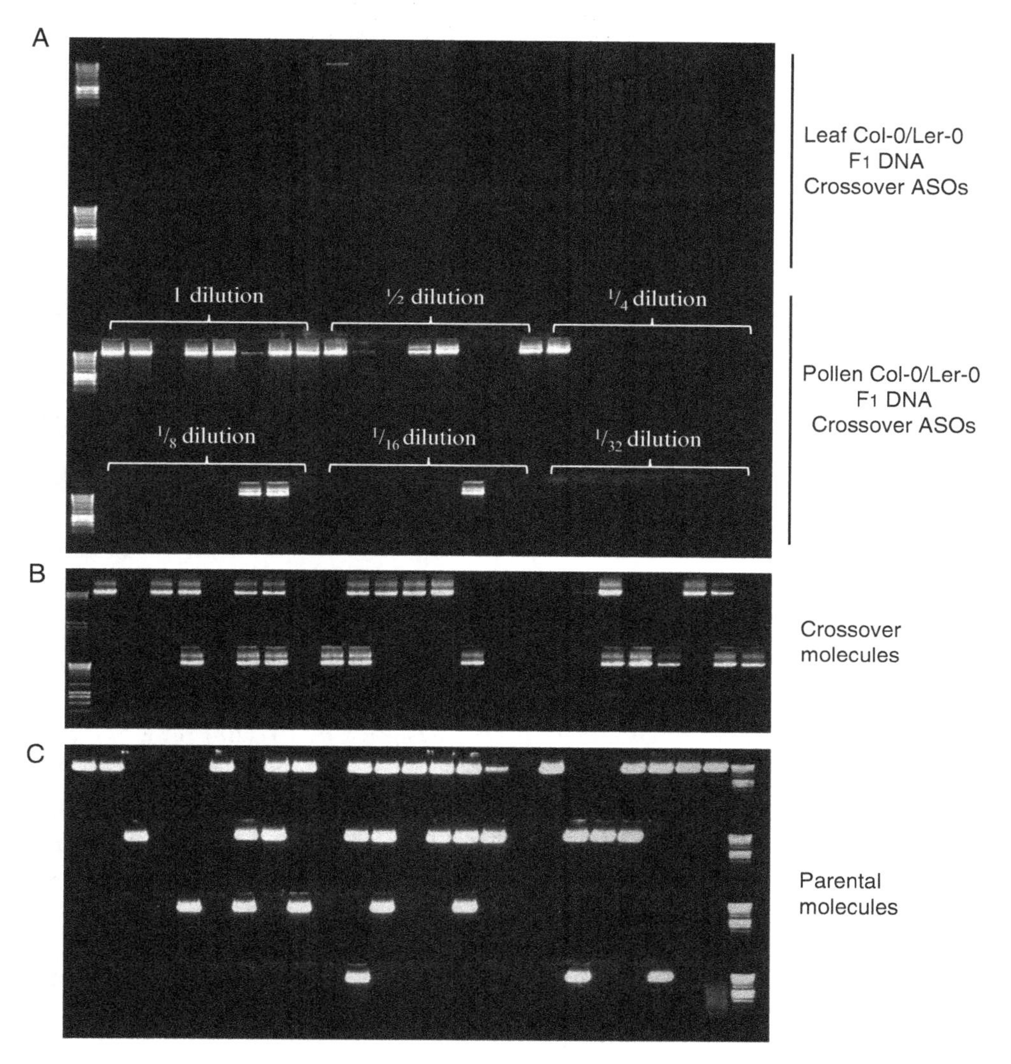

Fig. 4 Amplification and titration of single crossover and parental molecules from pollen genomic DNA. (**a**) Crossover PCR amplification products using nested pairs of ASOs were separated by gel electrophoresis using an agarose gel containing ethidium bromide. Amplification of crossover molecules is evident from Col-0/Ler-0 F_1 pollen genomic DNA, but not Col-0/Ler-0 F_1 leaf genomic DNA. 8 reactions were performed for each DNA dilution analyzed (Subheading 3.6, **steps 1–16**). (**b**) Representative 1 % agarose-ethidium bromide gel showing single crossover amplifications from 48 reactions, where approximately half (25 of 48) are negative (Subheading 3.6, **step 29**). (**c**) Representative 1 % agarose-ethidium bromide gel showing single parental molecule amplifications in a dilution series of 24 reactions. 6.5 μl of a 1:2000 dilution of 10 ng/μl pollen DNA solution was added to a PCR master mix. A dilution series of 1:1, 1:2 and 1:4 were made as described (Subheading 3.6, **step 34**). 24 PCR reactions were analyzed for each dilution

OligoCalc.html. Shorter ASOs (e.g., 17 nt) should have higher GC content (e.g., 45 %) and longer ASOs (e.g., 20 nt) should have a lower GC (30 %), in order to optimize annealing temperature.

4. Avoid ASOs predicted to form dimers or secondary structures, which can also be identified using the above oligo calculation website. Additional mismatches to the template can be added to ASOs to mitigate primer dimer formation and secondary structure, as required.
5. An additional mismatch (e.g., underlined bold T) close to the ASO 3′ end (-2 or -3 position) can be added in order to increase ASO specificity (e.g., ATCGAATTCGCGAC**T**A**G**). This can be useful for short ASOs with high GC content.
6. 3-6 poly-dC (e.g., underlined CCC) or poly-dG can optionally be added to the 5′-end of ASOs (e.g., **CCC**ACTAA-GACACAAATA**C**). This increases annealing temperature and can improve amplification efficiency, which is particularly useful for short ASOs with low GC content.

3.5 Optimisation of ASO Specificity and Efficiency

It is critical to experimentally test ASO template specificity and amplification efficiency using gradient PCR. These steps should be repeated until at least two nested pairs of high quality ASOs are obtained per hotspot (*see* Introduction to Subheading 3.4) (Fig. 1). For these assays universal, non-allele-specific oligonucleotides (UO) should also be designed in positions close to the ASOs. UOs should be designed with annealing temperatures above 65 °C, such that the T_m of the ASO has the greater effect on amplification efficiency.

1. Dilute leaf genomic DNA for the two *Arabidopsis thaliana* accessions (e.g., homozygous Col-0 and homozygous Ler-0), which are to be used in pollen typing analysis, to 5 ng/μl with TE.
2. Prepare a cocktail for 13 PCR reactions in a 1.5 ml Eppendorf tube, in order to test a single ASO using gradient PCR in 96-well plate, using DNA from the first accession (e.g., Col-0), as described below.

Components	Volume (μl)	13 Reactions
Genomic DNA (5 ng/μl)	0.2	2.6
10× Ex Taq buffer	1	13
dNTP, each 2.5 mM	1	13
ASO (5 μM)	1	13
UO (universal oligo) (5 μM)	1	13
Ex Taq (5 U/μl)	0.05	0.65
Nuclease-free water	5.75	74.75
Total	10	130

3. Prepare an additional 13 PCR reactions to test the same ASO with genomic DNA from the other *Arabidopsis thaliana* accession (e.g., Ler-0).
4. Transfer 10 μl of the reaction cocktail to each of 24 wells (e.g., A1-A12 for Col-0, B1-B12 for Ler-0) in a 96-well plate, using an automatic pipette.
5. Run a gradient PCR in a thermal cycler using the following parameters:

   1 cycle of,

   94 °C, 2 min;

   30 cycles of,

   94 °C, 30 s,

   55–65 °C (gradient), 30 s,

   68 °C (1 min/kb);

   1 cycle of 68 °C, 1 min/kb +1 min

   Hold at 4 °C.
6. Add 2.5 μl of 5× DNA loading buffer to each well and analyze the PCR products using gel electrophoresis with a 1.2% agarose gel and ethidium bromide staining (Fig. 3).
7. Assess whether ASOs are able to amplify PCR products efficiently and specifically from one parental template (e.g., Col-0) and not the other (e.g., Ler-0) (Fig. 3). A successful ASO will amplify a single band only from one parental template.
8. For ASOs that show PCR amplification specificity at higher temperatures (e.g., 59–65 °C), but produce weak or non-specific PCR bands at lower temperatures (e.g., 55–58 °C), further optimisation is required by either changing ASO length, adding mismatches and/or poly-dC (*see* Subheading 3.4.2), or redesigning ASOs at an alternative location.

3.6 Titration of Single Crossover and Parental Molecules

In order to estimate the recombination rate of a given hotspot it is necessary to estimate the concentration of crossover (recombinant) and parental (non-recombinant) molecules via titration (Fig. 4) [18, 23].

1. Prepare the first reaction cocktail using F_1 pollen genomic DNA (e.g., Col-0/Ler-0 F_1) for 20 PCR reactions and mix well by pipetting ten times. Depending on the amount of pollen extracted from, the concentration of F_1 pollen genomic DNA is typically between 10 and 50 ng/μl (*see* **Note 2**).
2. Prepare a second reaction cocktail without DNA template sufficient for 50 reactions and transfer 100 μl to each of five 1.5 ml tubes (1st, 2nd, 3rd, 4th, 5th), which will be used for serial twofold dilutions.

Components	Volume (μl)	20 Reactions	50 Reactions
Template	–	–	0
10× Ex Taq buffer	1	20	50
dNTP, each 2.5 mM	1	20	50
ASO 1st F (5 μM)	1	20	50
ASO 1st R (5 μM)	1	20	50
Ex Taq (5 U/μl)	0.05	1	2.5
Nuclease-free water	5.95	–	297.5
Total	10	200	Each 100 μl

3. Transfer 100 μl from the first reaction cocktail to the 1st dilution tube and mix in order to achieve a twofold dilution.
4. Transfer 100 μl from the 1st dilution tube to the 2nd dilution tube and mix well to achieve a twofold dilution.
5. In the same way dilute into the 3rd, 4th, and 5th tubes in order to achieve 8-, 16- and 32-fold dilutions, by transferring 100 μl from the 2nd, 3rd, and 4th dilution tubes respectively.
6. Transfer 10 μl from first reaction cocktail to eight wells (e.g., A1-H1) of a 96-well plate.
7. Transfer 10 μl from the 1st, 2nd, 3rd, and 4th dilution tubes to eight wells each (e.g., A2-H2, A3-H3, A4-H4, A5-H5 and A6-H6).
8. Repeat **steps 1–7**, but using the same amount of Col/Ler F_1 leaf genomic DNA in the remaining wells (e.g., A7-H12). In this DNA sample both parental allelic templates are present, but crossovers should be absent. This serves as a negative control for PCR artifacts, including template switching.
9. Run the first round of touchdown PCR in a thermal cycler using the following parameters (*see* **Note 3**):

   1 cycle of,

   94 °C for 2 min 30 s;

   8 cycles of,

   94 °C for 30 s, 4 °C higher than ASO annealing temperature (decrease by 0.5 °C per cycle), for 45 s,

   68 °C for 1 min/kb + 30 s;

   22–25 cycles of,

   94 °C for 20 s,

   ASO annealing temperature (°C) for 30 s,

   68 °C for 1 min/kb + 30 s;

1 cycle of,

68 °C 1 min/kb + 1 min;

Hold at 4 °C.

10. Dilute the PCR products 20-fold by adding 190 μl of 10 mM Tris–HCl (pH 8) to each well.
11. Prepare a reaction cocktail for a 96-well PCR plate for the second round of ASO PCR, as below.

Components	Volume (μl)	100 Reactions
Diluted first PCR product	1	Each
10× Ex Taq buffer	1	100
dNTP, each 2.5 mM	1	100
2nd ASO F (5 μM)	1	100
2nd ASO R (5 μM)	1	100
Ex Taq (5 U/μl)	0.05	5
Nuclease-free water	4.95	495
Total	10	Each 9

12. Aliquot 9 μl of the reaction cocktail to each well using an automatic pipette.
13. Transfer 1 μl of the diluted 1st round PCR plate to each well of the 2nd round PCR plate using an eight channel pipette.00
14. Run the second round of touchdown PCR in a thermal cycler using the following parameters:

1 cycle of,

94 °C for 2 min 30 s;

8 cycles of 94 °C for 30 s,

04 °C higher than ASO annealing temperature (decrease by 0.5 °C per cycle) for 45 s,

68 °C for 1 min/kb + 30 s;

22–25 cycles of,

94 °C for 20 s,

ASO annealing temperature (°C) for 30 s,

68 °C for 1 min/kb + 30 s,

1 cycle of,

68 °C 1 min/kb + 1 min,

Hold at 4 °C.

15. Add 2.5 μl of 5× DNA loading buffer to each well and analyze the PCR products by gel electrophoresis using a 1.2 % agarose gel and ethidium bromide staining (Fig. 4).

16. Repeat **steps 1–15** to test different ASO combinations and optimize annealing temperatures and cycle numbers in the first and second rounds of PCR. Repeat until strong amplification bands appear specifically when F_1 pollen genomic DNAs are used as a template, but which are absent when leaf genomic DNA is used (Fig. 4a). Optimizing this step is critical for the analysis of single crossover molecules (*see* **Note 4**). ASOs and PCR parameters are listed in Table 2 for analysis of different *Arabidopsis* hotspots.
17. Estimate the concentration of amplifiable crossover molecules, based on the dilution series. Identify a DNA concentration where approximately half of the reactions are negative and half are positive (Fig. 4a). The concentration of crossover molecules (c1) can be estimated as follows.

$$c1 = -\ln(N / T) / V,$$

where N is the number of negative reactions at a given dilution,

T is the total number of reactions at a given dilution,

V is the pollen DNA volume (μl) per reaction, and

ln is the natural logarithm.

For example, consider a case in which the 3rd dilution of eight wells (e.g., A4-H4) had four positive and four negative wells and 0.14 μl of input DNA was used for each undiluted reaction. In the 3rd dilution wells, there is 0.0175 μl of template DNA. Therefore, $c1 = -\ln(4/8)/(0.14/8) = 39.61$ crossovers/μl. If the 4th dilution of the same DNA had 1 positive band, then $c1 = -\ln(7/8)/(0.14/32) = 30.52$ crossovers/μl. Therefore, the input pollen DNA has roughly 35.07 crossovers/μl. Increasingly accurate estimates of crossover concentrations should then be determined by increasing the number of reactions with varying amounts of input DNA, as described below.

18. Estimate the volume of input DNA that produces approximately half negative wells from multiple independent PCR reactions (e.g., 24–96 wells), using the preliminary estimation of crossover concentration (c1) from **step 17**. The volume of DNA input in each reaction is $= -\ln(0.5) \times (1/c1)$. For example, if c1 from eight reactions is 32.88 crossovers/μl, then add $-\ln 0.5 \times (1/32.88)$, or 0.021 μl, to each of the 24 reactions.

Prepare four PCR reaction cocktails (a-d), each for 25 reactions, with varying DNA template input amounts, as below.

Table 2
Pollen typing PCR parameters for analysis of *Arabidopsis* hotspots

Hotspot	Molecules	First PCR	Annealing temperature	Second PCR	Annealing temperature
3a	Parental	6339—Col F 1st 6401—Col R 1st	59–55 °C (−0.5 °C/cycle)	6341—Col F 2nd 6399 – Col R 2nd	59–55 °C (−0.5 °C/cycle)
3a	Crossover	6339—Col F 1st 6401—Ler R 1st	59–55 °C (−0.5 °C/cycle)	6341—Col F 2nd 6399—Ler R 2nd	59–55 °C (−0.5 °C/cycle)
3b	Parental	KC167—Col F 1st KC166—Col R 1st	59–55 °C (−0.5 °C/cycle)	KC168—Col F 2nd KC152—Col R 2nd	57–53 °C (−0.5 °C/cycle)
3b	Crossover	KC156—Ler F 1st KC166—Col R 1st	59–55 °C (−0.5 °C/cycle)	KC160—Ler F 2nd KC152—Col R 2nd	57–53 °C (−0.5 °C/cycle)
RAC1	Parental	KC493—Col F 1st KC418—Col R 1st	65–63 °C (−0.4 °C/cycle)	KC495—Col F 2nd KC417—Col R 2nd	65–63 °C (−0.4 °C/cycle)
RAC1	Crossover	KC459—Ler F 1st KC418—Col R 1st	65–63 °C (−0.4 °C/cycle)	KC465—Ler F 2nd KC417—Col R 2nd	65–63 °C (−0.4 °C/cycle)

Components	Volume (μl)	25 Reactions
DNA template (*see* (a)–(d) below)	–	–
10× Ex Taq buffer	1	25
dNTP, each 2.5 mM	1	25
ASO 1st F (5 μM)	1	25
ASO 1st R (5 μM)	1	25
Ex Taq (5 U/μl)	0.05	1.25
Nuclease-free water	5.95	–
Total	10	250

(a) 1st reaction cocktail with volume of DNA template input according to **steps 17** and **18**

(b) 2nd reaction cocktail with $^1/_2$ DNA input volume of 1st reaction cocktail

(c) 3rd reaction cocktail with $^1/_4$ DNA input volume of 1st reaction cocktail

(d) 4th reaction cocktail with $^1/_8$ DNA input volume of 1st reaction cocktail

19. Transfer 10 μl from the 1st, 2nd, 3rd, 4th reaction cocktails to 24 wells (e.g., A1-H3, A4-H6, A7-H9 and A10-H12 respectively) of a 96-well plate.
20. Run the first round of allele-specific touchdown PCR (Table 2), as in **step 9**.
21. Dilute the PCR product from **step 21**, 20-fold by adding 190 μl of 10 mM Tris–HCl (pH 8) to each well.
22. Prepare a reaction cocktail for a 96-well PCR plate for the second round of allele-specific PCR, as in **step 11**.
23. Aliquot 9 μl of the reaction cocktail to each well, using an automatic pipette.
24. Transfer 1 μl of the diluted 1st round PCR plate (**step 22**) to each well of the 2nd round PCR plate, using an eight channel pipette.
25. Run the second round of allele-specific touchdown PCR as in **step 14**.
26. Add 2.5 μl of 5× DNA loading buffer to each well and analyze the PCR products using gel electrophoresis with a 1.2% agarose gel and ethidium bromide staining. Score the number of positive and negative amplifications (*see* **Note 4**).
27. Estimate the concentration (c2) of amplifiable crossover molecules, based on the dilution series of 24 reactions, as for **step 17**. For example, if there are 12 positive PCR bands in

the first set of 24 wells, when 0.019 µl input DNA was used for each reaction, $c2 = -\ln(12/24)/0.019 = 36.48$ crossovers/µl. If there were eight positive PCR bands in the second set of reactions, when 0.0095 µl input DNA were used for each reaction, $c2 = -\ln(16/24)/0.0095 = 42.68$ crossovers/µl. Here the mean c2 is 39.58 crossovers/µl.

28. Repeat **steps 18–27** with two sets of 48 reactions. For the first set, add the amount of input DNA which generates about half positive and half negative amplifications to each of the 48 wells, e.g., $-\ln(0.5) \times (1/c2) = 0.018$ µl. For the second set of reaction, use half this volume as input DNA, e.g., 0.009 µl, compared with the first 48 reactions.
29. Estimate the concentration (c3) of crossover molecules as in **step 17**.
30. Finally repeat **steps 18–27** with three sets of 96-well amplifications, using an input amount estimated from c3 to give half positive and half negative reactions.
31. Estimate the concentration (c4) of crossover molecules. The concentrations of crossover molecules from 48 and 3×96 reactions (c3 and c4) should be very similar between independent experiments. Whereas the c1 and c2 estimates from 8 and 24 reactions show greater variation.
32. Determine the final concentration of crossover molecules and standard deviation using the estimates c3 and/or c4 from multiple independent experiments with varying input amounts (Table 3).
33. Given that the genetic length of plant hotspots is typically between 0.1 to 0.5 cM [9, 11–13], the concentration of parental molecules is vastly higher than that of crossover molecules. Therefore, to estimate parental molecule concentrations it is necessary to repeat **steps 1–33** using 500 to 1000-fold diluted F_1 pollen genomic DNA input. Parental amplifications use two pairs of parental ASOs and/or UOs (Fig. 1 and Table 1).
34. Calculate crossover frequency (centiMorgans per megabase, cM/Mb) from the independent titration experiments (**steps 1–34**) using the following formula:

$$\text{cM} = (c / p) \times 100$$

c is the concentration of crossover molecules (**step 33**).

p is the concentration of parental molecules (**step 34**).

cM/Mb = cM/(bp of 2nd PCR amplicon/1,000,000)

3.7 Mapping Crossover Sites by Sanger Sequencing

1. Perform first and second rounds of allele-specific PCR in multiple (>3) 96-well plates to amplify single crossover molecules using F_1 pollen genomic DNA as a template. Estimate the amount of DNA template for each reaction

Table 3
Titration of crossover molecules. Average concentration of crossovers from c3a to c4b = 38.4 ± 2.65 crossovers/µl

Estimate	Number of positive wells (P)	Number of negative wells (N)	Number of total wells (T)	µl/well (V)	Crossovers/µl = −ln((N/T)/V)	Average crosssovers/µl	µl for half N/half P
c1a	4	4	8	0.0175	39.61		
c1b	2	6	8	0.00875	32.88	36.24	0.019
c2a	12	12	24	0.019	36.48		
c2b	8	16	24	0.0095	42.68	39.58	0.018
c3a	24	24	48	0.018	38.51		
c3a	15	34	48	0.009	38.32	38.41	0.018
c4a	45	51	96	0.018	35.14		
c4b	30	66	96	0.009	41.63	38.39	0.018

using 1/concentration of crossovers $\times$ -ln(0.5). This will lead to PCR amplifications where ~69 % are from single crossover molecules and ~31 % will be from two or more crossovers, $0.5 \times \ln(0.5)/-0.5 = 0.69$. This process should be repeated on a sufficient scale to isolate and sequence 100 or more independent crossovers per hotspot.

2. Add 2.5 μl of 5× DNA loading buffer directly to each well (total 12.5 μl) and run 7 μl of the PCR products using a 1.2 % agarose gel and ethidium bromide staining, in order to identify positive wells (Fig. 4b).
3. Transfer the remaining 5 μl of PCR product from the positive wells to a new 96-well plate.
4. Add 195 μl of 10 mM Tris–HCl (pH 8) to each well.
5. Amplify 1 μl of the diluted PCR products using the same conditions as for the second round of allele-specific PCR (Subheading 3.6, **step 14**).
6. Add 5 μl of the following cocktail (0.3 μl EXO I, 1.3 μl SAP, 3.4 μl D.W) to each well. Exonuclease (EXO) removes single-stranded DNA and shrimp alkaline phosphatase (SAP) dephosphorylates the PCR products to prepare for Sanger sequencing.
7. Incubate at 37 °C for 40 min and then 75 °C for 10 min.
8. Prepare Sanger sequencing reactions by transferring 2 μl of the exonuclease/SAP-treated DNA reactions to 8 μl of the following cocktail in a 96-well plate, or a 0.2 ml PCR tube as below.

Component	Amount per well (μl)
PCR product template	2
5× Sequencing buffer	2
BigDye (Applied Biosystems)	0.9
Primer (10 μM)	1
Nuclease-free water	4.1

9. Run the reactions in a thermal cycler with the following parameters:

   1 cycle of,

   96 °C for 30 s;

   27 cycles of,

   96 °C for 30 s,

   50 °C for 5 s,

   60 °C for 3 min;

   Hold at 4 °C.

10. Sequence the product using an ABI 3700 DNA sequencer, or similar machine, via a service provider.
11. Determine crossover sites by manually checking sequencing data chromatograms using visualization programs such as CLC Main Workbench and Chromas. If the PCR product is derived from two crossover molecules, double peaks will be evident in the chromatograms at polymorphic sites. This allows both crossover locations to be identified. However, if more than two crossover molecules are sequenced in a single reaction it is not generally possible to confidently identify recombination sites.
12. Sequence each crossover PCR product with additional primers until the recombination site is confirmed by observing a change in SNP genotype from one accession to the other (e.g., Col-0->Ler-0). Note this change should match the orientation of the ASOs used in the first and second rounds of allele-specific PCR (Fig. 1). Gene conversion associated with crossovers may also lead to multiple genotype switches in vicinity to a crossover.
13. Calculate the recombination activity in each SNP interval within the amplicon as follows:

$$\text{cM} / \text{Mb} = cM \times (N / T) / (L / 1{,}000{,}000)$$

cM: concentration of crossover molecules/concentration of parental molecules × 100 (Subheading 3.6., **step 35**).

L: Length of interval (bp).

N: Crossover numbers in the interval being considered.

T: Total number of mapped single crossovers.

Representative data for the *3a* hotspot are shown in Table 4.

3.8 Pollen Typing Quantitative PCR

Titration experiments are necessary for accurate hotspot crossover frequency estimation and to isolate single crossover molecules for Sanger sequencing. However, in many cases it is necessary only to estimate hotspot recombination rates during preliminary experiments, or when testing conditions that may influence crossover frequency. For these purposes we describe a pollen typing quantitative PCR (qPCR) assay that is useful to compare relative crossover rates between wild type and mutants, or different environmental conditions (Fig. 1). For these assays the input pollen genomic DNA amount and allele-specific PCR cycle number need to be optimized, according to the hotspot being studied. Crossover and parental amplification products are quantified from pollen and leaf F_1 genomic DNA, which either possess or lack crossovers respectively (Fig. 5).

Table 4
Crossover frequency in each interval within the 3a hotspot. The SNP position (bp) is relative to the TAIR10 Col-0 reference sequence. These data are reproduced from [12]

Chr 3	SNP genotype				
Position	Col-0	Ler-0	Length (bp)	Crossovers	cM/Mb
634,109	A	G	829	11	28.57
634,938	C	T	1,181	1	1.82
636,119	A	C	80	0	0
636,199	T	A	1,084	16	31.78
637,283	T	A	94	2	45.81
637,377	–	A	377	8	45.69
637,754	A	G	220	3	29.36
637,974	A	G	509	22	93.06
638,483	A	T	150	1	14.35
638,633	T	A	6	0	0
638,639	C	T	32	0	0
638,671	C	–	6	0	0
638,677	C	T	2	0	0
638,679	A	G	2	0	0
638,681	A	–	6	0	0
638,687	–	T	92	0	0
638,779	–	AT	885	26	63.25
639,664	C	A	210	6	61.52
639,874	A	G	60	2	71.77
639,934	A	T			
Total			5,825	98	36.22
cM					0.21

1. Prepare the first round of PCR reactions (a–f) using F_1 pollen genomic DNA equivalent to approximately 20 or 30 crossover molecules per reaction (estimated from Subheading 3.6. **Steps 1–34**), as below.

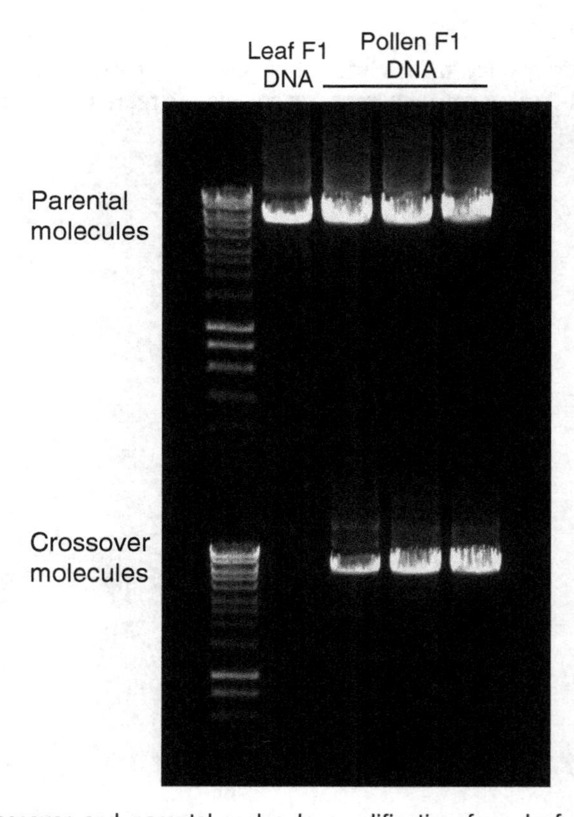

Fig. 5 Crossover and parental molecule amplification from leaf and pollen genomic DNA following qPCR analysis. A representative 1 % agarose ethidium bromide stained gel showing nested crossover and parental PCR products. Here the products of the first round of allele-specific PCR performed during qPCR analysis are used as templates for a second round of allele-specific PCR (Subheading 3.8). Crossover molecules are specifically amplified from pollen and not leaf F_1 genomic DNA

Components	Volume (μl)
F_1 pollen genomic DNA (20 or 30 crossovers) or F_1 leaf genomic DNA	–
10× Ex Taq buffer	2
dNTP, each 2.5 mM	2
1st ASO F (5 μM)	2
1st ASO R for COs or ASO/UO R for parentals (5 μM)	2
Ex Taq (5 U/μl)	0.1
Nuclease-free water	–
Total	20

(a) F_1 leaf DNA, ASO Col F/Col R or UR for parental.

(b) F_1 pollen DNA, ASO Col F/Col R or UR for parental.

(c) Mutant F_1 pollen DNA, ASO Col F/Col R or UR for parental.

(d) F_1 leaf DNA, ASO Col F/Ler R for crossover.

(e) F_1 pollen DNA, ASO Col F/Ler R for crossover.

(f) Mutant F_1 pollen DNA, ASO Col F/Ler R for crossover.

2. Run the first round of PCR as for single molecule amplifications (Subheading 3.6. **step 9**), but with a reduced number of PCR cycles (e.g., 16–25) (*see* **Note 5**).
3. Dilute the first PCR product 20-fold for crossover molecules and 1000–2000-fold for parental molecules, using 10 mM Tris–HCl (pH 8).
4. Use the diluted PCR products as templates for qPCR reactions with a pair of universal forward and reverse primers (UO F, UO R) (Table 5).

Components	Volume (μl)
Diluted first PCR products	2
10× SYBR green	2
10× DNA Taq buffer	2
dNTP, each 2.5 mM	2
UO F (5 μM)	2
UO R (5 μM)	2
Go Taq (5 U/μl)	0.2
Nuclease-free water	7.8
Total	20

Table 5
Sequences of universal oligonucleotides used for qPCR analysis

Hotspot	Orientation	Sequence (5'–3')
3a	Forward	GTTCAAGCTTAAAGGGAAATCG
3a	Reverse	GCCCATGACTCGGTGTAAAT
3b	Forward	GACGAGCTCGAAGGAATCAG
3b	Reverse	ATGCTGAGGCCTATCATTGC
RAC1/RPP9	Forward	GCCGATTTGATGCCTCACTT
RAC1/RPP9	Reverse	AAGGAATCCAGGATCGCTGT

5. Run real time thermal cycler analysis using the following parameters:

   1 cycle of,

   95 °C for 30 s;

   27 cycles of,

   95 °C for 30 s,

   60 °C for 20 s,

   72 °C for 30 s (detection);

   1 cycle of,

   72 °C for 1 min;

   Hold at 4 °C.

6. The diluted products of the first round of allele-specific PCR from **step 3** can also be used as templates for a second round of allele-specific PCR (as in Subheading 3.6. **steps 11–14**), in order to test specificity of crossover amplifications from pollen and leaf genomic DNA (Fig. 5).

7. Calculate the relative crossover amplification enrichment from F_1 pollen genomic DNA (sample) compared to F_1 leaf DNA (control) by using the $2^{-\Delta\Delta Ct}$ method, as below [45]. The relative fold change between sample and control is calculated as:

   $$\text{Relative fold change} = 2\text{^}-((C-P)-(\text{CL}-\text{PL})).$$

   C: Ct from crossover amplification from F_1 pollen DNA.

   P: Ct of parental amplification from F_1 pollen DNA.

   CL: Mean Ct of crossover amplification from F_1 leaf DNA.

   PL: Mean Ct of parental amplification from F_1 leaf DNA.

Here, Ct (the threshold cycle) reflects the PCR cycle number at which fluorescence from accumulating amplicon products within a reaction crosses the threshold above background signal [45]. Therefore, in Table 6 the fold change of wild type crossovers compared between pollen and leaf genomic DNA is $2\text{^}-((14.62-27.69)-(22.78-27.49)) = 328.56$.

3.9 High Throughput Crossover Sequencing Library Construction

In order to sample greater numbers of crossovers than is feasible from Sanger sequencing of single molecule amplification products, we have developed a method to mass amplify crossovers and generate high-throughput sequencing libraries (Fig. 1). Here, multiple independent PCR amplifications are performed, each containing an estimated 1–3 crossover template molecules, based on previous titrations experiments. Low numbers of estimated crossovers are used per reaction in order to evenly amplify from a large number of independent molecules.

Table 6
Crossover and parental molecule amplification from leaf and pollen genomic DNA measured using qPCR

ASO combination	Template	Code	Ct	Mean Ct	Fold change	Fold change mean	Fold change St. Dev.
Parentals	F_1 leaf	a	27.4	27.49			
	F_1 leaf		27.16				
	F_1 leaf		27.9				
	F_1 pollen wild type	b	27.69				
	F_1 pollen wild type		27.69				
	F_1 pollen wild type		27.67				
	F_1 pollen *arp6*	c	27.65				
	F_1 pollen *arp6*		27.66				
	F_1 pollen *arp6*		27.61				
Crossovers	F_1 leaf	d	23.03	22.78	0.79	1.04	0.36
	F_1 leaf		22.65		0.87		
	F_1 leaf		22.65		1.45		
	F_1 pollen wild type	e	14.62		328.56	363.65	52.99
	F_1 pollen wild type		14.58		337.79		
	F_1 pollen wild type		14.23		424.61		
	F_1 pollen *arp6*	f	15.53		170.07	195.86	23.27
	F_1 pollen *arp6*		15.29		202.25		
	F_1 pollen *arp6*		15.15		215.27		

Relative crossover enrichment from allele-specific amplication of leaf and pollen Col/Ler F_1genomic DNA, comparing between wild type and *arp6* mutants using the $2^{-\Delta\Delta Ct}$ method (Subheading 3.8, **steps 1**–7). a–f indicate the template and primer combination used, as described in Subheading 3.8, **step 1**. Three technical qPCR replicates were performed using first PCR products from reactions a-f as templates. These data are one of the biological replicates published in [12]

1. Prepare eight 96-well PCR plates for the first round of allele-specific PCR, as below. Based on the previous titration of crossover molecules (Subheading 3.6, **steps 1–35**), add the amount of pollen genomic DNA to each well, such that each contains an estimated 1–3 crossovers (e.g., 1/concentration of crossovers $\times -\ln(0.35)$).

Components	Volume (μl)	100 reactions
F_1 pollen genomic DNA (1–3 crossovers)	–	–
10× Ex Taq buffer	1	100
dNTP, each 2.5 mM	1	100
1st ASO F (5 μM)	1	100
1st ASO R (5 μM)	1	100
Ex Taq (5 U/μl)	0.05	5
Nuclease-free water	–	–
Total	10	Each 10

2. Run the first round of allele-specific PCR as for single crossover molecule amplifications (Subheading 3.6, **step 9**).
3. Dilute the PCR products 20-fold by adding 190 μl of 10 mM Tris–HCl (pH 8) to each well.
4. Prepare a reaction cocktail for eight 96-well PCR plates for the second round of allele-specific PCR, as below.

Components	Volume (μl)	100 Reactions
Diluted first PCR product	1	Each
10× Ex Taq buffer	1	100
dNTP, each 2.5 mM	1	100
2nd ASO F (5 μM)	1	100
2nd ASO R (5 μM)	1	100
Ex Taq (5 U/μl)	0.05	5
Nuclease-free water	4.95	495
Total	10	Each 9

5. Aliquot 9 μl to each well using an automatic pipette.
6. Add 1 μl of diluted 1st round PCR amplifications to each well of the 2nd round PCR plate, using an eight channel pipette.
7. Seal the plate and mix by centrifuging at 500 × *g* for 1 min.
8. Run the second round of PCR, as for single molecule crossover amplifications (Subheading 3.6, **step 14**).
9. Add 2.5 μl of 5× DNA loading buffer to 12 wells of each 96-well plate and analyze 5 μl of the PCR product using a 1.2 % agarose gel containing ethidium bromide, to confirm the presence of amplification products.

10. Pool the second round PCR products from each 96-well plate (900 μl) in a 2 ml tube.
11. Add 0.1 volume of 3 M sodium acetate and 1 vol of isopropanol.
12. Incubate at −20 °C for 2 h.
13. Centrifuge the samples at 16,000 ×*g* for 30 min at 4 °C.
14. Remove the supernatant, wash the pellet with 1 ml of 70% EtOH, centrifuge again for 5 min at 4 °C.
15. Remove the supernatant and dry the pellet at room temperature for 5 min.
16. Dissolve the pellets in 100 μl of TE and store at −80 °C.
17. Analyze the PCR products using gel electrophoresis with a 1.2% agarose and ethidium bromide staining.
18. Excise the amplification product band of expected size from the gel.
19. Purify the PCR products using a gel extraction kit.
20. Measure purified DNA amount using a fluorometer and dilute to 20 ng/μl.
21. Sonicate 2 μg of DNA in 100 μl of TE in a 1.5 ml tube to 300–500 bp by using a Bioruptor (high setting, 30 s ON–30 s OFF for 15 min). The size of DNA fragmentation can be varied according to distribution of polymorphisms. For example, if polymorphisms are relatively sparse in the analyzed region then using longer DNA fragments may be beneficial.
22. Analyze the PCR products by gel electrophoresis using a 2% agarose gel containing ethidium bromide.
23. Excise the 300–400 bp region from the gel containing the sonicated crossover PCR products.
24. Purify the PCR products using a gel extraction kit.
25. Measure the purified DNA amount using a fluorometer.
26. Generate a DNA sequencing library, according to the manufacturer's instructions.
27. Sequence the library using paired-end reads of 150 bp length, for example, using an Illumina miSeq or NextSeq500 instrument.

3.10 Analysis of Pollen Typing High Throughput Crossover Sequencing Data

1. It is first necessary to perform Sanger sequencing of the two parental haplotypes analyzed during pollen typing (e.g., Col-0 and Ler-0). Use these data to generate two FASTA files containing the sequence of the two parents within the amplified region. A suggested data analysis pipeline is described below.
2. Build a bowtie index for both sequences, or similar file for alternative alignment methods [46].

3. Trim adapter sequences from reads. Suitable programs include FastX or Trimmomatic (http://hannonlab.cshl.edu/fastx_toolkit/index.html) [47].
4. Paired-end reads should be separated into forward read and reverse read FASTQ files.
5. Align each read set to the Col-0 and Ler-0 amplicon sequences from **steps 1** and **2**, allowing only uniquely matching alignments. Suitable programs include BWA, bowtie or bowtie2 [46, 48].
6. This will result in four alignment BAM or SAM files (end1.col, end1.ler, end2.col, end2.ler), which should be sorted and indexed using samtools [49].
7. Use Rsamtools to input alignment data into R [50].
8. Filter for reads pairs, i.e., those with matching read names, where one read uniquely aligns to the Col-0 sequence and one uniquely aligns to the Ler-0 sequence. The read genotypes and alignment coordinates should match the ASO configuration used during allele-specific PCR amplification (Fig. 1).
9. Use the filtered read pairs for analysis, normalized by the total number of mapped, filtered reads. Crossover values can also be assigned to all intervening base pairs between the innermost polymorphisms present in paired crossover read alignments, weighted by interval sizes.

4 Notes

1. The quality of DNA obtained is strongly dependent on the vortexing step. Vortexing for longer periods increases the number of disrupted pollen grains, which increases DNA yield. However, vortexing also increases shearing of DNA. This is a problem due to the long length of pollen typing amplicons (5–10 kb). The vortexing step should be optimized for local laboratory equipment and conditions in order to maximize crossover amplification from pollen genomic DNA.
2. The final concentration of pollen genomic DNA is less important than the quantity of amplifiable crossover molecules. As crossover and parental amplicons are large, for example ~10 kb, it is important that DNA molecules remain as intact as possible. This is evident by a clear high molecular weight band when DNA is analyzed by agarose gel electrophoresis and ethidium bromide staining (Fig. 2). The number of amplifiable molecules is increased by careful manipulation of pollen genomic DNA. For example, never vortex or vigorously mix the DNA and resuspend pellets gently by hand.

3. Frequently ASOs show template specificity only at higher temperatures (e.g., >60.5) and not lower temperatures (e.g., <59.5) during gradient PCR optimisation (e.g., ASO_6 in Fig. 3). In this case ensure to use higher annealing temperatures during all amplifications.
4. PCR conditions, including cycle number, should be carefully optimized, as described in **steps 1–16** of Subheading 3.6, using pollen and leaf F_1 genomic DNA. Conditions should be identified where no amplification is observed from leaf F_1 DNA templates, after for example, 35 PCR cycles. Therefore, when amplifying crossover molecules from pollen F_1 DNA under the same conditions, any observed amplification is classified as positive and an absence of amplification is classified as negative.
5. It is important to use a minimal number of PCR cycles (e.g., 16–25), otherwise saturation of amplified PCR products can mask differences between samples. Varying first PCR cycle numbers should be tested until an approximately a 30–500-fold enrichment of crossovers is detected from F_1 pollen DNA, compared to F_1 leaf DNA (Subheading 3.8, **step 7** and Table 6).

References

1. Osman K, Higgins JD, Sanchez-Moran E et al (2011) Pathways to meiotic recombination in Arabidopsis thaliana. New Phytol 190:523–544
2. Kauppi L, Jeffreys AJ, Keeney S (2004) Where the crossovers are: recombination distributions in mammals. Nat Rev Genet 5:413–424
3. Villeneuve AM, Hillers KJ (2001) Whence meiosis? Cell 106:647–650
4. Barton NH, Charlesworth B (1998) Why sex and recombination? Science 281:1986–1990
5. Keeney S, Neale MJ (2006) Initiation of meiotic recombination by formation of DNA double-strand breaks: mechanism and regulation. Biochem Soc Trans 34:523–525
6. De Massy B (2013) Initiation of meiotic recombination: how and where? Conservation and specificities among eukaryotes. Annu Rev Genet 47:563–599
7. Mercier R, Mézard C, Jenczewski E et al (2014) The molecular biology of meiosis in plants. Annu Rev Plant Biol. doi:10.1146/annurev-arplant-050213-035923
8. Baudat F, Imai Y, de Massy B (2013) Meiotic recombination in mammals: localization and regulation. Nat Rev Genet 14:794–806
9. Choi K, Henderson IR (2015) Meiotic recombination hotspots – a comparative view. Plant J 83:52–61
10. Coop G, Przeworski M (2006) An evolutionary view of human recombination. Nat Rev Genet 8:23–34
11. Yelina NE, Choi K, Chelysheva L et al (2012) Epigenetic remodeling of meiotic crossover frequency in Arabidopsis thaliana DNA methyltransferase mutants. PLoS Genet 8, e1002844
12. Choi K, Zhao X, Kelly KA et al (2013) Arabidopsis meiotic crossover hot spots overlap with H2A.Z nucleosomes at gene promoters. Nat Genet 45:1327–1336
13. Drouaud J, Khademian H, Giraut L et al (2013) Contrasted patterns of crossover and non-crossover at Arabidopsis thaliana meiotic recombination hotspots. PLoS Genet 9:e1003922
14. Tiemann-Boege I, Calabrese P, Cochran DM et al (2006) High-resolution recombination patterns in a region of human chromosome 21 measured by sperm typing. PLoS Genet 2:e70
15. Baudat F, de Massy B (2007) Cis- and trans-acting elements regulate the mouse Psmb9 meiotic recombination hotspot. PLoS Genet 3:e100
16. Cole F, Keeney S, Jasin M (2010) Comprehensive, fine-scale dissection of homologous recombination outcomes at a hot spot in mouse meiosis. Mol Cell 39:700–710

17. Berg IL, Neumann R, Sarbajna S et al (2011) Variants of the protein PRDM9 differentially regulate a set of human meiotic recombination hotspots highly active in African populations. Proc Natl Acad Sci U S A 108:12378–12383
18. Kauppi L, May CA, Jeffreys AJ (2009) Analysis of meiotic recombination products from human sperm. Methods Mol Biol. doi:10.1007/978-1-59745-527-5
19. Jeffreys AJ, Kauppi L, Neumann R (2001) Intensely punctate meiotic recombination in the class II region of the major histocompatibility complex. Nat Genet 29:217–222
20. Arbeithuber B, Betancourt AJ, Ebner T, Tiemann-Boege I (2015) Crossovers are associated with mutation and biased gene conversion at recombination hotspots. Proc Natl Acad Sci U S A 112:2109–2114
21. De Boer E, Jasin M, Keeney S (2015) Local and sex-specific biases in crossover vs. noncrossover outcomes at meiotic recombination hot spots in mice. Genes Dev 29:1721–1733
22. De Boer E, Jasin M, Keeney S (2013) Analysis of recombinants in female mouse meiosis. Methods Mol Biol 957:19–45
23. Drouaud J, Mézard C (2011) Characterization of meiotic crossovers in pollen from Arabidopsis thaliana. Methods Mol Biol 745:223–249
24. Yelina NE, Lambing C, Hardcastle TJ et al (2015) DNA methylation epigenetically silences crossover hot spots and controls chromosomal domains of meiotic recombination in Arabidopsis. Genes Dev 29:2183–2202
25. Khademian H, Giraut L, Drouaud J, Mézard C (2013) Characterization of meiotic non-crossover molecules from Arabidopsis thaliana pollen. Methods Mol Biol 990:177–190
26. Auton A, McVean G (2012) Estimating recombination rates from genetic variation in humans. Methods Mol Biol 856:217–237
27. Charlesworth B, Charlesworth D (2010) Elements of evolutionary genetics. Roberts and Company, Englewood, CO
28. Weigel D, Nordborg M (2015) Population genomics for understanding adaptation in wild plant species. Annu Rev Genet 49:315–338
29. Auton A, McVean G (2007) Recombination rate estimation in the presence of hotspots. Genome Res 17:1219–1227
30. Fearnhead P (2006) SequenceLDhot: detecting recombination hotspots. Bioinformatics 22:3061–3066
31. Drouaud J, Mercier R, Chelysheva L et al (2007) Sex-specific crossover distributions and variations in interference level along Arabidopsis thaliana chromosome 4. PLoS Genet 3:12
32. Giraut L, Falque M, Drouaud J et al (2011) Genome-wide crossover distribution in Arabidopsis thaliana meiosis reveals sex-specific patterns along chromosomes. PLoS Genet 7:e1002354
33. Salomé PA, Bomblies K, Fitz J et al (2012) The recombination landscape in Arabidopsis thaliana F2 populations. Heredity (Edinb) 108:447–455
34. Wijnker E, Velikkakam James G, Ding J et al (2013) The genomic landscape of meiotic crossovers and gene conversions in Arabidopsis thaliana. Elife 2, e01426
35. Shilo S, Melamed-Bessudo C, Dorone Y et al (2015) DNA crossover motifs associated with epigenetic modifications delineate open chromatin regions in Arabidopsis. Plant Cell 27:tpc.15.00391. doi:10.1105/tpc.15.00391
36. Hellsten U, Wright KM, Jenkins J et al (2013) Fine-scale variation in meiotic recombination in Mimulus inferred from population shotgun sequencing. Proc Natl Acad Sci U S A 110:19478–19482
37. Dooner HK (1986) Genetic fine structure of the BRONZE Locus in maize. Genetics 113:1021–1036
38. Brown J, Sundaresan V (1991) A recombination hotspot in the maize A1 intragenic region. Theor Appl Genet 81:185–188
39. Saintenac C, Faure S, Remay A et al (2011) Variation in crossover rates across a 3-Mb contig of bread wheat (Triticum aestivum) reveals the presence of a meiotic recombination hotspot. Chromosoma 120:185–198
40. Chodavarapu RK, Feng S, Bernatavichute YV et al (2010) Relationship between nucleosome positioning and DNA methylation. Nature 466:388–392
41. Zilberman D, Coleman-Derr D, Ballinger T, Henikoff S (2008) Histone H2A.Z and DNA methylation are mutually antagonistic chromatin marks. Nature 456:125–129
42. Zhang X, Bernatavichute YV, Cokus S et al (2009) Genome-wide analysis of mono-, di- and trimethylation of histone H3 lysine 4 in Arabidopsis thaliana. Genome Biol 10:R62
43. Liu S, Yeh C-T, Ji T et al (2009) Mu transposon insertion sites and meiotic recombination events co-localize with epigenetic marks for open chromatin across the maize genome. PLoS Genet 5:13
44. Copenhaver GP, Nickel K, Kuromori T et al (1999) Genetic definition and sequence analysis of Arabidopsis centromeres. Science 286:2468–2474

45. Livak KJ, Schmittgen TD (2001) Analysis of relative gene expression data using real-time quantitative PCR and the 2(-Delta Delta C(T)) Method. Methods 25:402–408
46. Langmead B, Salzberg SL (2012) Fast gapped-read alignment with Bowtie 2. Nat Methods 9:357–359
47. Bolger AM, Lohse M, Usadel B (2014) Trimmomatic: a flexible trimmer for Illumina sequence data. Bioinformatics 30:2114–2120
48. Li H, Durbin R (2009) Fast and accurate short read alignment with Burrows-Wheeler transform. Bioinformatics 25:1754–1760
49. Li H, Handsaker B, Wysoker A et al (2009) The Sequence Alignment/Map format and SAMtools. Bioinformatics 25:2078–2079
50. R Development Core Team (2012) R: a language and environment for statistical computing. R Foundation for Statistical Computing, Vienna

Part II

Selective Sequencing of Gene Families (including the MHC)

Chapter 3

PacBio for Haplotyping in Gene Families

Wei Zhang and Joachim Messing

Abstract

The throughput and read length provided by Pacific Bioscience (PacBio) Single Molecule Real Time (SMRT) sequencing platform makes it feasible to construct contiguous, non-chimeric sequences. This is especially useful for genes with repetitive sequences in their gene bodies in gene families. We illustrate the use of PacBio to sequence and assemble hundreds of transcripts of gluten gene families from different cultivars of wheat using sequence from a single SMRT cell. To this end, we barcoded amplicons from different cultivars, then pooled these into one library for sequencing. Sequencing reads were later separated by the barcodes and further sorted into different gene groups by blast. The reads from each gene are then assembled by SeqmanNGen software. Given the length of 1 kb for each sequence derived from an initial molecule, the phase of the polymorphisms is not lost and can be used to infer also haplotype differences between different cultivars.

Key words PacBio RS, SMRT, Repetitive sequence, Gene family, Chimeric sequence, Haplotyping by sequencing

1 Introduction

1.1 Characteristics of Cereal Seed Storage Proteins

Major agricultural cereals wheat, maize, and rice belong to the grass family. Common cereals usually have prolamins as their source of major seed storage proteins. A primary characteristics of prolamins is the large number of tandem repeated genes in contiguous regions due to amplification, such as zeins in maize with over 20 genes in a few hundred kilobases [1–4] and glutens in wheat [5–9]. Another characteristic of prolamins is the presence of 6–15 repeated blocks of peptides rich in glutamine and proline in the central portion of their protein sequences [10–12]. These repeated blocks of peptides are problematic for successfully assembling non-chimeric genes with short sequencing reads. In order to sequence mRNAs that have internal repeats and a large set of gene copies at various levels, we need to sequence full-length cDNAs as single molecules.

1.2 Wheat Gluten Genes

Common wheat accumulates 8–15% protein in its seeds, with 80% being glutenins and gliadins proteins. Low-molecular-weight (LMW) glutenins and gliadins comprise gene copies in large numbers.

Irene Tiemann-Boege and Andrea Betancourt (eds.), *Haplotyping: Methods and Protocols*, Methods in Molecular Biology, vol. 1551, DOI 10.1007/978-1-4939-6750-6_3, © Springer Science+Business Media LLC 2017

LMW-glutenin genes are located at the Glu-3 locus on the short arm of chromosome 1 in the wheat and encode 30–50 KDa proteins. The estimated copy numbers of LMW glutenins are 22–40 [9, 13] averaging at the DNA level a size of 0.7 kb. The α/β-type gliadin genes are at the Gli-2 locus on the short arm of chromosome 6 [14]. The gene copy number of α/β-gliadins ranges from 25 to 35 copies at the low end and can reach up to 150 copies [5, 15, 16]. The γ-gliadin and ω-gliadin genes are located at the Gli-1 locus on the short arm of chromosome 1 [17]. The estimated gene copy number for γ-gliadin genes is 17–39, and for ω-gliadins is 15–18 [9].

1.3 PacBio RS Platform

The Pacific Bioscience RS (PacBio) third-generation sequencing system offers single molecule, Real-Time (SMRT) sequencing. Each SMRT cell is built with thousands of zero-mode waveguides (ZMWs). In PacBio sequencing reactions, single molecules of DNAs are immobilized on the bottom of the ZMWs and mixed with DNA polymerase. As the DNA replicates, ZMWs allow light to illuminate only the bottom of a well and phospholinked nucleotides in the cell allow real-time observation of the newly replicated DNA strand. PacBio offers two unique advantages to sequence gluten genes. First, it offers high throughput. On average, half of the 150,000 ZMWs are used to generate sequences, which translates to 50,000–70,000 single reads per SMRT cell [18]. Second, it offers long reads. The first version of the RS system gives read lengths over 3 kb and the RSII system produces even longer reads of about 10 kb on average [18]. The longer reads will prohibit chimeric joining of different gluten molecules at internal repeat regions.

Although PacBio produces a high error rate (median error rate 11%), the errors are random. With sufficient coverage, these errors will be canceled out and a high-consensus accuracy can be reached. Additionally, error rates can be partially corrected using circular consensus sequencing (CCS) reads. PacBio sequencing uses SMARTbell as templates which ligate hairpin adaptors to both ends of double-stranded DNA. Given a DNA polymerase that can replicate a few kb of DNA and a short template, the PacBio system is capable of sequencing the raw read a few times and generates a consensus sequence with higher accuracy [19]. The gliadins and glutenin genes that we have selected as examples are ~700 bp, and therefore an average CCS read of 3 kb yields four to fivefold coverage of these genes. Today, with the RSII system, this would be increased to 14-fold coverage.

2 Materials

2.1 Plant Materials

All plant materials (Chinese spring, PI 185357, PI 187165, PI 213833, PI 260900, PI 290911, PI 447403, PI 559679, PI 608017, PI 641165) were ordered through the USDA GRIN germplasm resource website http://www.ars-grin.gov.

2.2 Molecular Biology Reagents

1. TRIzol.
2. Chloroform.
3. Isopropanol.
4. 70% ethanol.
5. Nuclease-free water.
6. DNAse I and buffer from NEB.
7. 0.5 M EDTA.
8. Qiagen RNeasy Mini Kit.
9. Oligo$(dT)_{20}$.
10. 10 mM dNTP mix (10 mM each dATP, dGTP, dCTP, and dTTP at neutral pH).
11. Invitrogen SuperScript™ III Frist-Strand Synthesis System.
12. Promega Redtaq.
13. All purpose DNA ladder (50–10,000 bp) from Thermo Scientific.
14. Agarose (electrophoresis grade).
15. Qiagen PCR Purification Kit.

2.3 Supplies and Equipment

The Following supplies and equipments are used.

1. 2000 Geno/Grinder® and metal beads.
2. 1.5 ml microtubes, free of nuclease.
3. Pipettors and nuclease-free pipette tips.
4. Centrifuge to hold 1.5 ml and PCR reaction tubes.
5. Gel-running equipment including gel tank and power supply.
6. PCR tubes and lids.
7. PCR thermal cycler.
8. NanoDrop spectrophotometer.

2.4 Software

We used the following applications and online software/resources for our data analysis:

NCBI: http://www.ncbi.nlm.nih.gov

Galaxy: http://www.usegalaxy.org

Transeq: http://www.ebi.ac.uk/Tools/st/emboss_transeq/

Clustal Omega: http://www.ebi.ac.uk/Tools/msa/clustalo/

SeqmanNGen, an application under Lasergene of DNASTAR

EditSeq, an application under Lasergene of DNASTAR

Mega5, downloaded from http://www.megasoftware.net

3 Methods

3.1 RNA Extraction and cDNA Synthesis

If amplifying genes directly from DNA, this step can be replaced with DNA extraction.

1. Plant 10 seeds for each wheat line into 2 L plant pots with fertilizer. When the seedlings are about 5 cm in height, keep six plants per line per pot. Water regularly.
2. As the plant matures, take notes of the timing of anthesis for each line.
3. Collect immature seeds from wheat spikelets at 10–15 days after anthesis, put the collected seeds into 1.5 ml microtubes and keep on ice.
4. Put in two grinding beads per tube. Grind the collected kernels in genogrinder at 1600 strokes/min for 30 s.
5. Immediately put in 1 ml of TRIzol. Store at −80 °C if necessary (*see* **Note 1**).
6. Leave at room temperature (RT) for 1 min.
7. Add 200 μl of chloroform and invert the tubes 2–3 times to mix.
8. Leave at room temperature for 3 min.
9. Centrifuge at 12,000 rpm for 10 min.
10. Get supernatant and add equal amount of isopropanol.
11. Let it precipitate on ice for 10 min.
12. Centrifuge at 12,000 rpm for 10 min.
13. Remove supernatant and add 1.0 ml 70 % ethanol. Samples at this step can be left at -80 °C for a few months.
14. Centrifuge at 12,000 rpm for 10 min.
15. Pour off ethanol and dry the precipitate at room temperature by placing the tubes upside down onto a clean Kimwipes.
16. Put in Nuclease-free water and 10 μl 10× DNAse I buffer to make it up to 100 μl
17. Add 5 U of DNAse I and incubate at 37 °C for 10 min.
18. Put in 1 μl 0.5 M EDTA and heat inactivate at 75 °C for 10 min.
19. Clean up remaining starch and contaminants following RNA Cleanup protocol in Qiagen RNeasy Mini Kit following manufacturer' protocol.
20. Measure RNA concentration using NanoDrop spectrophotometer.
21. Take 1 μg of total RNA for each sample, put in 1 μl 50 μm oligo(dT) and 1 μl 10 mM dNTP mix then follow first-strand cDNA synthesis protocol from SuperScript III kit.

3.2 Barcodes and PCR Amplification

In order to mark PCR products from different lines of wheat, 10 unique barcodes were added to the 5′ end of primers of wheat gluten genes. The same barcode was added to the end of forward and reverse primers amplifying LMW-glutenins, α-gliadins, and γ-gliadins in the same wheat line (Fig. 1). The barcodes were added during primer synthesis. Barcode and primer sequences that amplify gluten genes are listed in Tables 1 and 2.

1. Add the following components to PCR reactions to amplify LMW-glutenins, α-gliadins, and γ-gliadins. A total of 30 different PCR reactions are performed for ten wheat lines.

   0.2–0.5 μl first-strand cDNA;

   1 mM forward primer with barcode;

   1 mM reverse primer with barcode;

   12.5 μl 2× Redtaq (*see* **Note 2**);

   Add water to a final volume of 25 μl.

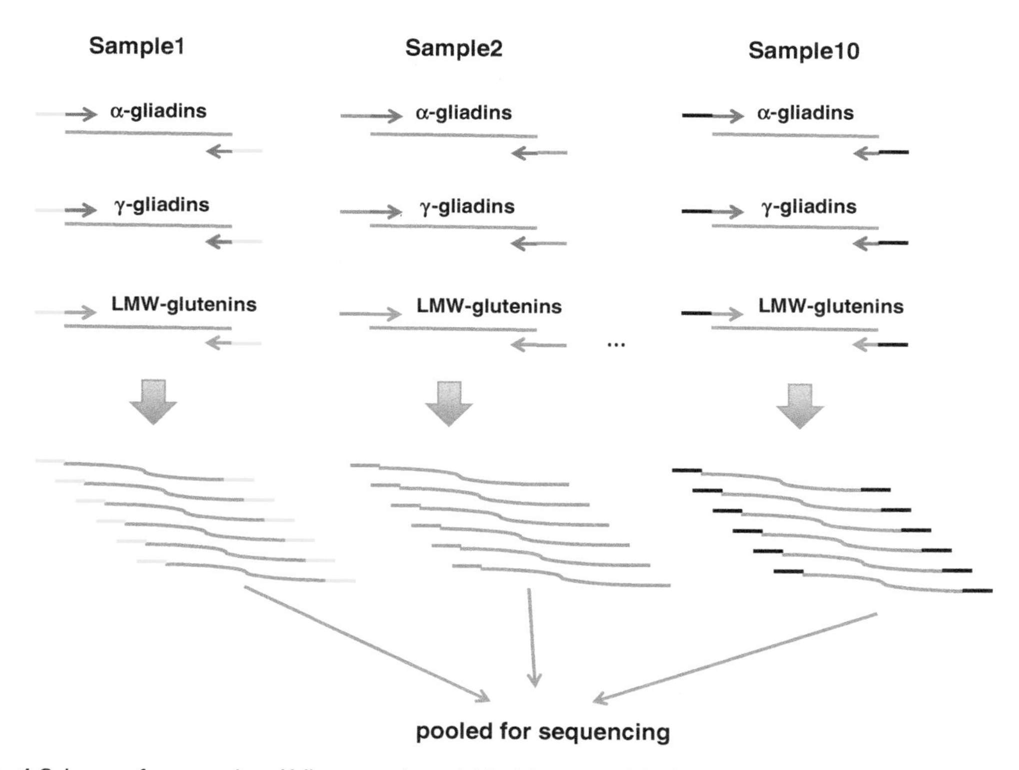

Fig. 1 Scheme of sequencing. *Yellow*, *purple*, and *black bars* stand for barcodes; *blue*, *green*, and *red arrows* represent primers for α-gliadins, γ-gliadins, and LMW-glutenins. . Samples 3–9 used in actual experiment are omitted in the figure

Table 1
Barcodes and their sequences

Barcodes	Sequences
Barcodes for Chinese Spring (BC1)	TTCTCCTTCA
Barcodes for PI 185357 (BC2)	ACCTTACCTT
Barcodes for PI 187165 (BC3)	CATTCCTCTA
Barcodes for PI 213833 (BC4)	TGTCATTCCT
Barcodes for PI 260900 (BC5)	CCATATGAAC
Barcodes for PI 290911 (BC6)	CGGAACTTAT
Barcodes for PI 447403 (BC7)	CCGGTGGAAT
Barcodes for PI 559679 (BC8)	CCGAACAGTG
Barcodes for PI 608017 (BC9)	GGAAGACCTC
Barcodes for PI 641165 (BC10)	GCCTTCAGGA

Table 2
Gliadin and glutenin primers for PCR amplification

alphaF	ATGAAGACCTTTCTCATCCTTG
alphaR	TCAGTTRGTACCGAAGATGCCA
gammaF	ATGAAGACCTTAYTCATCCT
gammaR1	TTTTCATTGKCCACTGATGCC
gammaR2	TTTTCATTGKCCACCAATGCC
gammaR3	TCATCGATATTGGCCACCAATG
LMWF	ATGAAGACCTTCCTCRTCTT
LMWR	TTATCAGTAGMVACCAACTCC

2. Let PCR run at the following reaction conditions.

   95 °C for 5 min;

   and 35 reactions of:

   95 °C for 30 s;

   58 °C for 30 s;

   72 °C for 1 min;

   and let extension for additional 8 min at 72 °C.
3. After the PCR reaction, purify PCR products with Qiagen PCR Purification Kit.
4. Measure the concentration of purified PCR products.

5. Take 200 ng for each of the PCR products and pool into one tube.
6. Send to facility with PacBio sequencing services for library preparation and sequencing.

3.3 Quality Control of Sequences and Assigning Sequences to Cultivars

PacBio sequences were filtered for read quality above 0.75 by the sequencing services. As a result, 77,462 reads with an average read length of 3050 bp passed quality control. Out of these reads, 32,863 CCS reads (>=2 full passes) were derived and used in our subsequent analysis. Using the following steps, 31234 reads were assigned to gluten genes and to wheat lines:

1. Go to usegalaxy.org and click Convert Formats.
2. Click FASTQ to FASTA converter.
3. Input fastaq file and execute. Now the fasta file containing all sequences is named ccs.fasta.
4. Click NGS: QC and manipulation.
5. Click Barcode Splitter.
6. Under 'Barcodes to use' input barcode sequences;

   Under 'library to split' input ccs.fasta;

   Under 'Barcodes found at' select Start of sequence (5′ end);

   Under 'Number of allowed mismatches' input 1;

   Under 'Number of allowed barcodes nucleotide deletions' input 1;

   Click 'Execute'.
7. Repeat **steps 4–6**, except under 'Barcodes found at' select End of sequence (3′ end).
8. Combine sequences from **steps 6** and **7** for the same cultivar and name them BCS1.fasta through BCS10.fasta for the different cultivars.

   Now that the sequences are separated in cultivars, we further assign them gene family (α-gliadins, γ-gliadins, or LMW-glutenins).
9. The following reference sequences from NCBI were used for classifications, JX141488 for α-gliadin, HQ404651 for γ-gliadin, and JF339162 for LMW-glutenin. Place sequences into text files and save them as template.fasta.
10. Run blastall -p blastn –i BCS1.fasta -d template.fasta –m 8 > BC1blast.text
11. In Excel, collapse sequences by name and group them by α-gliadins, γ-gliadins, and LMW-glutenins.
12. Collect names of sequences into alphaBC1.txt, gammaBC1.txt and LMWBC1.txt.

13. Extract sequences in fasta file by headers. This can be done using online tools: http://users-birc.au.dk/biopv/php/fabox/fasta_extractor.php. Input your fasta file containing all reads and a file containing the headers of sequences. The output file should contain reads that partially match alpha-gliadins and named BC1alpha.fasta (*see* **Note 3**).
14. Repeat **step 13** for gamma gliadins and LMW-glutenins in wheat line #1.
15. Repeat **step 10–14** for BCS2 to BCS10.

3.4 Sequence Assembly

We used SeqmanNGen to assemble our genes in two steps: (1) we used de novo assembly to generate initial contigs and then (2) we used these initial contigs for templated-assembly, to correct sequencing errors in and get final sequences of contigs.

1. Choose de novo assembly. This step is necessary to de novo generate contigs.
2. Input reads in BC1alpha.fasta.
3. Choose the following parameters: mer size = 50, min match percent = 90, gap penalty = 50, max gap = 3.
4. Save the contigs generated by **step 3** as BC1alphaContigTemp.
5. Choose templated-assembly. This step is necessary to correct errors in contigs generated from **step 3**.
6. Use contigs from **step 4** as templates.
7. Choose the following parameters: mer size = 35, min match percent = 80, gap penalty = 20 and max gap = 3.
8. Discard contigs with less than 20 assembled reads.
9. Check assembly to manually correct errors in contigs (Fig. 2 and **Note 4**).
10. Generate final contigs for BC1 alpha gliadins.
11. Repeat **steps 1–10** for BC1 gamma gliadins and BC1 LMW-glutenins.
12. Repeat **steps 1–11** for gluten genes in other wheat lines.

3.5 Phylogenetic Analysis

1. Transcribe DNA sequences into protein sequences using EBI online tool: http://www.ebi.ac.uk/Tools/st/emboss_transeq/dddd
2. Discard non-protein sequences or sequences with premature stop codons.
3. Align protein sequences. One way is to align protein sequences in Mega by ClustalW or Muscle – Data - Open A File/Session-select protein sequences in fasta file – select align – in alignment explorer select Alignment – Align by ClustalW – select all and use default parameters – save alignment as alpha.meg

```
Position: 1
Reference Coordinates           10        20        30        40        50        60        70        80        90        100
▷ Translate ▷ Consensus      ATGAA-GACCTTA-T-TCATCCTAACAATCCTTGCGATGGCAACAAC-TATCGC-CACCGCGAATATGCAGG-TCGACCC-AGCGGCC-AAGTACAATGGCC-
gamma_assemble-4(1>1016) →   ATGAA-GACCTTA-T-TCATCCTAACAATCCTTGCGATGGCAACAAC-TATCGC-CACCGCGAATATGCAGG-TCGACCC-AGCGGCC-AAGTACAATGGCC-
ID_697 →   ATGAA-GACCTTACT-TCATCCTAACAATCCT GCGATGGCAACAAC-TATCGC-CACCGCGAATATGCAGG-TCGACCC AGCGGCC-AAGTACAATGGCC-
ID_494 →   ATGAA-GACCTTA-C-TCATCCTAACAATCCTTGCGATGGCAACAAC-TATCGC-CACCGCGAATATGCAGG-TCGACCC AGCGGCC-AAGTACAATGGCC-
ID_478 →   ATGAA-GACCTTA-T-TCATCCTAACAATCCTTGCGATGGCAACAAC-TATCGC-CACCGCGAATATGCAGG-TCGACCC AGCGGCC-AAGTACAATGGCC-
ID_145 →   ATGAAAGACCTTA-C-TCATCCTAACAATCCTTGCGATGGCAACAAC-TATCGC-CACCGCGAATATGCAGG-TCGACCC AGCGGCC-AAGTACAATGGCC-
ID_863 →   ATGAA-GACCTTA-T-TCATCCTAACAATCCTTGCGATGGCAACAAC-TATCGC-CACCGCGAATATGCAGG-TCGACCC AGCGGCC-AAGTACAATG CC-
ID_801 →   ATGAA-GACCTTA-C-TCATCCTAACAATCCTTGCGATGGCAACAAC-TATCGC-CACCGCGAATATGCAGG-TCGACCC AGCGGCC-AAGTACAATGGCC-
ID_394 →   ATGAA-GACCTTA-C-TCATCCTAACAATCCTTGCGATGGCAACAAC-TATCGC-CACCGCGAATATGCAGG-TCGACCC AGCGGCC-AAGTACAATGGCC-
ID_275 →   ATGAA-GACCTTACT-TCATCCTAACAATCCTTGCGATGGCAACAAC-TATCGC-CACCGCGAATATGCAGG-TCGACCC AGCGGCC-AAGTACAATGGCC-
ID_955 →   ATGAA-GACCTTA-C-TCATCCTAACAATCCTTGCGATGGCAACAAC-TATCGC-CACCGCGAATATGCAGG-TCGACCC AGCGGC--AAGTACAATGGCC-
ID_728 →   ATGAA-GACCTTA-C-TCATCCTAACAATCCTTGCGATGGCAACAAC-TATCGC-CACCGCGAATATGCAGG-TCGACCC AGCGGCC-AAGTACAATGGCC-
ID_701 →   ATGAA-GACCTTA-T-TCATCCTAACAATCCTTGCGATGGCAACAAC-TATCGC-CACCGCGAATATGCAGG-TCGACCC AGCGGCC-AAGTACAATGGCC-
ID_856 →   ATGAA-GACCTTA-T-TCATCCTAACAATCCTTGCGATGGCAACAAC-TATCGC-CACCGCGAATATGCAGG-TCGACCC AGCGGCC-AAGTACAATGGCC-
ID_789 →   ATGAA-GACCTTA-T-TCATCCTAACAATCCTTGCGATGGCAACAAC-TATCGC-CACCGCGAATATGCAGG-TCGACCC AGCGGCC-AAGTACAATGGCC-
ID_826 →   ATGAA-GACCTTA-T-TCATCCTAACAATCCTTGCGATGGCAACAAC-TATCGC-CACCGCGAATATGCAGG-TCGACCC AGCGGCC-AAGTACAATGGCC-
ID_276 →   ATGAA-GACCTTA-T-TCATCCTAACAATCCTTGCGATGGCAACAAC-TATCGC-CACCGCGAATATGCAGG-TCGACCC-AGCGGCC-AAGTACAATGGCC-
ID_700 →   ATGAA-GACCTTA-T-TCATCCTAACAATCCTTGCGATGGCAACAAC-TATCGC-CAC GCGAATATGCAGG-TCGACCC AGCGGCC-AAGTACAATGGCC-
ID_622 →   ATGAA-GACCTTA-T-TCATCCTAACAATCCTTGCGATGGCAACAAC-TATCGC-CACCGCGAATATGCAGG-TCGACCC-AGCGGCC-AAGTACAATGGCC-
ID_611 →   ATGAA-GACCTTA-T-TCATCCTAACAATCCTTGCGATGGCAACAAC-TATCGC-CACCGCGAATATGCAGG-TCGACCC-AGCGGCC-AAGTACAATGGCC-
ID_889 →   ATGAA-GACCTTA-T-TCATCCTAACAATCCTTGCGATGGCAACAAC-TATCGC-CACCGCGAATATGCAGG-TCGACCC AGCGGCC-AAGTACAATGGCC-
ID_362 →   ATGAA-GACCTTA-T-TCATCCTAACAATCCTTACGATGGCAACAAC-TATCGC-CACCGCGAATATGCAGG-TCGACCC AGCGGCC-AAGTACAATGGCC-
ID_617 →   ATGAA-GACCTTA-T-TCATCCTAACAATCCTTGCGATGGCAACAAC-TATCGC-CACCGCGAATATGCAGG-TCGACCC-AGCGGCC-AAGTACAATGGCC-
ID_194 →   ATGAA-GACCTTA-T-TCATCCTAACAATCCTTGCGATGGCAACAAC-TATCGC-CACCGCGAATATGCAGG-TCGACCC AGCGGCC-AAGTACAATGGCC-
ID_783 →   ATGAA-GACCTTA-T-TCATCCTAACAATCCTTGCGATGGCAACAAC-TATCGC-CACCGCGA TATGCAGG-TCGACCC AGCGGCC-AAGTACAATGGCC-
ID_957 →   ATGAA-GACCTTA-T-TCATCCTAACAATCCTTGCGATGGCAACAACCTATCGC-CACCGCGAATATGCAGG-TCGACCC AGCGGCC-AAGTACAATGGCC-
ID_845 →   ATGAA-GACCTTA-T-TCATCCTAACAATCCTTGCGATGGCAACAAC-TATCGC-CACCGCGAATATGCAGG-TCGACCC-AGCG CC-AAGTACAATGGCC-
ID_305 →   ATGAA-GACCTTA-T-TCATCCTAACAATCCTTGCGATGGCAACAAC-TATCGC-CACCGCGAATATGCAGG-TCGACCC AGCGGCC-AAGTACAATGGCC-
ID_931 →   ATGAA-GACCTTA-T-TCATCCTAACAATCCTTGCGATGGCAACAAC-TATCGC-CACCGCGAATATGCAGG-TCGACCC AGCGGCC-AAGTACAATGGCC-
ID_977 →   ATGAA-GACCTTA-C-TCATCCTAACAATCCTTGCGATGGCAACAAC-TATCGC-CACCGCGAATATGCAGG-TCGACCC AGCGGC--AAGTACAATGGCC-
ID_341 →   ATGAA-GACCTTA-T-TCATCCTGACAATCCTTGCGATGGCAATAAC-CATCGC--ACCGC AATATGCAGG-TCGACCCTAGCGGCC-AAGTACAATGGCC-
ID_716 →   ATGAA-GACCTTA-T-TCATCCTAACAATCCTTGCGATGGCAACAAC-TATCGC-CACCGCGAATATGCAG--TCGACCC AGCGGCC-AAGTACAATGGCC-
ID_254 →   ATGAA-GACCTTA-T-TCATCCTAACAATCCTTGCGATGGCAACAAC-CATCGC-CAC-GC AATATGCAG--TCGACCCTAGCAGCCGAAGTACA▶
ID_502 →   ATGAA-GACCTTA-T-TCATCCTAACAATCCTTGCGATG CAACAAC-TATCGC-CACCGCGAATATGCAGGGTCGACCC AGCGGCC-A GTACAATGGCC-
ID_505 →   ATGAA-GACCTTA-C-TCATCCTAACA TCCTTGCGATGGCAACAAC-AATCGC-CACTGC AATATGCAGG-TCGACCT-AGCAGCC-GAGTACAATGGCC-
ID_811 →   ATGAA-GACCTTA-TCTCATCCTAACAATCCTTGCGATGGCA CAAC-TATCGC--ACCGCGAATATGCAGG-TCGACCC-AGCGGCC-AAGTACAATGGCC-
ID_558 →    ◀TGAA-GACCTTA-T-TCATCCTAACAATCCTTGCGATGGCAACAAC-TATCGCACACCGCGA-TGTGCAGG-TCGACCC AGCGGC--AAGTACAATGGCCC
ID_50  →    |TGAA-GACCTTACT-TCATCCTAACAATCCTTGCGATGGCAACAAC-TATCGC-CACCGCGAATATGCAGG-TCGACCC AGCGGCC-AAGTACAATGGCC-
ID_600 →        ◀A-GACCTTA-T-TCATCCTAACAATCCTTGCGATGGCAACAAC-TATCGC-CACCGCGAATATGCAG--TCGACCC AGCGGCC-AAGTACAATGGCC-
ID_625 →        ◀A-GACCTTA-T-TCATCCTAACAATCCTTGCGATGGCAACAAC-TATCGC-CACCGCGAATATGCAGG-TCGACCC-AGCGGCC-AAGTACAATGGCC-
ID_9   →        ◀A-GACCTTA-C-TCATCCTAACAATCCTTGCGATGGCAACAAC-AATCGC-CACTGC AATATGCAGG-TCGACCCTAGCAGCC-GAGTACAATG-CC-
ID_548 →          ◀GACCTTA-T-TCATC TAACAATCCT GCGATGGCAACAAC-TATCGC-CACCGCGAATATGCAGG-TCGACCC-AGCGGCC-AAGTACAATGGCC-
ID_770 →          ◀GACCTTA-T-TCATCCTAACAATC TTGCGATGGCAACGAC-TATCGC-CACCGCGAATATGCAGG-TCGACCC-AGCGGCC-AAGTACAATGGCC-
ID_72  →          ◀GACCTTA-T-TCATCCTAACA TCCTTGCGATGGCAACAAC-TATCGC-CACCGCGAATATGCAGG-TCGACCC AGCGGCC-AAGTACAATGGCC-
ID_610 →          ◀GACCTTA-T--CATCCTAACAATCCTTGCGATGGCAACAAC-TATCGC-CACCGCGAATATGCAGG-TCGACCC-AGCGGCC-AAGTACAATGGCC-
ID_112 →                   |CTAACAATCCTTGCGATGGCAACAAC-TATCGC-CACCGCGAATATGCAGG-TCGACCC AGCGGCC-AAGTACAATGGCC-
ID_800 →                                             ◀TATCGC-CACCGCGAATATGCAGG-TCGACCC-AGCGGCC-AAGTACAATGGCC-
ID_331 →                                             ◀TATCGC-CACCGCGAATATGCAGG-TCGACCC AGCGGCC-A GTACAATGGCC-
ID_49  →                                                   ◀CGCGAATATGCAGG-TCGACCC AGCG CC-AAGTACAATGGCC-
ID_839 →                                                                ◀G-TCGACCC AGCGGCC-AAGTACAATGGCC-
ID_139 →                                                                ◀G-TCGACCC-AGCGGCC-AAGTACAATGGCC-
ID_391 →                                                                               ◀AAGTACAATGGCC-
ID_73  →
```

Fig. 2 Screen shot of templated assembly. All sequencing reads matching the initial contig are aligned with the template. *First line* of sequence is the consensus sequence and *second line* of sequence is the template. Manual correction of initial contig sequences at position 81 is needed since C is missing in both the consensus sequence and the template

format. Another way is to use Clustal Omega: http://www.ebi.ac.uk/Tools/msa/clustalo/. In output page, click Phylogenetic Tree and save phylogenetic tree file as alpha.ph.

4. Generate phylogenetic tree. We prefer to generate phylogenetic tree by Mega using the alignment file in.meg format generated from **step 3**. Click Phylogeny in Mega—select Construct/Test Maximum Likelihood Tree or Construct/Test Neighbor-Joining Tree—in Analysis Preferences, choose Partial deletion under Gaps/Missing Data Treatment, 70% cutoff under Site Coverage Cutoff (%) and input 1,000 for No. of Bootstrap Replications. The phylogenetic tree can be saved in pdf format. Alternatively, if using the phylogenetic tree file generated by Clustal Omege online, click Data—Open a File/Session—select the alpha.ph tree file and the tree can be viewed and saved.

4 Notes

1. Immature seeds or endosperm samples usually contain large amounts of starch and sugars, which will give a sticky texture when put into TRIzol. It may therefore be necessary to put in an excess amount of TRIzol. Ground tissue samples can be stored in TRIzol at −80 °C for a month and extracted RNA can be stored in 70% ethanol for a year.
2. Redtaq mastermix contains DNA polymerase, buffer for DNA polymerase and dNTP.
3. Alternative ways to extract reads by header. One is under command line, use *cut -c 2 alphaBC1.TXT | xargs -n 1 samtools faidx alphaBC1.FA* for alpha gliadins. In order for this command line to work, you must have installed samtools (http://www.htslib.org/). Another simple way of doing this without writing custom code is to visit use galaxy.org. In that case, you will have to first convert the fasta file into a tabular file by using Convert Formats-FASTA to Tabular in galaxy. Then, from the "Join, Subtract and Group" menu, select Compare two Datasets.
4. Look for the lack of or excess of runs of the same nucleotide. In our case, we frequently found missing nucleotides in cases of homopolymers, typical for polymerase sequencing method. For example, in Fig. 2, nucleotide C is missing in the contig at position 81 of the assembly. This nucleotide is manually added in the contig by checking the templated-assembly in **step 6** under Subheading 3.4.

References

1. Llaca V, Messing J (1998) Amplicons of maize zein genes are conserved within genic but expanded and constricted in intergenic regions. Plant J 15(2):211–220
2. Miclaus M, Xu JH, Messing J (2010) Differential gene expression and epiregulation of alpha zein gene copies in maize haplotypes. PLoS Genet 7(6):e1002131. doi:10.1371/journal.pgen.1002131, PGENETICS-D-11-00531 [pii]
3. Song R, Llaca V, Linton E, Messing J (2001) Sequence, regulation, and evolution of the maize 22-kD alpha zein gene family. Genome Res 11(11):1817–1825. doi:10.1101/gr.197301
4. Song R, Messing J (2002) Contiguous genomic DNA sequence comprising the 19-kD zein gene family from maize. Plant Physiol 130(4):1626–1635. doi:10.1104/pp.012179
5. Harberd NP, Bartels D, Thompson RD (1985) Analysis of the gliadin multigene locus in bread wheat using nullisomic-tetrasomic lines. Mol Gen Genet 198:234–242
6. Huang XQ, Cloutier S (2008) Molecular characterization and genomic organization of low molecular weight glutenin subunit genes at the Glu-3 loci in hexaploid wheat (Triticum aestivum L.). Theor Appl Genet 116(7):953–966. doi:10.1007/s00122-008-0727-1
7. Marino CL, Tuleen NA, Hart GE, Nelson JC, Sorrells ME, Lu YH, Leroy P, Lopes CR (1996) Molecular genetic maps of the group 6 chromosomes of hexaploid wheat (Triticum aestivum L. em. Thell.). Genome 39(2):359–366, doi: g96-046 [pii]
8. Payne PI, Law CN, Mudd EE (1980) Control by homeologous group I chromosomes of the high molecular weight subunits of glutnin, a major protein of wheat endosperm. Theor Appl Genet 58:113–120

9. Sabelli P, Shewry PR (1991) Characterization and organization of gene families at the Gli-1 loci of bread and durum wheats by restriction fragment analysis. Theor Appl Genet 83:209–216
10. Geraghty D, Peifer MA, Rubenstein I, Messing J (1981) The primary structure of a plant storage protein: zein. Nucleic Acids Res 9(19): 5163–5174
11. Masci S, D'Ovidio R, Lafiandra D, Kasarda DD (1998) Characterization of a low-molecular-weight glutenin subunit gene from bread wheat and the corresponding protein that represents a major subunit of the glutenin polymer. Plant Physiol 118(4):1147–1158
12. Sugiyama T, Rafalski A, Peterson D, Soll D (1985) A wheat HMW glutenin subunit gene reveals a highly repeated structure. Nucleic Acids Res 13(24):8729–8737
13. Cassidy BG, Dvorak J, Anderson OD (1998) The wheat low-molecular-weight glutenin genes: characterization of six new genes and progress in understand gene family structure. Theor Appl Genet 96:743–750
14. Payne IP (1987) Genetics of wheat storage proteins and the effect of allelic variation on bread-making quality. Annu Rev Plant Physiol 38:141–153
15. Anderson OD, Litts JC, Greene FC (1997) The α-gliadin gene family. I. Characterization of ten new wheat α-gliadin genomic clones, evidence for limited sequence conservation of flanking DNA, and southern analysis of the gene family. Theor Appl Genet 95: 50–58
16. Okita TW, Cheesbrough V, Reeves CD (1985) Evolution and heterogeneity of the alpha-/beta-type and gamma-type gliadin DNA sequences. J Biol Chem 260(13):8203–8213
17. Payne PI, Jackson EA, Holt LM, Law CN (1984) Genetic linkage between endosperm storage protein genes on each of the short arms of chromosomes 1A and 1B in wheat. Theor Appl Genet 67:235–243
18. Revolutionize genomics with SMRT sequencing. Pacbio brochure. http://www.pacb.com/wp-content/uploads/2015/09/PacBio_RS_II_Brochure.pdf
19. Pacific Biosciences (2013). Specifics of SMRT sequencing data. https://speakerdeck.com/pacbio/specifics-of-smrt-sequencing-data

Chapter 4

High Molecular Weight DNA Enrichment with Peptide Nucleic Acid Probes

Nicholas M. Murphy, Colin W. Pouton, and Helen R. Irving

Abstract

Here, we describe how peptide nucleic acid (PNA) probes can be used to enrich genomic DNA fractions to facilitate downstream analysis, such as the haplotype phasing of the isolated genomic pieces. This method enriches for polymorphic regions of fragmented chromosomes by physically separating the desired sequence and flanking regions. The PNA probes used for enrichment are novel synthetic nucleic acids with highly specific targeting and hybridization properties. Using a enrichment technique, we capture high molecular weight genomic DNA using nothing more than a simple modification to standard genomic DNA extraction from blood.

Key words Peptide nucleic acid probes, Haplotype enrichment, Major histocompatibility complex, Human leukocyte antigen system alleles, High molecular weight DNA

1 Introduction

In this chapter, we describe how we use peptide nucleic acid (PNA) probes to enrich haplotype sequences within the major histocompatibility complex (MHC). PNAs are synthetic molecules based on DNA, but containing an achiral uncharged polyamide backbone instead of the nucleic acid deoxyribose phosphate backbone [1]. Sequencing the major histocompatibility complex (MHC) at 6p22.1 to 6p21.3 is one of the most challenging tasks in genomics. The inherent complexity of the immune system, paralogous nature of the genes, large number of alleles and diploid copy number means that special molecular and bioinformatics techniques are required to resolve MHC gene sequences [2]. High-throughput next generation sequencing has facilitated the sequencing of human genomes; however, the MHC still remains an elusive target, unless supplementary techniques are included in the workflow [3, 4]. The diversity of the human leukocyte antigen (HLA), which directly presents antigens for the immune system

Irene Tiemann-Boege and Andrea Betancourt (eds.), *Haplotyping: Methods and Protocols*, Methods in Molecular Biology, vol. 1551, DOI 10.1007/978-1-4939-6750-6_4, © Springer Science+Business Media LLC 2017

provides researchers with an unique group of alleles that can be targeted for enrichment strategies. The importance of precise HLA typing is highlighted by the breadth of fields in which HLA alleles and haplotypes are relevant, ranging from cancer [5], autoimmunity [6], pharmacogenetics [7], and vaccinomics [8]. To demonstrate how peptide nucleic acids could assist in this process, we previously described how two disparate alleles (in terms of base content) of the *HLA-DRB1* gene were used to enrich specific haplotypes [9]. However due to the high specificity of PNA hybridization, polymorphic variations throughout the genome could be used.

The targeting and enrichment of specific alleles prior to sequencing simplifies the workflow by broadening the available options for amplification and sequencing. DNA enrichment utilizing peptide nucleic acid probes provides unique advantages over traditional DNA based enrichments due to increased thermal stability, strand invasion and limited interference with downstream PCR amplification [10–12]. The procedures described below explain how we used PNA probes to enrich specific haplotypes of the *HLA-DRB1* gene and Fig. 1 shows the basic workflow followed.

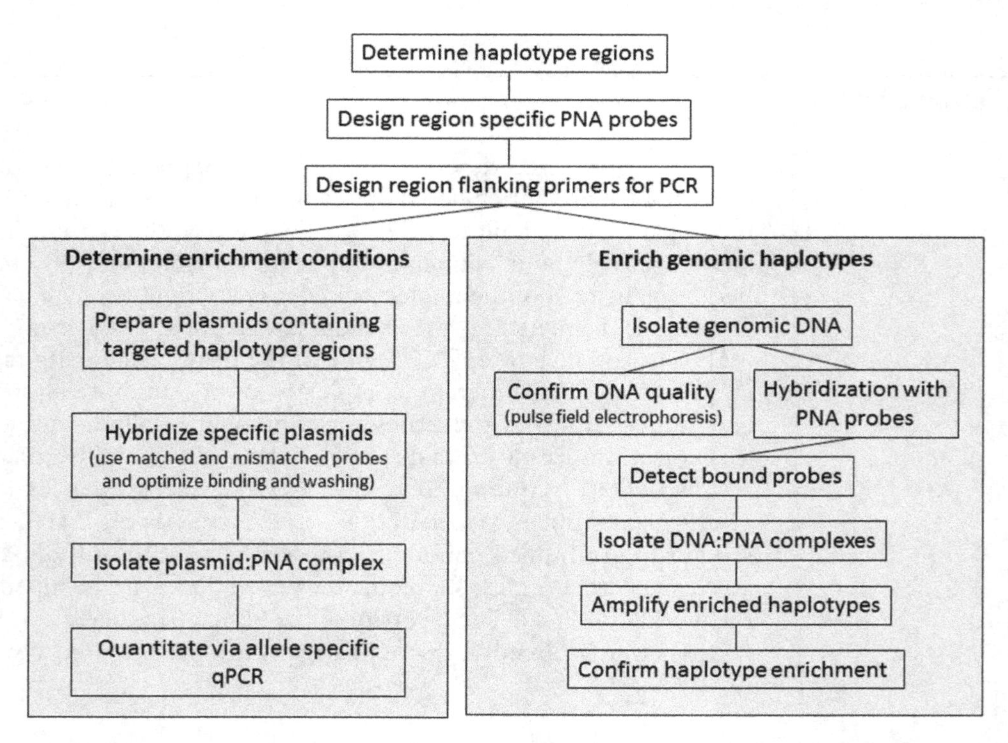

Fig. 1 Schematic outline showing the experimental methodologies using PNAs to enrich genomic DNA haplotypes

2 Materials

Reagents should be of analytical (or molecular biology) grade and must be prepared in ultra-pure water unless otherwise stated. Prepare and store all reagents at room temperature (20–26 °C) unless otherwise stated. If using human or animal samples, first obtain the suitable ethics approval following the regulations and requirements of your institution and country. Dispose of all waste materials following the regulations and recommendations of your institution.

2.1 Cloning and Analysis of Test Plasmid Components

1. Basic molecular biology reagents and equipment for agarose gel electrophoresis, *E. coli* strains, suitable growth media (e.g., LB broth) and appropriate antibiotics.
2. Shaking incubator.
3. Thermocycler.
4. Probe sonicator.
5. Horizontal rocking platform
6. Quantitative PCR machine (e.g., Rotor-Gene (Qiagen) or a C1000™ Thermal Cycler and CFX96™ Real-Time System (Bio-Rad Laboratories)).
7. Microcentrifuges capable of spinning at >12,000 × *g* and a bench top centrifuge with rotors capable of holding 15 or 50 mL tubes and spinning at >12,000 × *g*.
8. A suitable cloning vector (e.g., pDONR201 or pDONR207 vector, Invitrogen) into which your target gene region can be inserted.
9. Allele-specific forward and reverse primers (e.g., Table 1) for your locus of interest or gene of interest (*see* **Note 1**) for cloning and sequencing purposes.
10. Additional cloning primers with cloning vector specific sequences incorporating relevant gateway recombination sites (i.e., attB sites in our case, Table 1) (or specific restriction endonucleases for traditional cloning).
11. High fidelity DNA polymerase such as Phusion High Fidelity DNA polymerase (New England Biolabs) or AmpliTaq Gold DNA polymerase (Thermo Fisher Scientific).
12. PCR Clean up Kit such as UltraClean® PCR Clean-Up kit (Mo Bio, USA).
13. Gateway® BP Clonase enzyme mix (Invitrogen).
14. Media: Luria Broth (LB) medium. Dissolve 20 g of powdered LB in 800 ml of distilled water and adjust the solution to a pH of 7.5. Make up the solution to 1 L with distilled water and autoclave at 121 °C for 15 min. LB plates. Weigh out 10 g of

Table 1
PCR primers used to generate and confirm plasmid insert sequences

Primer name	Sequence 5′–3′
HLA-DRB1*01 Forward	TGGCAGCTTAAGTTTGAA
HLA-DRB1*01 Reverse	GTGTCCACCGCGGCCCGCC
HLA-DRB1*03 Forward	GGAGTACTCTACGTCTGAG
HLA-DRB1*03 Reverse	TAGTTGTCCACCCGGCCCCGCT
Gateway BP _HLA-DRB1*01_ Forward	GGGGACAAGTTTGTACAAAAAAGCAGGCTCGTATCTTTCTTGTGGCAGCTTAAGTTTGAA
Gateway BP _HLA-DRB1*01_ Reverse	GGGGACCACTTTGTACAAGAAAGCTGGGTGGTGTCCACCGCGGCCCGCC
Gateway BP _HLA-DRB1*03_ Forward	GGGGACAAGTTTGTACAAAAAAGCAGGCTCGTCGAGTTTCTTGGAGTACTCACGTCTGAG
Gateway BP _HLA-DRB1*03_ Reverse	GGGGACAAGTTTGTACAAAAAAGCAGGCTCTCGATAGTTGTCCACCCGGCCCCGCT
201 Forward (for sequencing)	TCGCGTTAACGCTAGCATGGATCT C
201 Reverse (for sequencing)	GTAACATCAGAGATTTTGAGACAC

The cloning primers were originally reported in Table S1 in [9]

LB powder and dissolve in 500 ml of distilled water before adding 15 g bacterial agar. Autoclave and cool to ~55 °C and add and thoroughly mix in suitable antibiotic before pouring out in Petri dishes.

15. Solutions. Hybridization buffer stock solution: 10 mM sodium phosphate buffer (pH 7.5) with 1 mM EDTA. Wash buffer: Tris buffered saline 25 mM Tris, 150 mM NaCl (pH 7.2).
16. Plasmid isolation kit such as PureLink HiPure Plasmid Miniprep and PureLink HiPure Plasmid Filter Maxiprep kits (ThermoFisher Scientific).
17. Reacti-Bind Neutravidin Coated 96-Well Plate (Pierce, USA). Streptavidin-coated plates would also be suitable.
18. A fluorescence microplate reader capable of measuring fluorescence at excitation 488 and 520 nm and emission wavelengths of 530 nm and 560 nm, respectively. (The EnVision™ 2101 plate reader from PerkinElmer® is suitable for this purpose).
19. Fluorescent moiety (we selected fluorophores from the Alexa Fluor range, however the commonly used cyanine dyes are acceptable) coupled to peptide nucleic acid probes in accordance to standard recommended design specifications [13] coupled to biotin and non-biotinylated probes (Table 2) obtained from a commercial supplier such as Panagene Inc. (South Korea) (*see* **Note 2**).

2.2 Genomic DNA Isolation and Analysis Components

1. A source of genomic DNA. We use blood and collect 10–15 mL in a heparin Vacutainer (BD Diagnostics). It is important to note that the blood needs to be processed on the day of collection (*see* **Note 3**).
2. Genomic DNA Extraction Kit such as the Wizard Genomic DNA Extraction Kit (Promega).
3. Solutions: Proteinase K (20 mg/mL), 3 M sodium acetate, 70% isopropanol, nuclease-free water.

Table 2
Sequence of PNAs designed to target the *HLA-DRB1*01* and *HLA-DRB1*03* alleles

PNA target	Sequence
[Alexa Fluor 488]-PNA*01	TG TGG CAG CTT AAG TTT GAA TG-Lys-AlexaFluor488
[Alexa Fluor 488]-PNA*01-Biotin	Biotin-OOO-E-TGT GGC AGC TTA AGT TTG AAT G-E-Lys-AlexaFluor488
[Alexa Fluor 532]-PNA*03	Lys-GAG TAC TCT ACG TCT G-Lys-AlexaFluor532
[Alexa Fluor 532]-PNA*03-Biotin	Biotin-OOO-GAG TAC TCT ACG TCT G-Lys-AlexaFluor532

"T," "G," "C," "A" each denote standard PNA bases, "O" denotes a spacer, "Lys" denotes a lysine residue, and "E" the standard Panagene Inc. PNA linker
This table was originally published as Table 1 in [9]

4. A suitable gel electrophoresis system for pulse field electrophoresis (e.g., CHEF PFGE system (Bio-Rad Laboratories)). Low melting point agarose (Bio-Rad Laboratories), 0.5× TBE (Tris–boric acid–EDTA, *see* [14]).
5. High molecular weight markers for pulse field electrophoresis such as *Saccharomyces cerevisiae* DNA ladder (New England Biolabs) or CHEF DNA size marker and 4.5–5.7 Mb DNA Ladder (Bio-Rad Laboratories) and 12 kb DNA ladder (1 Kb DNA Ladder Invitrogen).
6. Reacti-Bind Neutravidin Coated 96-Well Plate (Pierce, USA). Streptavidin-coated plates would also be suitable.
7. Fluorescently coupled peptide nucleic acid probes coupled to biotin and non-biotinylated probes obtained from a commercial supplier such as Panagene Inc. (South Korea) (*see* **Note 2**).
8. PCR Clean up Kit such as UltraClean® PCR Clean-Up kit (Mo Bio, USA).

3 Methods

3.1 Optimization of PNA Hybridization Using Control Plasmids

The researcher should be familiar with standard molecular biology techniques [14].

3.1.1 Plasmid Generation

1. Design primers to amplify the targeted region of interest to provide a suitable platform for generating a hybridization standard. We insert our regions of interest into DONR201 vectors using the Gateway (Invitrogen, USA) cloning system. However any cloning vector would be suitable (*see* **Note 1**). Since we use Gateway system, additional primers are designed with the Gateway recombination sites incorporated (Table 1). We have found if amplifying from genomic DNA, it is better to undertake the amplification in two steps, with the primary primers spanning ~30–100 bp either side of the targeted sequence as it is easier to incorporate the recombinant sites in the second PCR step. The ensuing product is used as a template for incorporation of the Gateway recombination sites [15]. Use no more than 25 amplification cycles with a high fidelity DNA polymerase such as Phusion High Fidelity DNA polymerase (New England Biolabs) for both PCR steps. Purify the resulting product with a PCR clean-up kit. Recombine with Gateway® BP Clonase enzyme mix (Invitrogen) to incorporate your sequence into pDONR201 or pDONR207 (Invitrogen) using standard molecular biology techniques.
2. Confirm the resulting plasmid by sequencing in order to verify that your gene segment of interest is inserted into the cloning vector. If using pDONR201 or pDONR207, you can use the 201 forward and 201 reverse primers (Table 1).

3. Amplify the vector in a suitable host bacterial strain such as Stbl3 *E. coli* and isolate it as a pure plasmid using commercial plasmid preparation kit such as PureLink HiPure Plasmid Maxiprep kit (Invitrogen) to concentrate pure plasmid in the order of 10 μg/μl. Bulk plasmid preparations can be used to generate sufficient plasmid stocks for the enrichment protocol optimization.

3.1.2 PNA Probe Preparation

1. Design the PNA probes. The PNA probes that we used include non-biotinylated PNA probes that are used as negative controls in conjunction with biotinylated PNAs as assay probes in order to optimize the procedure. In the example we describe, we used *HLA-DRB1*01* and *HLA-DRB1*03* plasmid constructs. The *HLA-DRB1*01* specific PNAs were coupled covalently at the N-terminus to Alexa Fluor 488 and the *HLA-DRB1*03* specific PNAs were coupled to Alexa Fluor 532 (Table 2) in order to optimize the enrichment parameters of plate binding and washing.
2. Solubilize the PNA probes. Our probes were obtained from the commercial supplier Panagene Inc. (South Korea) and made up in a solution of 10:2:1 dimethylformaldehyde–H_2O–trifluoroacetic acid (TFA) due to the low solubility of the probes (*see* **Note 4**).

3.1.3 Plasmid Enrichment

1. To achieve this, plasmid constructs were hybridized separately to both perfectly matched (PM) or mismatched (MM) fluorescently labeled PNAs (*see* **Note 5**). 4.0 μg of each plasmid DNA was incubated with 40 pmol of the specific PNA probe (e.g., perfectly matched (PM) or mismatched (MM) probes) in 100 μl reaction PCR tubes in 10 mM sodium phosphate buffer (pH 7.5) with 1 mM EDTA in a total volume of 100 μl.
2. Hybridize the probes to the plasmids by heating to 95 °C for 10 min, 85 °C for 5 min, 75 °C for 5 min, 65 °C for 10 min, 55 °C for 5 min, 45 °C for 5 min, 35 °C for 10 min, and 4 °C for 10 min to create a controlled temperature decrement. Use a thermocycler to control the hybridization temperatures.
3. Preblock and wash in triplicate a Reacti-Bind Neutravidin Coated 96-Well Plate (Pierce) with Tris buffered saline.
4. Transfer the hybridization reaction to designated Reacti-Bind microwells using a wide-bore pipette tip.
5. Incubate overnight at 4 °C on a horizontal rocker at 50 rpm.
6. Wash between 3 and 7 times with Tris-buffered saline, monitoring fluorescence of the hybridized probes with a fluorescent microplate reader (*see* **Note 6** for troubleshooting approaches).
7. Remove bound PNA–DNA product from the microplate by aspirating with boiling water three times into a 1.5 mL

Table 3
Primers used for allele-specific PCR

Primer Name	Sequence 5′–3′	Reference
I1-RB1 (*HLA-DRB1*01* Forward)	TCCCAGTGCCCGCTCCCT	[17]
I2-RB2 (*HLA-DRB1*01* Reverse)	ACACACTCAGATTCTCCGCTT	[17]
I1-RB9 (*HLA-DRB1*03* Forward)	TGGTGGGCGTTGGGGCG	[17]
I2-RB28 (*HLA-DRB1*03* Reverse)	ACACACACACTCAGATTCCCA	

These are *HLA-DRB* specific primers [17] that were reported in Table S1 in [9]

microcentrifuge tube and subsequently precipitate using 3 M sodium acetate. Then resuspend the PNA–DNA complex in 20 μl dH_2O.

8. Quantification of the enrichment assay can be performed using allele-specific primers and quantitative PCR. In our example, we used *HLA-DRB1*01* and *HLA-DRB1*03* allele-specific real-time quantitative PCR (Table 3) with standard curves of the neat plasmid samples (*see* **Note 7**).

3.2 Genomic DNA Isolation and Analysis

3.2.1 Blood Collection and Genomic DNA Isolation

1. Genomic DNA is isolated from lymphocytes. Collect 10–15 mL blood in a heparin Vacutainer via venipuncture, storing at room temperature and perform DNA extraction on the same day (*see* **Note 3**).
2. Place 3 mL whole blood in 15 mL centrifuge tube and follow the procedures of a standard Genomic DNA Extraction Kit such as the Wizard Genomic DNA Extraction Kit (Promega). However, since the aim is to optimize the extent of haplotyping, large regions of contiguous genomic DNA need to be obtained, the genomic DNA extraction protocols need to be followed with specific qualifications (*see* **Note 8**).
3. We used the Wizard Genomic DNA Extraction Kit (Promega) and used the accompanying reagents. Add 9 mL of Cell Lysis Solution provided. Mix by inverting the tube 2–3 times. Incubate for 10 min at room temperature, inverting and centrifuge 2–3 times during incubation.
4. Centrifuge at 2000 × g for 5 min. Aspirate supernatant leaving ~20 μl of residual buffer/plasma with pellet. If excessive red blood cells remain, repeat **steps 3** and **4**.
5. Resuspend white blood cells in remaining liquid via vigorous repeated pipetting. Add 3 mL of Nuclei Lysis Solution and gently mix via flicking 5–6 times. Add 15 μL of RNase solution.
6. At this stage, we modify the protocol to ensure that we obtain high molecular weight genomic DNA (*see* **Note 9**). Place the solution on a horizontal (circular) rocker for approximately 15 min at ~100 rpm.

7. Add 1 mL of Protein Precipitation and gently mix on a horizontal rocker for 10 min at ~30 rpm. Then centrifuge the sample at 500 × *g*.
8. Transfer the maximum volume of liquid possible into a new 50 ml centrifuge tube containing 25 ml of 70 % isopropanol without disturbing and/or collecting the lower protein component.
9. Place on a vertical rocker (~30 rpm) for several minutes or until high molecular weight DNA precipitates.
10. Finally, use a glass rod to stir around the precipitated DNA with a swirling motion and transfer captured DNA to 5–10 ml of pure distilled H_2O. The goal is to obtain genomic DNA with fragments >70 kbp in length (*see* **Note 9**).

3.2.2 Confirmation of High Molecular Weight Genomic DNA

1. The size range of the high-molecular weight DNA can be determined with pulse field gel electrophoresis (PFGE). 10 μl aliquots of DNA samples obtained from the genomic extraction are mixed with low-melting point agarose (Bio-Rad Laboratories) and proteinase K by slicing each agarose block into quarters prior to mixing with agarose as per typical PFGE preparation.
2. A 0.8 % agarose gel in 0.5× TBE is used and a range of ladders are run simultaneously with the samples to span a full range of genomic DNA lengths. We utilized a 500 bp to ~12 kb DNA ladder (1 Kb DNA Ladder Invitrogen); 225–1900 kb *Yeast DNA Ladder* (New England Biolabs) and; a 4.5–5.7 Mb DNA Ladder (Bio-Rad Laboratories).
3. Run the DNA on the gel for approximately 24 h at 14 °C, rotating 120° at 6 V/cm, every 60 s.
4. The smears and/or bands on the gel can be visualized by staining with ethidium bromide or alternative DNA dyes such as GelRed (Biotium) and indicate the range of molecular weights of DNA that has been isolated.

3.2.3 Enrichment of High Molecular Weight Genomic DNA

1. Enrichment with high-molecular-weight genomic DNA should be performed as per the plasmid enrichment (Subheading 3.1.3), using a pre-blocked and washed neutravidin-coated microplate. Genomic DNA is considerably more fragile than plasmid DNA. For aspirating and aliquoting the high-molecular weight genomic DNA, the choice of pipette is important due to the possibility of DNA shearing (*see* **Note 10**)
2. Genomic DNA will contain significantly reduced target sites relative to the plasmids used to initially adjust the probe conditions and thus the initial optimization steps should account for this (*see* **Note 11**).

3. The enrichment of high molecular weight DNA will produce long DNA fragments, the quantity can vary depending on the region of interest and the stringency of washing required by the PNA–DNA hybridization specificity and off-target binding. A nonessential but informative practice is to amplify the enriched product of the protocol by quantitative real-time PCR. If long DNA fragments are successfully enriched, it is feasible to quantitate with alternative specific alleles as we previously described using the *HLA-DQA1* gene which is downstream of the HLA-DRB1 locus that we probed [9]. Alternatively, standard low-copy number whole-genome amplification steps can be used in order to determine nonspecific enrichment.
4. Sequencing (Sanger and/or next generation) is useful to determine if haplotypes are being successfully isolated (*see* **Note 12**).

4 Notes

1. You can use an insert that is confined to your locus of interest or a larger insert containing more of the gene or the full gene sequence. We used homozygote DNA regions from within exon 2 to generate PCR fragments using allele-specific primers, for the HLA region, using the IMGT/HLA database which is a powerful tool for primer design [16] The gene insert can be size-compatible with the plasmid system that you are using (e.g., 100–5000 bp).
2. Non-biotinylated PNA probes can be used as negative controls in conjunction with biotinylated PNAs as assay probes in order to optimize the procedures for specific genes of interest.
3. Source of genomic DNA. If using blood, when possible process the high-molecular weight DNA extraction on the day of collection in order to minimize cell death and pro-apoptosis shearing of genomic DNA.
4. Due to the hydrophobic backbone, PNAs can be challenging to place into solution. The addition of a strong acid may be necessary along with vigorous agitation to improve solubility with a sonicator. Care should be taken limit the amount of TFA to the minimum required to improve solubility as addition of excessive TFA can denature the probe, indicated by evidence of precipitation of the probe.
5. It is important to ensure that the enrichment procedures for each probe are robust and independent. To achieve this, plasmid constructs were hybridized separately to both perfectly matched (PM) or mismatched (MM) fluorescently labeled PNAs. In this instance, the detection of Alexa Fluor 488 can be performed using a 488/8 nm excitation filter and a

520/8 nm emission filter, and in a separate assay for Alexa Fluor 532 labeled PNAs using a 530/8 nm excitation filter and 560/10 nm emission filter using a fluorescent pate reader such as Envision (Perkin Elmer).

6. If troubleshooting is required, monitor wash solutions to determine amount of non-microplate bound PNA. Using real-time PCR to generate a standard curve of the degree of DNA enrichment against input will guide the optimization of the enrichment. Alternatively, if enrichment is successful at the targeted sequence but not over flanking loci, the DNA may not be of sufficiently high molecular weight for the washing procedure and the number of washes needs to be decreased.
7. The quantification of the PNA–DNA complex can be influenced by the specific hybridization interaction, so the amount of washes, wash solutions and/or incubation times may need to be optimized. That is a lack of enrichment at the target site may be overcome by increasing stringency of wash stages by changing wash temperatures, buffer selection and/or wash number.
8. Extraction of high molecular weight genomic DNA from blood can be universally applied with standard DNA extraction kits, with a modified latter stage of the protocol. Columns are not recommended due to the increased shearing of DNA to low molecular weight. Typical DNA extraction strategies directly from blood which involve the separation of red blood cells (RBSs) via hypotonic lysis and buffy coat isolation allows for robust handling of samples until the actual DNA extraction from isolated lymphocytes.
9. Genomic DNA extracted by traditional means methods will produce sizes of DNA ranging between 10 and 50 kb in length. However, for the haplotype spanning *HLA-DRB1* to *HLA-DQB1*, enrichment of fragments >70 kb is essential since the aim of the protocol is to obtain significantly higher molecular weights than this value. This is achieved by eliminating steps that involve the utilization of purification columns and/or pelleting in exchange for glass-rod extraction and minimizing the steps that could involve brief agitation via increasing the duration at little to no cost in overall purity.
10. When handling genomic DNA it is particularly important to minimize shearing of the DNA strands. Wide pore pipette tips should be used and this can easily be achieved cutting the tips of disposable pipette tips and blunting with a naked flame.
11. Optimization of the protocol to the conditions of the users' laboratory setting and conditions using bulk amounts of plasmid is a critical component for the next stage in acquiring enrichment using genomic DNA. Successful optimization should factor a significantly decreased number of target sites,

lower quantities of DNA and increased fragility of linear genomic DNA with respect to shearing in comparison to circularized plasmid DNA. Optimization may include the PNA–DNA hybridization parameters (i.e., denaturation temperatures, hybridization buffer, incubation temperature, duration) wash optimization (numbers, adding and alternating wash detergents, buffers) and optimization of the elution from the solid-support microplate.

12. Sanger or next-generation sequencing of the enriched genomic DNA product can be informative for indicating if the hybridization or wash parameters require further optimization. That is if mismatched (MM) hybridization (i.e., the undesired allele) or entirely off-target hybridization is limiting the enrichment. In the case of MM hybridization, optimizing PNA–DNA hybridization conditions may be fruitful in increasing specificity, which should be followed by re-optimization of the washing and elution procedure. However, even a small enrichment of 5–10 % can often make it easier to resolve haplotypes.

Acknowledgment

N.M.M. was supported by a scholarship from the Monash Institute of Pharmaceutical Sciences, Monash University.

References

1. Nielsen PE, Egholm M, Berg RH, Buchardt O (1991) Sequence-selective recognition of DNA by strand displacement with a thymine-substituted polyamide. Science 254:1497–1500
2. Tewhey R, Bansal V, Torkamani A, Topol EJ, Schork NJ (2011) The importance of phase information for human genomics. Nat Rev Genet 12:215–223
3. Hosomichi K, Shiina T, Tajima A, Inoue I (2015) The impact of next-generation sequencing technologies on HLA research. J Hum Genet 60(11):665–673
4. Wittig M, Anmarkrud JA, Kassens JC, Koch S, Forster M, Ellinghaus E, Hov JR, Sauer S, Schimmler M, Ziemann M, Gorg S, Jacob F, Karlsen TH, Franke A (2015) Development of a high-resolution NGS-based HLA-typing and analysis pipeline. Nucleic Acids Res 43(11), e70
5. Boon T, Coulie PG, Van den Eynde B (1997) Tumor antigens recognized by T cells. Immunol Today 18(6):267–268
6. Gough SCL, Simmonds MJ (2007) The HLA region and autoimmune disease: associations and mechanisms of action. Curr Genom 8(7):453–465
7. Pavlos R, Mallal S, Phillips E (2012) HLA and pharmacogenetics of drug hypersensitivity. Pharmacogenomics 13(11):1285–1306
8. Ovsyannikova IG, Jacobson RM, Vierkant RA, Pankratz VS, Poland GA (2007) HLA supertypes and immune responses to measles–mumps–rubella viral vaccine: findings and implications for vaccine design. Vaccine 25(16):3090–3100
9. Murphy NM, Pouton CW, Irving HR (2014) Human leukocyte antigen haplotype phasing by allele-specific enrichment with peptide nucleic acid probes. Mol Genet Genomic Med 2:245–253
10. Mamanova L, Coffey AJ, Scott CE, Kozarewa I, Turner EH, Kumar A, Howard E, Shendure J, Turner DJ (2010) Target-enrichment strategies for next-generation sequencing. Nat Methods 7:111–118
11. Ray A, Nordén B (2000) Peptide nucleic acid (PNA): its medical and biotechnical applications

and promise for the future. FASEB J 14: 1041–1060
12. Siddiquee S, Rovina K, Azriah A (2015) A review of peptide nucleic acid. Adv Tech Biol Med 3:131
13. Paulasova P, Pellestor F (2004) The peptide nucleic acids (PNAs): a new generation of probes for genetic and cytogenetic analyses. Ann Genet 47:349–358
14. Ausubel FM, Roger Brent R, Kingston RE, Moore DD, Seidman JG, Smith JA, Struhl K (eds) (2002) Short protocols in molecular biology. Wiley, Hoboken, NJ
15. Muleya V, Wheeler JI, Irving HR (2013) Structural and functional characterization of receptor kinases with nucleotide cyclase activity. In: Gehring C (ed) Cyclic nucleotide signaling in plants: methods and protocols, vol 1016, Methods in molecular biology. Springer, New York, pp 175–194
16. Robinson J, Halliwell JA, Hayhurst JD, Flicek P, Parham P, Marsh SGE (2015) The IPD and IMGT/HLA database: allele variant databases. Nucleic Acids Res 43: D423–D431
17. Kotsch K, Wehling J, Blasczyk R (1999) Sequencing of HLA class II genes based on the conserved diversity of the non-coding regions: sequencing based typing of HLA-DRB genes. Tissue Antigens 53:486–497

Chapter 5

High-Throughput Sequencing of the Major Histocompatibility Complex following Targeted Sequence Capture

Johannes Pröll, Carina Fischer, Gabriele Michelitsch, Martin Danzer, and Norbert Niklas

Abstract

The Human Major Histocompatibility Complex (MHC) is a highly polymorphic region full of immuno-regulatory genes. The MHC codes for the human leukocyte antigens (HLA), proteins that present on the cellular surface and that are involved in self-non-self recognition. For matching donors and recipients for organ and stem-cell transplants it is important to know an individual's HLA haplotype determinable in this region. Now, as next-generation sequencing (NGS) platforms mature and become more and more accepted as a standard method, NGS applications have spread from research laboratories to the clinic, where they provide valid genetic insights. Here, we describe a cost-effective microarray-based sequence capture, enrichment, and NGS sequencing approach to characterize MHC haplotypes. Using this approach, ~4 MB of MHC sequence for four DNA samples (donor, recipient and the parents of the recipient) were sequenced in parallel in one NGS instrument run. We complemented this approach using microarray-based genome-wide SNP analysis. Taken together, the use of recently developed tools and protocols for sequence capture and massively parallel sequencing allows for detailed MHC analysis and donor-recipient matching.

Key words Next-generation sequencing, Targeted sequence capture, Major histocompatibility complex

1 Introduction

Constant improvements in next-generation sequencing will drive haematopoietic stem cell transplantation (HSCT) to produce significant better outcomes in clinical practice constantly.

The basis for these improved outcomes is a carful match of donors and recipients using detailed characterization of the highly polymorphic, multigene region called the Major Histocompatibility Complex (MHC). The MHC region is a ~4 Mb long region on chromosome 6 (6p21.3) [1–3]. Although it depicts a relatively

Electronic supplementary material: The online version of this chapter (doi:10.1007/978-1-4939-6750-6_5) contains supplementary material, which is available to authorized users.

Irene Tiemann-Boege and Andrea Betancourt (eds.), *Haplotyping: Methods and Protocols*, Methods in Molecular Biology, vol. 1551, DOI 10.1007/978-1-4939-6750-6_5, small part of the genome, it provides an interesting application for targeted sequencing approaches.

Sequence capture (or targeted enrichment) depends on two technologies, developed in parallel. Invention of massively parallel sequencing was triggered by the development of pyrosequencing in the late 1990 [4], emulsion PCR [5], and polony or bridge PCR [6, 7], which later led to the development of massively parallel sequencing, or next-generation sequencing (NGS). Independent of the development of NGS, DNA microarrays [8] led to the evolution of sequence capture techniques. In sequence capture, DNA fragments are used to physically isolate the region of interest that can then be sequenced using an NGS method. Sequence capture targeting the MHC refines the picture of HLA variation in addition to the known and well-studied HLA transplantation determinants. In addition, targeted NGS can be used for detecting novel MHC-related inflammatory, infectious or autoimmune pathologies. Moreover, targeted sequencing offers a sophisticated, modern method of HLA typing. HLA typing has been of interest since the beginning for several NGS companies, including major players like Roche [9, 10], Illumina [11], Ion Torrent [12], or Pacific Biosciences [13]. These companies have developed laboratory protocols and about 15 software products so far [14]. Among these, only one relies on a targeted enrichment strategy for library preparation (Wittig et al. [15]). In that context, soon after invention and first description by Albert et al. 2007 [16], we used targeted sequence capture and NGS for novel MHC analysis. This concept was then utilized by other groups to fine-tune the use of NGS for HLA typing.

The presented protocol describes the use of NimbleGen sequence capture technology on solid surface for the MHC and subsequent Roche/454 Life Sciences sequencing as published in Pröll et al. [17]. To assess the accuracy of this approach, we analyzed approximately 1500 SNP markers mapping along the sequenced MHC region using genome-wide SNP arrays (Affymetrix Genome Wide Human SNP Nsp/Sty 6.0). In our specific experimental setting, we concentrated on a patient who underwent HLA-matched allogeneic stem cell transplantation, the parents of the patient and the fully matched unrelated donor. Despite perfect HLA match established with traditional methods (Sanger sequencing-based HLA typing (SBT)), the patient developed acute graft-versus-host disease after transplantation and died one and a half month later. Using NGS, we describe 3025 single nucleotide variations, insertions, and deletions between recipient and donor in the investigated MHC region. Thus, we showed that MHC resequencing can identify a large number of genetic differences and variants potentially linked to onset of graft-versus-host disease (Figs. 1 and 2), although HLA regions were typed identically using traditional methods. In contrast to microarray-based SNP analysis or haplotype association studies, the sequence capture and

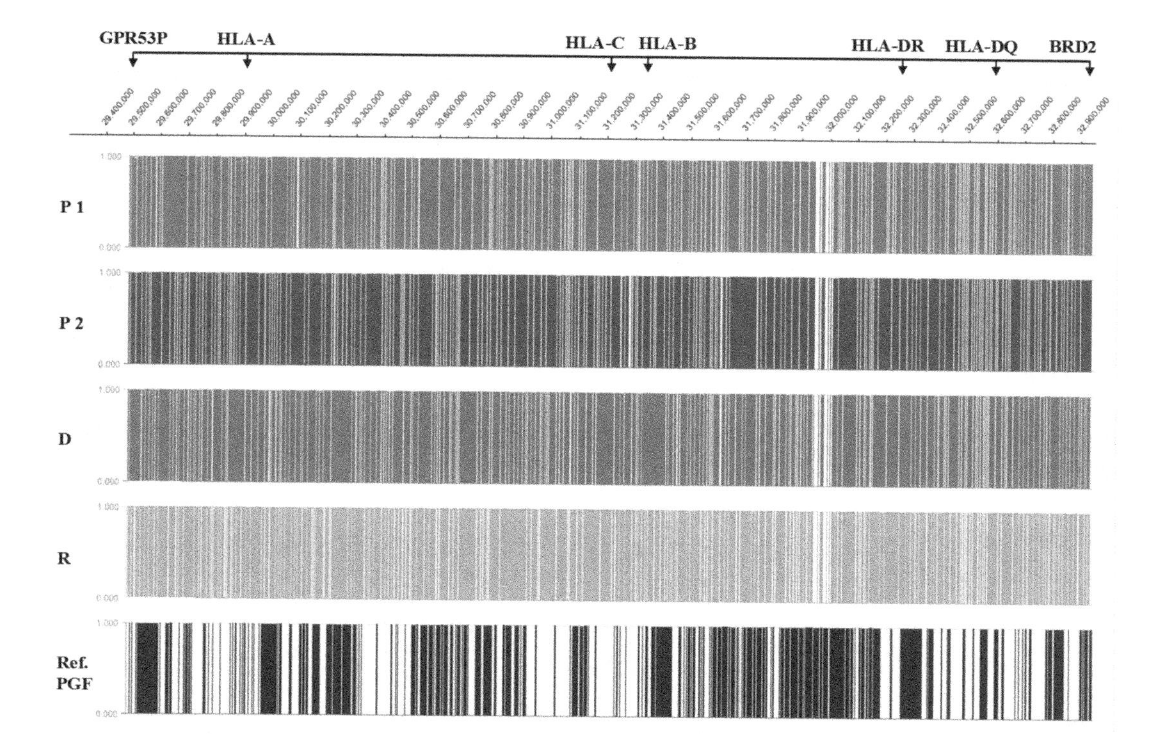

Fig. 1 MHC resequencing results for parents, donor, and recipient compared to annotated gene loci. *Graphs* are representing at least threefold coverage (present = 1) or gaps (absent = 0) for the complete 3.5 Mb spanning MHC region of the four sequenced DNA samples (*P1*, parent 1), (*P2*, parent2), (*D*, donor), and (*R*, recipient). The *graph* for the reference cell line PGF shows known annotated gene loci in *black* (gene annotation = 1)

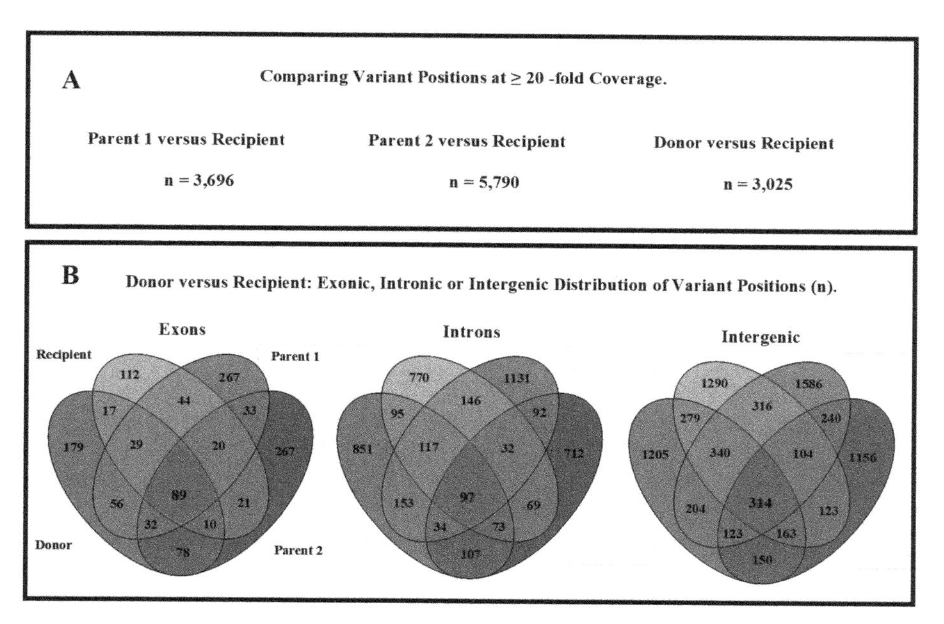

Fig. 2 Summary of MHC resequencing results representing variant positions between parents, donor, and recipients. Variant correlation to visualize the degree of sequence match among parents, donor, and recipient representing all variant positions in percentage of reads per position for two individuals (**A**). A count of variant positions is displayed in (**B**)

sequencing approach offers the near-complete information on genotypic differences. Important transplantation outcome determinants can be revealed at the single nucleotide resolution level. In the protocol described in this chapter, we have included all steps necessary for differentiated MHC analysis, enabling high-throughput even in a larger sample cohort.

2 Materials

2.1 Overview on Protocols, Reagents, and Equipment

1. Library Preparation (454 GS FLX Titanium General Library Preparation Kit).
   (a) GS Titanium DNA Library Prep Kit Nebulizers, Roche.
   (b) GS Titanium DNA Library Prep Kit Buffers, Roche.
   (c) MinElute PCR Purification Kit, Qiagen.
   (d) Ampure Beads, Agencourt.
   (e) 70% EtOH.
   (f) GS Titanium Library Reagent/Adaptors General, Roche 454 Life Sciences.
   (g) GS Titanium DNA Library Prep Kit Immobilization Beads, Roche 454 Life Sciences.
   (h) 3 M sodium acetate pH 5.2.
   (i) 10 N NaOH (Melt Solution).
   (j) BioAnalyzer DNA 7500 LabChip, Agilent.
   (k) Bioanalyzer 2100, Agilent.
2. Library Quality and Quantity Assessment.
   (a) Bioanalyzer RNApico 6000 LabChip.
3. LM-PCR of Pre-Captured Library.
   (a) GC-RICH PCR System, dNTPack, Roche.
   (b) QIAquick PCR Purification Kit, Qiagen.
   (c) Thermocycler, ABI.
4. Microarray Hybridization, Assembly of Elution Chamber, Microarray Washing and Elution of Captured Samples.
   (a) Sequence Capture 385 k Arrays, NimbleGen.
   (b) Sequence Capture Elution Champer ES1, NimbleGen.
   (c) Sequence Capture Wash&Elution Kit, NimbleGen.
   (d) Sequence Capture Hybridization Kit, NimbleGen.
   (e) Hybridization Oligo 454 Ti-A, NimbleGen.
   (f) Hybridization Oligo 454 Ti-B, NimbleGen.
   (g) MinElute PCR Purification Kit, Qiagen.
   (h) COT Human DNA, fluorometric grade, Invitrogen.

5. LM-PCR of Post-Captured Library.
    (a) GC-RICH PCR System, dNTPack, Roche.
    (b) QIAquick PCR Purification Kit, Qiagen.
6. qPCR of Pre-Captured and Post-captured Libraries.
    (a) LightCycler® 480 SYBR Green I Master, Roche.
    (b) Primers, NimbleGen.
    (c) LightCycler® 480 Instrument II, Roche.
7. emPCR (GS FLX Tit.LV or SV emPCR Kits).
    (a) GS Titanium emPCR Reagents (Lib-L), Roche 454 Life Sciences.
    (b) GS Titanium emPCR Emulsion Oil SV, Roche 454 Life Sciences.
8. Sequencing: Genome Sequencer FLX Operation (GS FLX Titanium Sequencing Kit XLR70); all Roche 454 Life Sciences.
    (a) GS Titanium Sequencing Reagent XLR70.
    (b) GS Titanium Sequencing Packing Beads 70.
    (c) GS Titanium Sequencing Buffer CB.
    (d) GS Titanium Sequencing Supplement CB.
    (e) GS Titanium Sequencing Bead/Wash Buffers.
    (f) GS FLX Sipper tubes.
    (g) GS Sequencing Pre-Wash Tubes.
    (h) Sodium Chlorite tablet.
    (i) Genome Sequencer FLX instrument.
9. Data Processing and Analysis (454 GS Reference Mapper Software, Roche 454 Life Sciences).

2.2 Genomic DNA

Following the instruction of the manufacturer, genomic DNA was isolated and purified from venous blood samples using magnetic bead extracting on a MagNaPure Compact instrument (Roche Diagnostics, Mannheim, Germany). The DNA concentration and quality was photometrically determined (BioPhotometer, Eppendorf, Hamburg, Germany). Quality control of successful DNA fragmentation via gel or Bioanalyzer (Agilent) is optional. Any other instrument for genomic DNA extraction would be applicable as well.

2.3 Sequence Capture and Enrichment

1. NimbleGen Sequence Capture Arrays specifically enrich regions of interest prior to sequencing: long oligos (55–100 mers) complementary to user-defined genomic regions are directly synthesized on an array (Supplement File).
2. NimbleGen supplies appropriate arrays, mixers, mixer port seals, and reusable elution chambers (Fig. 3).
3. NimbleGen user's guide and design files (in gff, general feature format) and bed format together with basic visualization software.

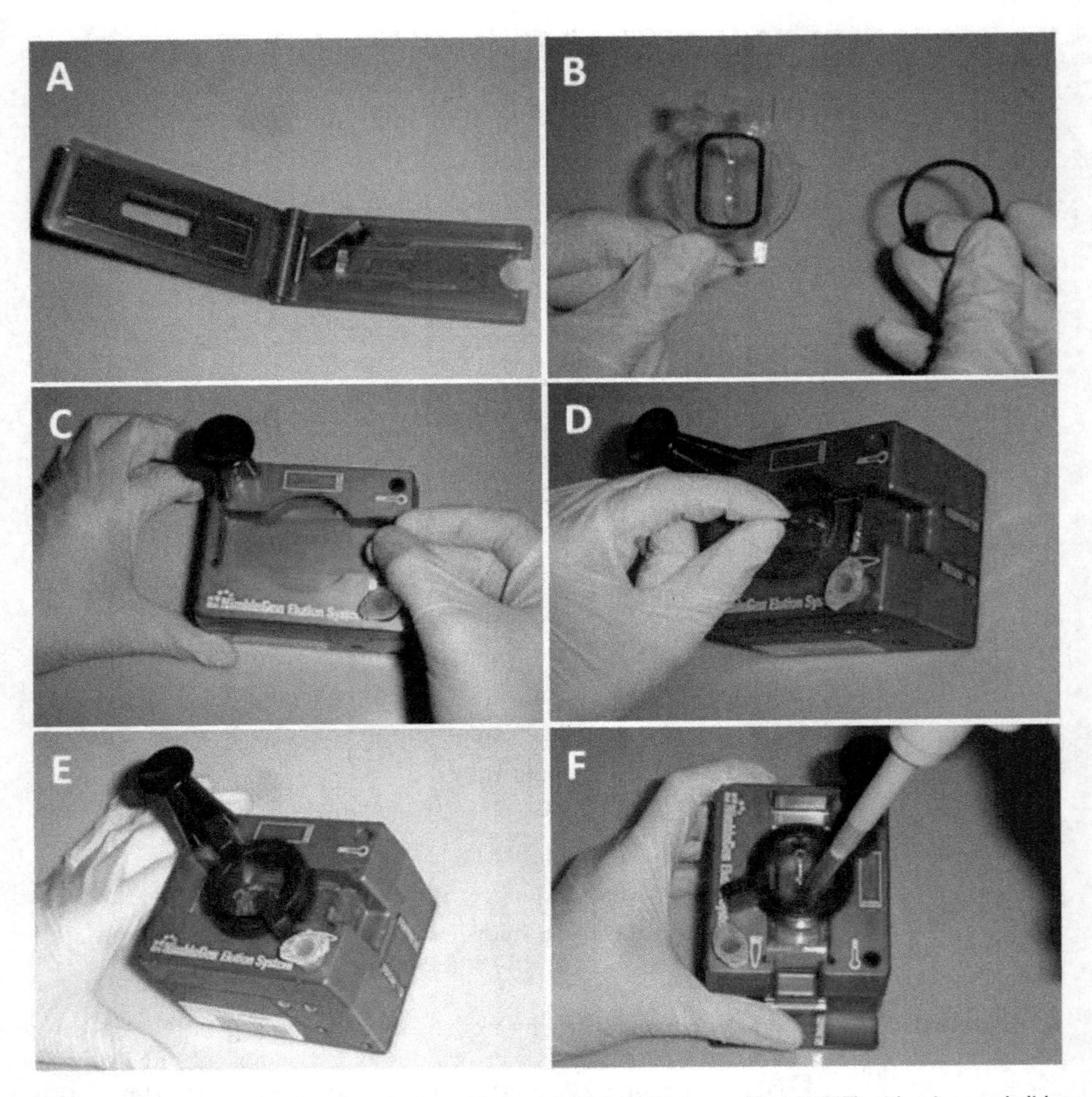

Fig. 3 Overview of sequence capture devices. (**a**) Precision Mixer Alignment Tool (PMAT) with mixer and slide, (**b**) Elution chamber, (**c**) Placement of a slide into the elution system, (**d**) Placement of the elution chamber in the elution system, (**e**) Placement and locking the elution system retaining ring, (**f**) Tilting the elution system and loading sample into the elution chamber

2.4 Next-Generation Sequencing

In this protocol the Genome Sequencer FLX Instrument and GS FLX Titanium reagents are used for performing next-generation sequencing, both Roche/454 Life Sciences owned products (*see* **Note 1**). The pre-capture LM-PCR (ligation mediated PCR), hybridization and washing protocol of the NimbleGen Sequence Capture array is followed by elution of the captured library, LM-PCR of captured samples, and quantitative PCR (qPCR) to assess the capture success. All steps are carried out exactly as indicated in the provided protocols. The NGS process is started using the 454 GS FLX Titanium General Library Preparation Kit (Roche Applied Science, Cat. No. 05233747001) for emulsion PCR). The process is continued with the purification and quantification of DNA-carrying beads, loading of the PicoTiterPlate (PTP), and supplying the NGS-Sequencer with sequencing reagents, and wash buffer.

2.5 NGS-Data Analysis

Computing performance depends on the total number of reads generated. Using 454 sequencing an average desktop computer is sufficient, subsequent analysis can be performed using a series of Perl (https://www.perl.org/) scripts or R [19]. The GS Reference Mapper can be replaced by any other mapping and variant detection tool, e.g., Genomics Workbench (Qiagen/CLCbio, Aarhus, Denmark).

3 Methods

It is a major advantage that the protocol does not require an additional construction of a sequencing library after the sample is enriched using the NimbleGen Sequence Capture array. This reduces time and reagent costs significantly. After fragmentation by nebulization, the genomic DNA library is captured, eluted, and amplified. After these steps, it is directly ready for emulsion PCR (emPCR amplification) using GS FLX Titanium emPCR Kits (Roche Applied Science).

3.1 Sequence Capture and Enrichment

1. The long oligonucleotides of the NimbleGen Sequence Capture Arrays complementary to the region of interest are synthesized on an array. These oligonucleotides hybridize their targets of a genomic DNA sample in a typical hybridization reaction performed in the NimbleGen sequence capture array (*see* Fig. 3). A specific feature is built in that enables the assessment of successful linker to the probes and the determination of an enrichment factor by quantitative PCR. These additions are extremely helpful during the progress of the protocol, as described later when referring to LM-PCR (linker-mediated PCR). The NimbleGen protocol leads to a sequencing ready library in a minimum of 3–4 days; *see* Fig. 4 for a complete hands-on and total time needed (*see* **Note 2**).
2. In the first step, it is necessary to define the region-of-interest (ROI) in terms of chromosomal coordinates. Using ENSEMBL v49 release (HG 18 in UCSC nomenclature) as a reference genome, we identified the target bases covering the complete MHC-class I, II, and III regions. Capture sequences for a ROI of 3,451,791 bp in total were designed, resulting in 2,922,962 target bases (84.7%) being covered physically by

Fig. 4 Overview of time loads for sequence capture and sequencing

capture oligonucleotides. This design was achieved using NimbleGen's default settings for probe selection, resulting in the omission of 528,829 bp (15.3 %) of the initial target region due to, e.g., hybridization conditions and repeat status. Capture sequence coordinates are provided in supplements.

3. Design and production of a sequence capture array (385 k) was done at NimbleGen (Roche NimbleGen, Madison, WI) targeting 3.5 Mb reference sequence spanning the MHC complex on chromosome 6 (NCBI: chr6 NGS primary_target_region 29,594,756-33,046,546).

3.2 Targeted Sequence Capture Protocol

This protocol was performed according to the manufacturer's instructions, including pre-capture ligation-mediated-PCR (LM-PCR), hybridization and washing of the NimbleGen sequence capture array, elution of the captured library, LM-PCR of captured samples, and quantitative PCR (qPCR) assays to assess capture success. Preparation of a GS FLX Titanium General DNA Library requires 3–5 μg of genomic DNA for the double Solid Phase Reversible Immobilization (SPRI) method we used or 5–10 μg for the alternative gel cut method (*see* **Note 3**).

The targeted sequence capture protocol starts with 5 μg DNA for nebulization and consists of the following steps:

1. Library Preparation (454 GS FLX Titanium General Library Preparation Kit).
2. Small Fragment Removal with Ampure Beads and magnetic particle concentrator (MPC).
3. DNA Sample Quality Assessment with DNA 7500 LabChip on the BioAnalyzer (Fig. 2).
4. Fragment End Polishing protocol with T4 DNA Polymerase Kit.
5. Adaptor ligation with adaptors and ligases.
6. Small Fragment Removal with Ampure Beads.
7. Library Immobilization.
8. Fill-In Reaction.
9. ssDNA (single strand) Library Isolation.

3.2.1 Library Preparation (454 GS FLX Titanium General Library Preparation Kit)

1. Press the pierced rubber stopper firmly into the Nebulizer top and insert the Millipore filter unit into the stopper hole (Fig. 3).
2. Affix a nebulizer condensor tube around the aspiration tube. To ensure proper function, make sure to push the condensor tube all the way down around the base of the aspiration tube, being careful not to rotate the aspiration tube. Set the assembled nebulizer top on the bench, with the aspiration tube pointing upward. Make sure that the inside parts do not contact any contaminated surfaces (counter top, hands).
3. *Double SPRI size with* 5 μg of sample DNA (in TE), also *see* **Note 4**.
4. Add TE to a final volume of 100 μl.
5. Add 500 μl of nebulization buffer; add the nebulization condenser tube on the flexible hose of the closer head of the beaker.
6. Transfer the nebulizer to the externally vented nebulization hood.
7. Connect the loose end of the nebulizer tubing to the nitrogen tank.
8. Direct 2.1 bar of nitrogen through the nebulizer for 1 min, centrifuge and check sample volume (should be >300 μl); do

not collect any material that may have lodged outside the nebulizer. This material may not have been completely fragmented and could cause problems later on.

9. Elution step (MinElute):
10. Add 2.5 ml of Buffer PBI to nebulized sample, transfer 750 μl each into two spin columns, spin down for 1 min. 18,000 rpm, discard flow through, add the second half of your sample approx. 750 μl to the column, spin again, discard flow through, add 750 μl PE, spin, discard the tube, take a new one 1.5 ml tube and spin (leave the lid opened), turn the tube 180°, spin, discard tube and use a new one 1.5 ml tube, incubate at 37 °C for 5 min), elute with 15 μl of Buffer EB, incubate at RT for 2 min and spin, reapply, incubate for another 2 min and spin again, pool the eluates of the two columns, for a total volume of ~20 μl, 10) take out 2 μl for check on BioAnalyzer.

3.2.2 Small Fragment Removal

Allow Ampure Beads to equilibrate to RT for 30 min before use.

1. Measure the volume of the pooled eluates (e.g., using a pipettor).
2. Add EB Buffer to a final volume of 100 μl. Always vortex the Ampure beads vigorously, optionally aliquot the beads; add 65 μl of Ampure beads as determined per the calibration, vortex.
3. Incubate 5 min at room temperature (let sit in tube rack).
4. Pellet the beads using an MPC, (wait 2 min) (*see* **Note 5**).
5. Remove the supernatant and discard.
6. Wash the beads twice with 500 μl of 70% Ethanol (do not vortex!), incubating for 30 s each time, leave the tubes in the MPC, remove the magnetic slide.
7. Remove the supernatant and discard.
8. Allow the Ampure beads to air dry completely (5 min. 37 °C), assure that the pellet is completely dry.
9. Remove the tube from the MPC.
10. Add 24 μl of Buffer EB, vortex to resuspend the beads.
11. Pellet the beads using an MPC.
12. Transfer the supernatant to a fresh tube, store sample(s) at −15 to −25 °C.

3.2.3 DNA Sample Quality Assessment (Fig. 5)

Run 1 μl aliquot on a BioAnalyzer DNA 7500 LabChip, sample 1: after Nebulization Sample, sample 2: after Ampure purification.

3.2.4 Fragment End Polishing

1. Prepare the following mix:

   ~23 μl purified, nebulized DNA fragments (add EB Buffer to a final volume of 23 μl), 5 μl 10× Polishing Buffer, 5 μl BSA, 5 μl ATP, 2 μl dNTPs, 5 μl T4 PNK, 5 μl T4 DNA Polymerase, 50 μl final volume, mix well.

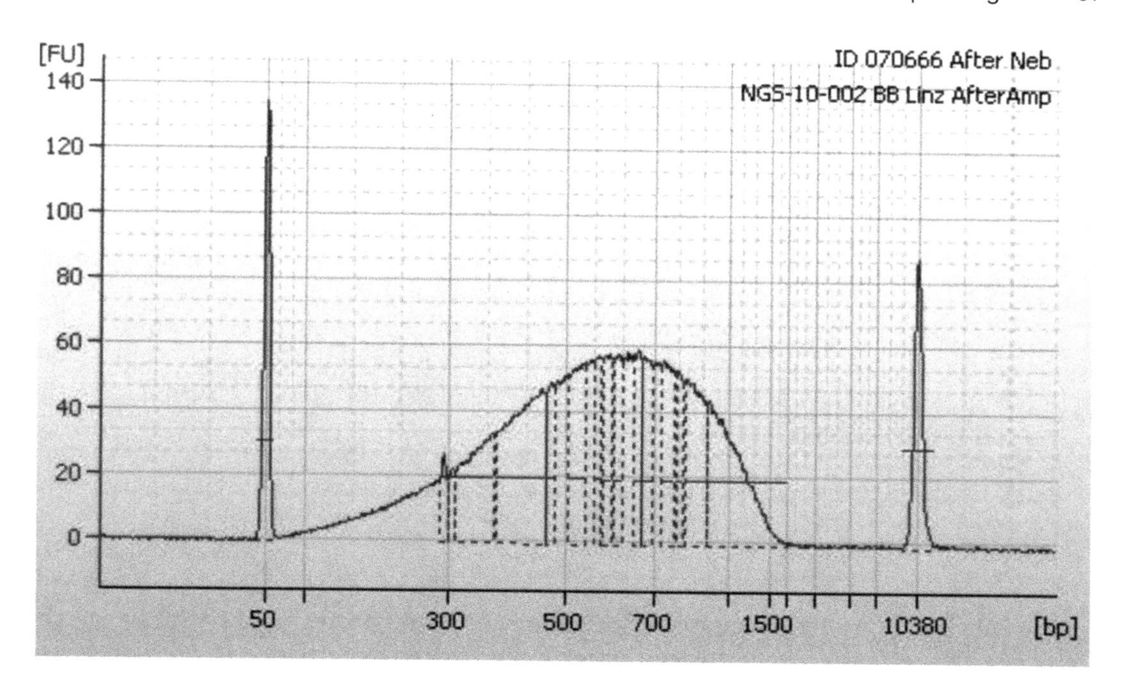

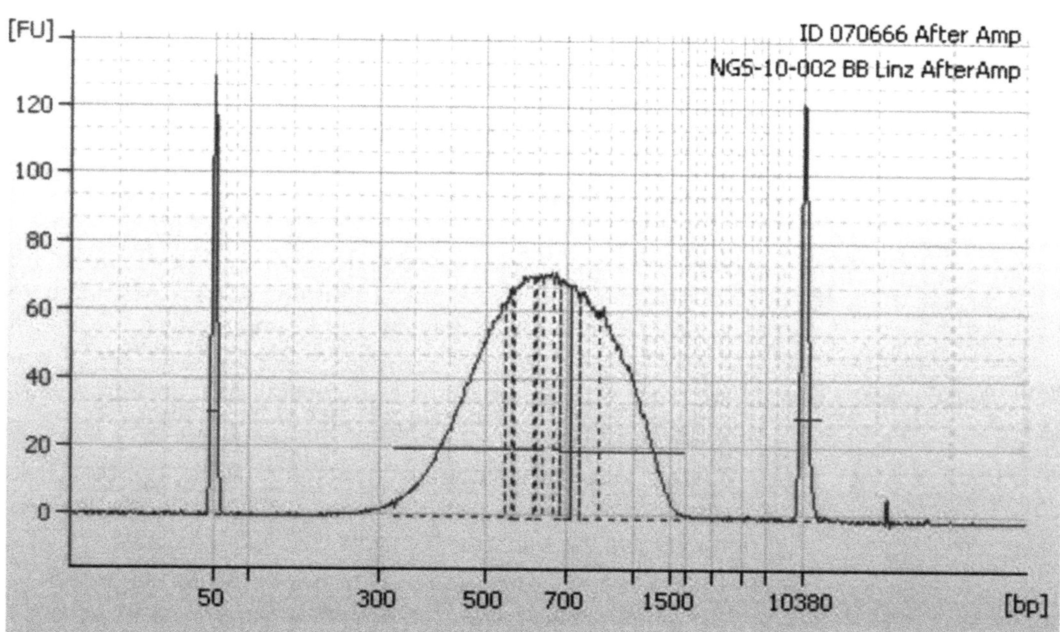

Fig. 5 DNA sample quality assessment after nebulization and after purification using the BioAnalyzer DNA 7500 LabChip. The mean size (*top of peak*) is between 500 and 800 bp, and not more than ~10 % of the material is below 350 or above 900 bp

2. Incubate for 15 min at 12 °C (Thermomixer).
3. Incubate for an additional 15 min at 25 °C (Thermomixer).
4. Purify the polished fragments on one column as follows (MinElute): to 1 sample/column add 5 volumes (250 μl) of Buffer PBI, transfer to one spin column, spin down, discard the flow-through, add 750 μl PE, spin again, discard the tube, take a new one 1.5 ml tube and spin (leave the lid opened), turn the tube 180°, spin, discard tube and use a new one 1.5 ml tube, incubate at 37 °C for 5 min, elute with 15 μl of Buffer EB, incubate at RT for 2 min and spin, reapply, incubate for another 2 min and spin again.

3.2.5 Adaptor Ligation

1. Prepare following mix: ~10 μl nebulized, polished DNA fragments, 20 μl 2× Ligase Buffer, 5 μl Adaptors, 5 μl Ligase, 40 μl total, mix well and spin briefly.
2. Incubate for 15 min at 25 °C (Thermomixer). Start with washing the library immobilization beads, use 50 μl of Library Immobilization Beads. In a new tube, use MPC to pellet the beads and remove the buffer, wash two times with 100 μl of 2× Library Binding Buffer, using the MPC, resuspend the beads in 25 μl of 2× Library Binding Buffer, store on ice.
3. Purify the ligation products from **step 2** as follows (MinElute): to 1 sample/column add 5 volumes (200 μl) of Buffer PBI, transfer to one spin column, spin down, discard the flow-through, add 750 μl PE, spin again, discard the tube, take a new one 1.5 ml tube and spin (leave the lid opened), turn the tube 180°, spin, discard tube and use a new one 1.5 ml tube, incubate at 37 °C for 5 min, elute with 50 μl of Buffer EB, incubate at RT for 2 min and spin, reapply, incubate for another 2 min and spin again, store sample on ice.

3.2.6 Small Fragment Removal

1. Measure the volume of the eluate, using a pipettor.
2. Add EB Buffer to a final volume of 100 μl.
3. Always vortex the Ampure Beads vigorously, optionally aliquot the beads, add the amount of Ampure beads as determined per the calibration, vortex.
4. Incubate 5 min at room temperature (let sit in tube rack).
5. Pellet the beads using an MPC, (wait 2 min) (*see* **Note 5**).
6. Remove the supernatant and discard.
7. Wash the beads twice with 500 μl of 70% Ethanol, do not vortex! incubating for 30 s each time, leave the tubes in the MPC, remove the magnetic slide.
8. Remove the supernatant and discard.
9. Allow the Ampure beads to air dry completely (min. 5 min. 37 °C), assure that the pellet is completely dry.

10. Remove the tube from the MPC.
11. Add 25 μl of Buffer EB, vortex to resuspend the beads.
12. Pellet the beads using an MPC.
13. Transfer the supernatant to a fresh microcentrifuge tube.

3.2.7 Library Immobilization

1. Vortex washed immobilized. Beads before adding ligated DNA.
2. Add the 25 μl of the ligated DNA sample to the beads, mix well.
3. Rotate sample for 20 min at +15 °C to +25 °C, 30×, Do not vortex your sample after this step.
4. Wash the immobilized Library with 100 μl of Library Wash Buffer, using the MPC.
5. Resuspend in 100 μl of Library Wash Buffer.

3.2.8 Fill-In Reaction

1. Prepare following mix: 40 μl Molecular Biology Grade water, 5 μl 10× Fill-in Polymerase Buffer, 2 μl dNTPs, 3 μl Fill-in Polymerase, 50 μl total.
2. Using the MPC, remove 100 μl of Library Wash Buffer from the library-carrying beads.
3. Add the 50 μl of the fill-in Reaction Mix to beads.
4. Mix gently, do not vortex and without pipetting in.
5. Incubate for 20 min at 37 °C (Thermomixer).
6. Wash the immobilized library with 100 μl of Library Wash Buffer, using the MPC.
7. Resuspend in 100 μl of Library Wash Buffer.

3.2.9 ssDNA Library Isolation

1. Prepare the neutralization solution by mixing 500 μl of PBI and 10 μl of 3 M Sodium acetate (pH 5.2); prepare the melt solution by adding 125 μl of 10 N NaOH to 9.875 ml of PCR grade water.
2. Using the MPC, remove the 100 μl of Library Wash Buffer from the Library-carrying beads.
3. Add 50 μl of Melt solution to the beads; vortex.
4. Using the MPC, pellet the beads and transfer the supernatant to the neutralization solution repeat for a total of two 50 μl Melt solution washes of the beads (pooled together into the same tube of neutralization solution) after the addition of each Melt Solution wash and mixing.
5. Purify neutralized ssDNA library as follows (MinElute): 1 sample/column, transfer melted sample to one spin column, spin down, discard the flow-through, add 750 μl PE, spin again, discard the flow-through, add 750 μl PE again, spin again, discard the tube, take a new one 1.5 ml tube and spin (leave the lid opened), turn the tube 180°, spin, discard tube and use a

new one 1.5 ml tube, wait for a minimum of 2 min at 37 °C, elute with 20 μl of Buffer TE, incubate at RT for 2 min and spin, reapply, incubate for another 2 min and spin again.

6. Run 1 μl of library on a RNA Pico 6000 LabChip to assess the quality of the library run, program: mRNA Pico.
7. Optional: Quantitate the DNA library (1 μl, in triplicate) by fluorometry. Use the RiboGreen method (Invitrogen).
8. Assess the quality of the DNA library for the characteristics as follows: Average fragment length: between 500 and 800 bp, Lower size cutoff: <10% below 350 bp, upper size cutoff: <10% above 1000 bp, DNA yield: ≥5 ng, adaptor dimer peak: <5% of library peak height.
9. For each of the five reactions in the Pre-Capture Library amplification, 3 ng of template is required.

3.3 Pre-Capture Library Amplification by LM-PCR

Mastermix for 1 PCR reaction with 3 ng each sample:

Sample	1	2	3	4
DNA (μl)	1.33	1.42	1.45	5.48
Water (μl)	23.67	23.58	23.55	19.51

MM for 5 PCR reactions:

Sample	1	2	3	4
DNA (μl)	7.31	7.81	7.96	20.00
Water (μl)	130.19	129.69	129.54	105.00

1. Prepare five PCR tubes each containing about 3 ng of template in a total volume of 25 μl of PCR-grade water.
2. Prepare LM-PCR Master Mix in a 1.5 ml tube, the amount of each reagent needed for one capture (five reactions) is listed below:

Mastermix: (125 μl)	
5× GC-RICH PCR reaction buffer (vial 2):	50 μl
25 mM $MgCl_2$ (vial 4):	10 μl
PCR grade Nucleotide Mix (vial 6):	5 μl
25 μM LM-PCR 454 Ti-A Oligo:	5 μl
25 μM LM-PCR 454 Ti-B Oligo:	5 μl
PCR-grade water (vial 5):	45 μl
GC-RICH Enzyme Mix (vial 1):	5 μl

3. Pipette 25 µl of the LM-PCR Master Mix into each of the five tubes containing 25 µl of template, mix well by pipetting.
4. Perform PCR amplification in a thermocycler using the following cycling conditions:

94 °C	4 min	
94 °C 58 °C 68 °C	30 s 1 min 1.5 min	11 cycles
68 °C	3 min	
4 °C	hold	

Store samples at 4 °C until ready for cleanup

5. Clean up with QIAquick PCR Purification Kit.

   Combine the five reactions for each sample into one 1.5 ml tube (~250 µl), add 1250 µl (5 volumes) of Buffer PB to each tube and mix well, transfer 750 µl of the sample to the spin column and centrifuge for 1 min, discard the flow through, load the rest of the sample into the spin column and centrifuge for 1 min, add 750 µl of Buffer PE to the column, centrifuge and discard the flow-through, take a new one 1.5 ml tube and spin (leave the lid opened), turn the tube 180°, spin, discard tube and use a new one 1.5 ml tube, incubate at 37 °C for 5 min, elute with 50 µl of PCR-grade water, incubate at RT for 2 min and spin, reapply, incubate for another 2 min and spin again,
6. Measure the volume and the DNA concentration on a photometer.

Sample	1	2	3	4	Negative control
ng/µl	0.204	0.2069	0.1963	0.1208	0.0145
260/280	1.85	1.83	1.76	1.77	1.5

7. For quality check, run 1 µl of pre-capture LM-PCR product on a BioAnaylzer DNA 7500 chip (*see* **Note 6**). The required sample quality should be A260/A280: 1.7–2.0, pre-capture LM-PCR yield: > 3 µg, library fragments: 350–1000 bp, average fragment length: 500–800 bp.

3.4 Microarray Hybridization, Assembly of Elution Chamber, Microarray Washing and Elution of Captured Samples

3.4.1 Prepare Hybridization System

1. Prepare the Hybridization System: set the Hybridization System to 42 °C with the cover closed, allow at least 3 h for the heat block temperature to stabilize at 42 °C (*see* **Note 7**).
2. Turn on one heat block to 95 °C and another to 70 °C and let equilibrate.
3. Prepare the Hybridization Cocktail. Add 100 µg of COT Human DNA to 3 µg of pre-capture LM-PCR amplified library, keep the tube lid closed and poke one hole on the top

with a 18–20 G needle, dry the sample in a SpeedVac on high heat (60 °C).

4. Add 3.5 μl of water to each sample to rehydrate, place a port seal on the top of the hole on the 1.5 ml tube, vortex the sample, and centrifuge at maximum speed for 30 s.
5. Place the sample in 70 °C heatblock for 10 min.
6. Vortex the sample and centrifuge at maximum speed for 30 s.
7. Add to each sample: Reagent 385 K, 0.65 μl of Hybridization Enhancing 454 Ti-A, 0.65 μl Hybridization Enhancing 454 Ti-B and vortex the sample and centrifuge at maximum speed for 30 s.
8. Add to each sample: Reagent 385 K, 8 μl of 2× Hybridization buffer, 3.2 μl of Hybridization Component A. Vortex the sample and centrifuge at maximum speed for 30 s.
9. Place each sample in a 95 °C heat block for 10 min to denature the DNA.
10. Centrifuge samples at maximum speed for 30 s and store at 42 °C until ready for hybridization, hybridize within 15 min of denaturation.

3.4.2 Prepare Mixers

1. Remove the NimbleGen X1 (for 385 K arrays) mixer from its package. For best results blow compressed gas across the mixer and slide to remove any dust or debris. Load samples within 30 min of opening the vacuum-packaged mixer to prevent the formation of bubbles during loading and/or hybridization.
2. Position the Precision Mixer Alignment Tool (PMAT) with the hinge on the left and then open it.
3. Snap the mixer onto the two alignment pins on the lid of the PMAT with the tab end of the mixer toward the inside hinge and the mixers adhesive gasket facing outward.
4. Place the slide in the base for the PMAT while pushing back the plastic spring so that the barcode is on the right and facing up and the corner of the slide sits against the plastic spring. Remove your thumb and make sure the spring is engaging the corner of the slide and the entire slide is registered to the edge of the PMAT to the rightmost and closest to you. In addition, be sure that the slide is lying flat against the PMAT. Optional: gently blow compressed gas across the mixer.
5. Remove the backing from the adhesive gasket using forceps and close the lid of the PMAT so that the gasket makes contact with the slide.
6. Lift the lid by grasping the long edges of the PMAT while simultaneously applying pressure with two fingers through the window in the lid of the PMAT to free the mixer-slide assembly from the pins.

7. Remove the mixer-slide assembly from the PMAT. Place the mixer-slide assembly on the backside of the PMAT.
8. Rub the mixer brayer over the mixer with moderate pressure to adhere the adhesive gasket and remove any bubbles. Start in the center of the array and rub outward. The adhesive gasket will become clear when fully adhered to both surfaces.
9. Place the mixer-slide assembly in the slide bay of the Hybridization System.
10. Repeat **steps 1–9** for all slides to hybridize.

3.4.3 Load and Hybridize Samples

1. Load the sample into the mixer chamber without introducing any bubbles, if there are, gently massage bubbles to either of the ends of the mixer chamber or to the sides, pipette any extra sample onto the port (no bubbles) and remove bubbles through the opposite port sample volume: 385 K array + X1 mixer (15–16 μl).
2. Dry any expose sample from the fill and vent ports with a clean tissue or cotton swab.
3. Adhere port seals to cover the fill and vent ports on the mixer, press on both the vent and fill port stickers at the same time to ensure equal pressure.
4. Close the bay clamp until locked (you should hear a click).
5. Turn on the mix mode and select mode B on the Hybridization Station, confirm the system recognizes the slide in each occupied bay.
6. Approximately 5–10 min after starting the Hybridization System ensure that the mix mode is set to B, ensure that the green light is displayed for all occupied stations, ensure that the temperature reads 42 °C.
7. Hybridize samples at 42 °C for at least 64 h (not more than 72 h).

3.4.4 Assembly Elution Chamber (See Note 8*)*

1. Remove one clear polycarbonate Elution Insert from its plastic bag.
2. Remove one Elution System Gasket for each Elution Chamber and make sure it is intact.
3. Insert the Elution System Gasket into the oblong channel of the molded Elution Insert, starting with the two straight portions of the channel. Fit the Elution System Gasket into the remaining rounded part of the channel while still holding down the center portion of the Elution Insert.
4. Gently press the Elution System Gasket down into the channel to ensure proper, uniform Elution Chamber assembly (in plastic bag).

3.4.5 Microarray Washing

Prepare the water bath temperature at 47.5 °C; equilibrate buffers at 47.5 °C at least 2 h before washing microarray (*see* **Note 9**).

1. Prepare the following solutions: (appropriate for single array): 125 mM NaOH (one 1.5 ml tube per sample!) (*see* **Note 10**), 20% acetic acid; vortex and spin down briefly, store at room temperature until use.
2. Dilute 10× SC Wash Buffers I, II, and III and 2× Stringent Wash Buffer to 1× working solution. Prepare the wash buffers for each slide:

Wash Buffer	Volume (ml)	Temperature
1× Stringent wash buffer	240	47.5 °C (±1 °C) (2 tubes, à 120 ml)
1× Wash buffer I	120	RT
1× Wash buffer II	240	1×RT; 1×47.5 °C (2 tubes, à 120 ml
1× Wash buffer III	120	RT

3. Remove the slide from the Hybridization System and place slide into disassembly basin containing 100 ml Wash II heated to 47.5 °C, wait about 10 s for the temperature to equilibrate.
4. Carefully peel the mixer off of the slide, handle the slide only by the edges and portions exposed before mixer removal, and never touch the surface of the slide inside the mixer area.
5. Transfer the slide to wash tube containing 47.5 °C Wash Buffer II; wash the array as indicated below (*see* **Note 11**):

Wash buffer	Inversions/incubation	Temperature (°C)
1× Wash buffer II	10×	47.5
1× Stringent wash buffer	10×/5 min	47.5
1× Stringent wash buffer	10×/5 min	47.5
1× Wash buffer I	1 inversion per s/2 min	RT

6. Remove the slide from the wash tube using your fingers and gently tap the slide on a paper towel and wipe the back side with a clean tissue (the front side of the slide is the side on which you can read the barcode numbers in the correct orientation).
7. Place the slide within the room temperature Elution System.
8. Place the disposable Elution Chamber on the top of the slide with the Elution System Gasket facing down.
9. Place an Elution System Retaining ring over the chamber and lock into place, tilt the top of the Elution System until it locks into the upright position.

10. Slowly pipet 425 µl of the freshly made 125 mM NaOH into the bottom filling port of the Elution Chamber, return the Elution System to the horizontal position. Return any excess 125 mM NaOH that does not fit in the Elution Chamber to the original 1.5 ml tube (*see* **Note 12**).
11. Incubate at room temperature for 10 min.
12. Add 20% Acetic Acid (16 µl) solution to Buffer PBI (500 µl) in a 1.5 ml tube for each array, pipette up and down ten times to mix.
13. Pipette eluted DNA out to the Elution Chamber and place into a 1.5 ml tube containing 20% Acetic Acid/Buffer PBI. Any extra volume of 125 mM NaOH that did not fit into the Elution Chamber transfer also to the neutralization solution, pipette up and down ten times to mix (*see* **Note 13**).
14. Disassemble and discard the Elution Chamber.

3.4.6 Purification with MinElute (max. Binding Capacity: 5 µg)

1. Transfer the neutralization solution into one spin column, spin down for 1 min 13,000 rpm, discard flow through, add the remaining volume of the sample to the column, spin again, discard flow through.
2. Add 750 µl PE, spin, discard the tube, take a new one 1.5 ml tube and spin (leave the lid opened), turn the tube 180°, spin, discard tube and use a new one 1.5 ml tube, incubate at 37 °C for 5 min.
3. Elute with 50 µl Buffer EB, incubate at RT for 1 min, spin.
4. Reapply, incubate for another 1 min and spin again.
5. Proceed to captured LM-PCR or keep the eluate at –20 °C until use.

3.5 LM-PCR of Post-Captured Library

LM-PCR captured samples:

Amplification of each captured sample	Ten reactions
Amplification of negative control	Two reactions

1. Prepare the LM-PCR Master Mix. The amount of each reagent needed for 1 captured sample (= 10 reactions) is listed below:

5× GC-RICH PCR reaction buffer (vial 2)	100 µl
25 mM MgCl2 (vial 4)	20 µl
PCR grade Nucleotide Mix (vial 6)	10 µl
25 µM LM-PCR 454 Ti-A Oligo	10 µl
25 µM LM-PCR 454 Ti-B Oligo	10 µl
PCR-grade water (vial 5)	300 µl
GC-RICH Enzyme Mix (vial 1)	10 µl
	460 µl

2. Prepare ten PCR reactions, each containing 4 µl of captured eluted DNA.
3. Pipette 46 µl of LM-PCR Master Mix into each of 10 reactions, mix by pipetting up and down.
4. Place 0.2 ml tubes in the thermocycler und use following program:

94 °C	4 min	
94 °C 58 °C 68 °C	30 s 1 min 1 min 30 s	19 cycles
68 °C	3 min	
4 °C	hold	

5. Purification with QIAquick PCR Purification Kit, combine the ten reactions for each sample into two 1.5 ml tube, five reactions per tube (~250 µl), add 1250 µl (5 volumes) of Buffer PB to each tube and mix well, transfer 750 µl of the sample to the spin column and centrifuge for 1 min, discard the flow through, load the rest of the sample into the spin column and centrifuge for 1 min, add 750 µl of Buffer PE to the column, centrifuge and discard the tube, take a new one 1.5 ml tube and spin (leave the lid opened), turn the tube 180°, spin, discard tube and use a new one 1.5 ml tube, incubate at 37 °C for 5 min, elute with 50 µl of TE-buffer, incubate at RT for 1 min and spin, reapply, incubate for another 1 min and spin again.
6. Combine the two 50 µl eluates per sample (100 µl total).
7. Analyze 1 µl of the sample on BioAnalyzer (DNA 7500 Chip) and measure on the photometer the required sample quality: A260/A280: 1.7–2.0, LM-PCR yield: > 1 µg, no visible primer dimer peak, average fragment length: 500–800 bp.

Sample	1	2	3	4	Negative control
ng/µl	76.6	28.7	40.2	54.2	11.1
260/280	1.77	1.77	1.73	1.69	1.6

If samples meet the sample requirements proceed to qPCR on LM-PCR amplified samples. If samples do not meet the ratio 260/280 requirement purify a second time with QIAquick. If a primer dimer peak is observed, adjust sample volume to 100 µl and add the amount of AMPure beads appropriate for the double SPRI method, follow the protocol of Purification with AMPure beads (see GS FLX Titanium General Library Preparation Manual) elute the sample instead of 24 µl in 100 µl TE buffer; Reanalyze 1 µl of purified sample on a BioAnalyzer to ensure no primer dimers are evident.

8. qPCR of Pre-Captured and Post captured Libraries using the LightCycler® 480 SYBR Green I Master, Roche and Primers.

3.6 qPCR on LM-PCR Amplified Samples

1. Dilute the pre-captured and the captured samples to 5 ng/μl.
2. Determine the number of DNA samples to be analyzed, aliquot Master Mix for duplicates: 28 μl MM into 0.5 ml tube, + 2 μl DNA to every tube, vortex and spin down, pipette in a 96-well plate 15 μl of the DNA-MM-Mix; without air bubbles, seal the plate with a LC480 sealing foil, centrifuge the plates for 30 s with 900 × *g* to eliminate the air-bubbles, program LC480: detection format SYBR Green, reaction volume 15 μl.

Pre-incubation	No analysis mode
95 °C/10 min/4.8 ramp rate	1 cycle
Amplification:	Quantification analysis mode
95 °C/10 s/4.8 ramp rate	40 cycles
60 °C/1 min/2.5 ramp rate	
Melting curve:	Melting analysis mode
95 °C/10 s/4.8 ramp rate	1 cycle
65 °C/1 min/2.5 ramp rate	
95 °C/5 acquisitions per °C	
Cooling:	No analysis mode
40/10 s/2 ramp rate	1 cycle

Perform Absolute Quantification and Tm Calling on LC480

3.7 Next-Generation Sequencing (454 Genome Sequencer FLX)

1. Using the Titanium-optimized sequence capture protocol provides the captured DNA ready for use in 454 emPCR, as required adaptor sequences are already integrated during the LM-PCR step. Approximately 500,000 molecules of DNA (quantification done with Pico Green) are loaded into a emPCR setup, whereas a tenfold amount of DNA carrying beads is used to ensure a DNA-copy to bead ratio of 1:1 without risking to many mixed reads (e.g., two DNA molecules coupled on one DNA carrying bead). The total library preparation protocol consists of emPCR (GS FLX Tit. emPCR Kits), the preparation of the Emulsion Oil, the process of Emulsification, and the FLX Shotgun Bead Recovery and Enrichment process (Lib L LV).

   Program: emPCR Titanium Shotgun, volume: 100 μl.

Place 0.2 ml 96-well plate in the thermocycler and use the following program:

94 °C	4 min	
94 °C 58 °C 68 °C	30 s 4 min 30 s 30 s	40 cycles
68 °C	30 min	
4 °C		

2. The four DNA-sequencing libraries were prepared separately, and each loaded onto one of the 4-lane gasket PicoTiterPlate device (PTP; 70 × 75 mm; Roche/454) respectively. In our approach each sample was physically separated, so no use of sample-identifying tags is necessary. With a coverage of 20× we were able to describe the SNPs and haplotypes.
3. Each library was quantified by PicoGreen® (Quant-iT™, Molecular Probes, Invitrogen), diluted to 1×10^5 molecules/μl, and used for emPCR at a ratio of 0.1 copies of library fragments per DNA capture bead (*see* **Note 14**).
4. After the DNA capture bead recovery enrichment (to separate DNA-carrying beads from non-DNA carrying beads), 490,000–690,000 single strand DNA carrying beads were loaded on the PTP. For enrichment and recovery of the DNA carrying beads streptavidin-coated magnetic beads are used (use the magnetic particle concentrator: Dynal MPC-S). These form a biotin-streptavidin bond and so the DNA carrying beads can be collected. Finally, after many washing steps, beads are counted using standard particle counters, calibrated for a particle size of 20 μm (e.g., use the handheld device Scepter from Millipore) (*see* **Note 15**).
5. The sequencing itself was performed on a Genome Sequencer FLX system using Titanium chemistry and standard protocols (Lib-L SV; Rev. Jan 2010). To assure that the sequencing by synthesis reaction runs properly add the sequencing primers and the different types of enzyme beads (Bead layer 1/Enzyme Beads pre-layer, Bead layer 2/DNA and Packing Beads layer, Bead layer 3/Enzyme Beads post-layer, Bead layer 4/PPiase Bead layer) on the PicoTiterPlate device and load together with the Sequencing Reagents on the instrument GS-FLX and start sequencing (*see* **Notes 16** and **17**).

3.8 Data Analysis

Bioinformatics is still the key for a successful sequencing project. When sequencing is complete, raw sequences, files in FASTA or fastq format, undergo several rounds of analysis including removal of primer sequences and low-quality reads, mapping, variant detection, and annotation (*see* **Note 18**).

1. Image processing and base calling was performed with GS FLX software. The GS Reference Mapper (Roche/454 Life Sciences) was used for mapping and variant detection. For the analysis of the eight MHC haplotypes sequenced here, the MHC Haplotype Project provided an adequate reference sequence database.
2. Haplotype (A3-B7-DR15) of the cell line PGF was chosen as the MHC reference sequence, representing the reference sequence used for capture oligonucleotide design. We encourage using an ethnically matched reference haplotype, when studying different human populations.
3. Genes were annotated according to the Vertebrate Genome Annotation database (http://www.VEGA.sanger.ac.uk) on the basis of the referenced PGF cell line.
4. Perl scripts were applied to extract type (substitute, insertion, deletion, synonymous or nonsynonymous) and quantity (number of reads per variant) of variation between the four samples and the reference cell line and between donor and recipient. A simple, but clear-cut ratio with 20× coverage of the number of matched or mismatched variants in relation to the total number of sequenced base pairs defined the similarity of actual or potential donor and recipient MHC, reliably enabling fast qualification of possible donor-recipient matches with a single quality factor over and above the obligatory HLA-match. Parental information clearly shows the match between donor and recipient while clearly different to parents.
5. Additional preparation of variants allowed to search and compare for known disease associations or other valued medical information. Parameters like intergenic, intronic, coding/noncoding, synonymous/nonsynonymous and coverage as well as prevalence were used to define a list of SNPs to start batch analysis in NCBI's OMIM or related databases. The application of these data analysis programs allowed for immediate report generation with complete variant and disease association information available some hours after the Genome Sequencer instrument run has finished.

4 Notes

1. All Genome Sequencer instruments (GS-FLX, GS JUNIOR) and reagents from Roche 454 are phased out mid-2016, but not the NimbleGen enrichment protocol. However, the above-described protocol can easily be adapted with the change in adapter sequences to any other sequencing or enrichment technology.

2. All reagents used should be strictly stored or handled on the conditions needed, being either room temperature, +4, −4, or −20 °C. 454 NGS on the GS-FLX or the GS-Junior needs a controlled room temperature not higher than 25 °C, whereas NimbleGen Sequence Capture microarrays can be stored at room temperature until use.
3. To estimate the efficiency of the enrichment process, the four different samples (included in the sequence capture assay) were quantified by qPCR before and after the enrichment process. This method revealed that approximately 337-fold enrichment had been achieved (ranging from 164- to 851-fold). The usage of microarray-based sequence capture systems demands an enrichment higher than 100-fold as a quality standard that was obviously achieved by the enrichment of the four samples. Make sure that the RNA area of the RNA ladder falls within the normal range of 100–150. If it is out of the range, run another RNA 6000 Pico chip or estimate the sample concentration by dividing the RNA area of the sample by 137. A typical RNA area for a 1 μl ladder (1000 pg/μl) is 137. The maximum elution efficiency is between pH 7.0 and 8.5. It is critical that the amplified library is eluted with PCR-grade water rather than Buffer EB or 1× TE-buffer.
4. Preparation of a GS FLX Titanium General DNA Library requires 3–5 μg of genomic DNA for the double SPRI method or 5–10 μg for the alternative gel cut method. Pipette DNA to the bottom (cup) of the Nebulizer.
5. Leave the tube of beads in the MPC during all wash steps.
6. If the sample meets these requirements proceed to hybridization; if not, repeat LM-PCR or reconstruct the library.
7. This step may take 30 min or longer, to minimize drying time, dry the COT human DNA ahead of time. Denaturation with high heat is not problematic after adaptor ligation because the hybridization utilizes ssDNA.
8. Carefully select the correct Elution Chamber for the type of microarray: 385 K arrays use ES1 Elution Chambers; 2.1 M arrays use EL1 Elution Chambers.
9. It is extremely important that the water bath temperature is closely monitored.
10. Prepare fresh 125 mM NaOH and 20% acetic acid solutions each time prior to use.
11. After last wash step, it is recommended to work as quickly as possible.
12. Do not remove the pipette tip from the filling port until you have returned the Elution System to the horizontal position or

elution reagent will leak out. Pipette the 125 mM NaOH solution slowly so it does not come out the top (vent) port.

13. The elution station can be tilted to an upright position and a pipette can be used to remove any remaining volume from the slide and transfer it to the 1.5 ml tube containing 20% Acetic Acid/Buffer PBI. The color of the eluate/Acetic Acid/PBI should be yellow, if the color is purple add 20% Acetic Acid, 1 μl at a time to the DNA mixture until the color changes to yellow.
14. If possible, use Qubit 3.0 Fluorometer (Thermofisher; Q33216) for the library quantification, as sensitivity in the ng/μl range is needed.
15. Specificity of the enrichment process. After mapping 908,152 high-quality reads to the PGF reference, only 706,431 reads were located inside the specific MHC target region, these are 78% of reads. 61,455 reads (8.7%) were rejected because of short length (<50 bp), repetitive or chimeric characteristics.
16. 454-pyrosequencing. Libraries 1–4 (MHC-parent 1, MHC-parent 2, MHC-donor, MHC-recipient) showed a post emPCR enrichment efficiency of 2.5, 4.5, 5.7, and 5.8%. As a consequence, 490,450, 544,500, 682,950, and 690,900 DNA-capture beads per lane were loaded onto the PTP regions 1–4. NGS resulted in 223,005, 225,385, 226,501, and 234,017 individual read sequences with a mean read length of 350 bp that passed quality control filters of base calling and that were subsequently used for data analysis.
17. Error rates. In our amplicon sequencing approach 0.18% of total bases used for HLA typing were erroneous [18]. Therefore, 454-sequencing features a limit of detection (LOD) of ~1% for the minor alleles. When using Illumina sequencing, the PhiX spike-in should give a per-run estimate about the applicable LOD.
18. Computer power. Using Illumina sequencing an appropriate high-performance computer (min. 16 GB RAM) is recommended to reduce computing time. The 454 protocol requires a Linux environment with the GS Reference Mapper (Roche/454 Life Sciences) installed.

Acknowledgment

This project was supported by the Austrian Research Promotion Agency (FFG grant: Nr. 815468 B1) and Roche Austria.

References

1. Dausset J (1958) Iso-leuko-antibodies. Acta Haematol 20(1–4):156–166
2. Trowsdale J (2011) The MHC, disease and selection. Immunol Lett 137(1-2):1–8
3. Trowsdale J, Knight JC (2014) Major histocompatibility complex genomics and human disease. Annu Rev Genomics Hum Genet 14:301–323
4. Ronaghi M, Uhlén M, Nyrén P (1998) A sequencing method based on real-time pyrophosphate. Science 281(5375):363, 365
5. Dressman D, Yan H, Traverso G, Kinzler KW, Vogelstein B (2003) Transforming single DNA molecules into fluorescent magnetic particles for detection and enumeration of genetic variations. Proc Natl Acad Sci U S A 100(15): 8817–8822
6. Mitra RD, Church GM (1999) In situ localized amplification and contact replication of many individual DNA molecules. Nucleic Acids Res 27(24), e34
7. Fedurco M, Romieu A, Williams S, Lawrence I, Turcatti G (2006) BTA, a novel reagent for DNA attachment on glass and efficient generation of solid-phase amplified DNA colonies. Nucleic Acids Res 34(3), e22
8. Southern EM, Maskos U, Elder JK (1992) Analyzing and comparing nucleic acid sequences by hybridization to arrays of oligonucleotides—evaluation using experimental models. Genomics 13:1008–1017
9. Holcomb CL, Höglund B, Anderson MW, Blake LA, Böhme I, Egholm M, Ferriola D, Gabriel C, Gelber SE, Goodridge D, Hawbecker S, Klein R, Ladner M, Lind C, Monos D, Pando MJ, Pröll J, Sayer DC, Schmitz-Agheguian G, Simen BB, Thiele B, Trachtenberg EA, Tyan DB, Wassmuth R, White S, Erlich HA (2011) A multi-site study using high-resolution HLA genotyping by next generation sequencing. Tissue Antigens 77(3):206–217
10. Danzer M, Niklas N, Stabentheiner S, Hofer K, Pröll J, Stückler C, Raml E, Polin H, Gabriel C (2013) Rapid, scalable and highly automated HLA genotyping using next-generation sequencing: a transition from research to diagnostics. BMC Genomics 14:221
11. Lange V, Böhme I, Hofmann J, Lang K, Sauter J, Schöne B, Paul P, Albrecht V, Andreas JM, Baier DM, Nething J, Ehninger U, Schwarzelt C, Pingel J, Ehninger G, Schmidt AH (2014) Cost-efficient high-throughput HLA typing by MiSeq amplicon sequencing. BMC Genomics 24(15):63
12. Barone JC, Saito K, Beutner K, Campo M, Dong W, Goswami CP, Johnson ES, Wang ZX, Hsu S (2015) HLA-genotyping of clinical specimens using Ion Torrent-based NGS. Hum Immunol 28. pii: S0198-8859(15)00451-6
13. Mayor NP, Robinson J, McWhinnie AJM, Ranade S, Eng K, Midwinter W, Bultitude WP, Chin CS, Bowman B, Marks P, Braund H, Madrigal JA, Latham K, Marsh SG (2015) HLA typing for the next generation. PLoS ONE 10(5):e0127153
14. Hosomichi K, Shiina T, Tajima A, Inoue I (2015) The impact of next-generation sequencing technologies on HLA research. J Hum Genet 60(11):665–673
15. Wittig M, Anmarkrud JA, Kässens JC, Koch S, Forster M, Ellinghaus E, Hov JR, Sauer S, Schimmler M, Ziemann M, Görg S, Jacob F, Karlsen TH, Franke A (2015) Development of a high-resolution NGS-based HLA-typing and analysis pipeline. Nucleic Acids Res 43(11):e70
16. Albert TJ, Molla MN, Muzny DM, Nazareth L, Wheeler D, Song X, Richmond TA, Middle CM, Rodesch MJ, Packard CJ, Weinstock GM, Gibbs RA (2007) Direct selection of human genomic loci by microarray hybridization. Nat Methods 4(11):903–905
17. Pröll J, Danzer M, Stabentheiner S, Niklas N, Hackl C, Hofer K, Atzmüller S, Hufnagl P, Gülly C, Hauser H, Krieger O, Gabriel C (2011) Sequence capture and next generation resequencing of the MHC region highlights potential transplantation determinants in HLA identical haematopoietic stem cell transplantation. DNA Res 18(4):201–210
18. Westbrook CJ, Karl JA, Wiseman RW, Mate S, Koroleva G, Garcia K, Sanchez-Lockhart M, O'Connor DH, Palacios G (2015) No assembly required: full-length MHC class I allele discovery by PacBio circular consensus sequencing. Hum Immunol 76(12):891–896
19. R Development Core Team (2012) R: a language and environment for statistical computing. R Foundation for Statistical Computing, Vienna, Austria

Chapter 6

Pedigree-Defined Haplotypes and Their Applications to Genetic Studies

Chester A. Alper and Charles E. Larsen

Abstract

A haplotype is a string of nucleotides or alleles at nearby loci on one chromosome, usually inherited as a unit. Within the major histocompatibility complex (MHC) region on human chromosome 6p, independent population studies of multiple families have identified conserved extended haplotypes (CEHs) that segregate as long stretches (≥1 megabase) of essentially identical DNA sequence at relatively high (≥0.5%) population frequency ("genetic fixity"). CEHs were first identified through segregation analysis in the early 1980s. In European Caucasian populations, the most frequent 30 CEHs account for at least one-third of all MHC haplotypes. These CEHs provide all of the known individual MHC susceptibility and protective genetic markers within those populations for several complex genetic diseases. Haplotypes are rigorously determined directly by sequencing single chromosomes or by Mendelian segregation analysis using families with informative genotypes. Four parental haplotypes are assigned unambiguously using genotypes from the two parents and from two of their haploidentical (to each other) children. However, the most common current technique to phase haplotypes is probabilistic statistical imputation, using unrelated subjects. Such probabilistic techniques have failed to detect CEHs and are thus of questionable value in identifying long-range haplotype structure and, consequently, genetic structure–function relationships. Finally, with haplotypes rigorously defined, association studies can determine frequencies of alleles among unrelated patient haplotypes vs. those among only unaffected family members (i.e., control alleles/haplotypes). Such studies reduce, as much as possible, the confounding effects of population stratification common to all genetic studies.

Key words Allele, Disease, Haplotype, HLA, Immunogenetics, MHC, Pedigree, Polymorphism, Sequence, Whole-genome sequencing

1 Introduction: Haplotypes and the MHC

It is no accident that the word "haplotype" was coined by an immunogeneticist studying the major histocompatibility complex (MHC). Ceppellini et al. [1] used the term, derived from "haploid genotype," to designate the string of nucleotides or alleles on one chromosome usually inherited as a unit. Initially, interest in the MHC was due to the role of the genes it contained in determining whether organs or tissues could be successfully transplanted from one individual to another. Because tissue-compatible individuals

Irene Tiemann-Boege and Andrea Betancourt (eds.), *Haplotyping: Methods and Protocols*, Methods in Molecular Biology, vol. 1551, DOI 10.1007/978-1-4939-6750-6_6,

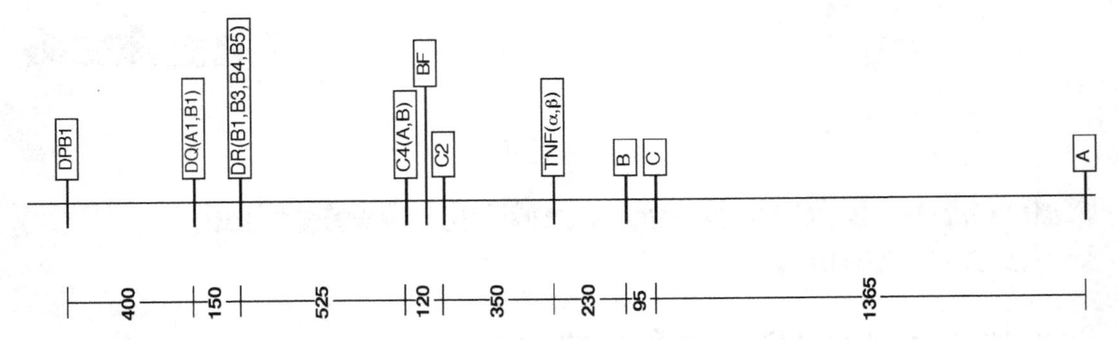

Fig. 1 Map of the human major histocompatibility complex (MHC) region showing the physical distances (in kilobases) between specific loci. Distances are approximate and vary at many locations in different haplotypes as a result of various DNA insertions, deletions, and gene duplications. The region shown is from near the centromeric end (*HLA-DPB1*, at 6p21.32) to near the telomeric end (*HLA-A*, at 6p22.1) of the human MHC. The human leukocyte genes ("HLA group") shown are: the class I genes *HLA-A*, *HLA-C*, and *HLA-B* and the class II genes *HLA-DPB1*, two of the HLA-DQ genes (*HLA-DQA1* and *HLA-DQB1*) and four of the HLA-DR genes (*HLA-DRB1*, *-DRB3*, *-DRB4*, and *-DRB5*). The four complotype genes (*C2*, *BF* (now named *CFB*), *C4A*, and *C4B*) are in the central (so-called class III) region of the human MHC as are the two cytokine genes shown (*TNFα* (now named *TNF*) and *TNFβ* (now named *LTA*)). (Reprinted slightly modified with permission from ref. 34, John Wiley & Sons, Hoboken, NJ.)

are most often found among family members, particularly siblings, histocompatibility gene studies have been carried out in families from the beginning. It quickly became evident that the MHC region, in the middle of the short arm of chromosome 6 (6p21.32–6p22.1), is highly gene-rich and that many of these genes are very polymorphic, i.e., have many alleles. Figure 1 shows a map of the human MHC, depicting several of the major genes.

Haplotypes provide far more information than individual single nucleotide polymorphisms (SNPs). For example, only phased or homozygous data contain information about *cis* and *trans* molecular interactions affecting gene expression [2]. Therefore, a number of methods for determining haplotype structure (or "phasing") have been developed. These methods are not always needed for haplotyping the MHC region. The International Histocompatibility Workshop (IHW) homozygous cell lines do not usually require phasing, and they are useful in themselves for haplotype structural analysis. Traditional methods use family studies in which patterns of Mendelian inheritance reveal the four parental chromosomes (except in rare cases of null alleles, novel mutations, or meiotic crossovers). In this analysis, any two family members that share a haplotype ("haploidentical"; e.g., a parent and a child) usually can provide at least partially phased haplotypes. While conceptually straightforward, pedigree analyses are not without challenges. Positions where both the parent and child are heterozygous identical cannot be phased. ("Heterozygous identical" refers to a situation where both subjects have the same two differing alleles at a given locus (e.g., *HLA-A*02:01/HLA-A*03:01*), a common occurrence for SNPs with high minor allele frequencies.) While most sites can be phased with trios (both parents and one child), sites where trios are heterozygous identical require both parents and two haploidentical children for full phasing. Special phasing problems in short chromosomal segments often require additional family members for confirmation. Even the two-parent, two-child design may be insufficient for particularly difficult problems, such as distinguishing homozygous identical parents (with all four chromosomes identical) from heterozygous identical parents with each parent having one chromosome with a null allele at the given locus. Further, while meiotic recombination, affecting one chromosome inherited by a child, and polymorphic sequence deletions or duplications are rare at individual loci, they are common in whole-chromosome phasing. In these complicated cases, additional family members (grandparents, parental siblings, additional children, or grandchildren) can be necessary to provide additional information from which to decipher the four (or more) individual family chromosomes.

1.1 Association Studies

The goal of association studies is the identification of genes involved in specific traits or in disease pathogenesis. Several decades ago, case/control association studies [3] revealed highly significant associations between certain HLA alleles and a variety of diseases, many of which are autoimmune diseases, such as type 1 diabetes (T1D) [4], celiac disease or gluten-sensitive enteropathy (GSE) [5], multiple sclerosis (MS) [6], and pemphigus vulgaris (PV) [7]. These case/control association studies have since been expanded to genome-wide association studies (GWAS) [8]. In GWAS, SNP genotypes are determined throughout the genome at various intervals, and those that show statistically different frequencies between patients and controls are considered to be associated with an increased risk of disease. Most analyses assume these SNPs occur in or near genes with small additive effects that contribute to the disease.

In general, the use of isolated SNPs to identify genes of interest has a number of inherent problems. How can one distinguish a gene that is directly involved in the disease from one that is merely a marker for such a gene because of proximity and linkage disequilibrium (LD)? What guarantees that SNPs in or near relevant genes have been included in the analysis? To some extent, the use of multiple-SNP-defined haplotypes in family studies overcomes or helps solve these problems.

A fundamental question that needs to be answered in all association studies is which genes are causal susceptibility genes and which genes are merely markers. In particular, some alleles may be only markers of the differing subpopulations from which the patients originated, compared with unrelated controls [9, 10]. A means to minimize subpopulation differences is a family-based association study [11, 12], in which alleles and/or haplotypes occurring in patients with the disease are compared to those only in unaffected family members (family controls or FCs). Importantly, this method allows the use of haplotypes defined by multiple genetic markers (including SNPs) that define many alleles [13] instead of SNPs themselves with only two (or, rarely, three) alleles.

1.2 Linkage Determined from Allele Sharing by Affected Sib Pairs

A different approach to unraveling the genetics of complex disease is to estimate linkage between disease and alleles (or haplotypes) in families with, for example, two disease-affected siblings [14]. The affected sib pairs (ASPs) are analyzed to detect sharing of one or two alleles/haplotypes that is significantly different from the null expectation (e.g., for siblings 50%, and 25%, respectively), usually summarized with a logarithm of the odds (LOD) score [15]. An excess of haplotype sharing among affected siblings implicates the shared haplotypes in the genetic disease; significant linkage among ASPs is manifested by LOD scores equal to or greater than 3.6 [16]. For example, whole genome ASP studies in T1D showed chromosome 6p (MHC) allele/haplotype LOD scores in excess of 200 [17, 18], whereas all other regions (with one exception in one of the reports [18]) showed no significant linkage.

Family studies not only define haplotypes, provide evidence of genetic linkage, and minimize population stratification, but also provide important information about genomic structure. Until about 30 years ago, the only picture of genomic structure was that defined by LD [19]. LD is a statistical measure of the extent of the coinheritance of genome markers. Normalized LD (D') is highest when calculated pairwise for two markers, is lower for three markers, and loses statistical significance for more than three markers [20]. When applied to the human genome, statistical analysis of D' values determined by pairwise analysis of neighboring SNPs found the median size of regions showing little to no population-level recombination (so-called LD blocks) was about 25 kb, although occasional larger blocks of up to 200 kb or so were sometimes detected [21]. Here, we will describe another level of genomic structure, involving long-range haplotypes, which appears particularly important for the MHC, but may also play a role elsewhere in the genome.

2 Definition and Methods: Genetic Fixity and Conserved Extended Haplotypes

It is vital that any analytical approach to genetic questions incorporates an accurate picture of genomic structure. Family MHC studies paint a very different picture than that based on pairwise SNP LD. Unlike pairwise SNP LD, which typically declines over a kilobase scale, conserved extended haplotypes (CEHs) [22] (termed later by some as ancestral haplotypes (AHs) [23]) are megabase (Mb)-length haplotypes at relatively high frequency in a population (e.g., ≥0.5 %), with virtually identical sequence [22–28]. CEHs can be thought of as structural genetic units (on the scale of 1–5 Mb) that are composed of several to many smaller LD blocks. We refer to such relatively common haplotypes that expanded to relatively high frequency in a given population without apparent recombination as "frozen" or "fixed" population haplotypes. Thus, we define genetic haplotype or sequence "fixity" as sequence identity and conservation of a large stretch of genomic sequence in a population, shared by a relatively large number of apparently unrelated individuals, without apparent recombination from an ancestral sequence [22, 23, 28, 29]. By "identity" we mean "essential identity," thus allowing for minor private mutation or microvariation within individual haplotype sequences comprising a particular CEH (or CEH fragment or block). The evidence for CEH sequence conservation (with minor microvariation) has increased whenever loci were defined at higher resolution (please *see* below) or at intervening locations. What others simply call haplotype "LD blocks" (shorter sequences at high frequency in a population, with little to no apparent historical recombination), we call "fixed" blocks. These are often shared by more than one CEH [28]. Note that we do *not* mean to imply a lack of polymorphism (i.e., a single variant fixed at unity) within a human population for the genomic region(s) in which these blocks and CEHs exist.

A short delineation of MHC genes and the technologies used to identify their variants is helpful to understand the structure of MHC CEHs. Over 300 genetic loci (and over 160 protein-coding genes) exist in the MHC region [30]. The histocompatibility-determining genes, those in the HLA group (*see* Fig. 1), are the best-known genes of the region. HLA alleles were initially defined serologically [31]. These serological "alleles" are now known to be "specificities" (e.g., HLA-A1, HLA-A2, HLA-A3 are three different *HLA-A* specificities). HLA specificities are usually not italicized, whereas genes and alleles usually are. With the introduction of practical and relatively inexpensive nucleotide sequencing and SNP detection, these previously defined specificities have been resolved at the nucleotide scale, and are now known to be composed of groups of related alleles with the same serological specificity [32]. For example, the HLA-A2 specificity (the most common throughout the world) includes, among its relatively common alleles (defined here only at 4-digit resolution), *HLA-A*02:01*, *HLA-A*02:06*, and *HLA-A*02:10*.

In addition to the HLA genes, the MHC region also includes four genes of the complement cascade that also exist very close to one another in the central MHC (sometimes called "class III" HLA). In the early 1980s, alleles at these four loci were found to exist in the population as fixed haplotype blocks [29], with a limited number of variants, far fewer than expected with frequent recombination between LD blocks. A "complotype" is a small

(75–140 kb) fixed genomic block, and is defined and named by its *CFB*, *C2*, *C4A*, and *C4B* alleles, in that arbitrary order [29]. Null or *Q0* alleles are simply designated 0. Thus, the complotype named FC31 indicates the complotype *CFB*F*, *C2*C*, *C4A*3*, *C4B*1*.

Where and how long are CEHs? MHC CEHs were originally defined by their *HLA-B*, complotype, and *HLA-DRB1* alleles [22]. The *HLA-B* to *-DRB1* distance is approximately 1.2 Mb (Fig. 1). Later CEH reports expanded the core region from *HLA-C* to *HLA-DQB1* (a distance of 1.4 Mb), and some reports describe full-length MHC CEHs as extending from *HLA-A* to *HLA-DPB1* (a distance of 3.1 Mb).

The CEHs from *HLA-B* to *HLA-DRB1* constitute one-third to one-half of all MHC haplotypes in European populations [20, 22–28]. That is, large fractions of a European Caucasian population's MHC haplotypes are comprised of a limited number (30 or fewer) relatively frequent (≥0.5 %) long-range haplotypes of essentially identical sequence (identical within each CEH). This genomic structure is undetected by standard SNP-LD analysis. When such analysis was applied to the MHC [33], only small blocks were detected, not entire CEHs. It is of interest that significant three-point LD between HLA-B, complotype, and HLA-DR alleles was observed in the original description of CEHs [22]. Although it is not known whether CEHs occur elsewhere in the genome, it seems likely that they do [34]. Suggestive evidence for non-MHC CEHs has been presented [35]. Thus, at least for the MHC and probably for the genome in general, GWAS are based on a faulty concept of the population-level structure of the human genome.

Identifying the genetic elements responsible for complex diseases requires knowing the genomic haplotype architecture of the population(s) in which the diseases exist. The Wellcome Trust Sanger Institute MHC Haplotype Project (MHP) made a major advance toward that goal 10–15 years ago, when it determined complete (or nearly complete) sequences of eight "common" European Caucasian MHC haplotypes [30, 36–38]. When these were related by partial resequencing to known CEHs [28], many, but not all, were found to be either complete or partial examples of CEHs.

These considerations lead to the inevitable conclusion that the reality of the haplotype structure of the MHC is the directly observed and phased (i.e., pedigree-determined) long stretches of DNA fixity in a multitude of CEHs. This structure is poorly reflected in the statistical constructs of small SNP/LD blocks between which there is thought (incorrectly) to be random historical crossing over [21, 33]. Put another way, the reality of the haplotype structure of the MHC is extensive long-range DNA fixity [22–26, 28], which has been directly observed by pedigree-analysis and only poorly and incorrectly approximated by typical measures of LD [33].

Using imputation [39] of individual SNP and sequencing data to guess at haplotypes based on LD blocks is tempting as it lowers cost, but we argue that these methods result in loss of valuable information that can only be definitively obtained through phasing of family sequence data or sequencing of isolated chromosomal DNA [2, 40, 41]. As discussed above, the pedigree structure optimal for such studies is that containing several haploidentical individuals.

3 The Genetics of Human Disease: The MHC and Complex Disease

Inheritance patterns of MHC gene alleles in early family studies of T1D [42] suggested that the disease is caused by Mendelian inheritance of a single recessive susceptibility gene. While it is now clear that T1D is not monogenic, there remains strong evidence that the MHC-linked T1D gene(s) exhibit(s) recessive inheritance. The recessive allele model is also consistent with the fact that MHC alleles appear to be at Hardy–Weinberg equilibrium in patients with T1D in some European subpopulations [43, 44]. In many other European populations, there is instead an excess of HLA-DR3/DR4 heterozygotes observed among patients [45], but this is not *a priori* evidence against the recessive nature of MHC susceptibility, as has been argued [46]. An alternative explanation [47, 48] is that, because selection against *any* required causal gene in a polygenic disease would reduce susceptible individual frequency, separate subpopulations could have selected against different required T1D susceptibility genes. The excess of HLA-DR3/DR4 heterozygotes might then reflect an increased incidence of T1D among offspring of parents who come from different, previously isolated, subpopulations (with one subpopulation identity marked by HLA-DR3 and the other by HLA-DR4). As a reduction in polygenic disease incidence could be achieved through selection against *any* of the *required* susceptibility loci involved in the disease, it may be that selection acted more effectively against *different* required genes in the different isolated subpopulations. The offspring of these parents from these separate subpopulations would have higher disease susceptibility than either parent, as the children would have a higher likelihood of having disease alleles at all required loci. In keeping with this interpretation, only specific CEHs are involved in the HLA-DR3/DR4 excess, and there are also excess heterozygotes for specific HLA-DR3/DRX (where HLA-DRX is neither HLA-DR3 nor HLA-DR4), HLA-DR4/DRX, and HLA-DRX/DRX genotypes (unpublished observations).

3.1 MHC Conserved Extended Haplotypes and Disease

In the first published family association study of a complex disease [12], all MHC markers of T1D detected by case/control studies belonged to specific CEHs. While this underlying haplotype structure facilitates association of the MHC with the disease, it makes identification of specific susceptibility genes far harder. In our view, specific causal susceptibility genes for the great majority of complex diseases are yet to be discovered [49], and complex genomic structure such as CEHs may obscure associations with specific genes. The example of hemochromatosis illustrates this point: hemochromatosis is a monogenic disorder shown to be associated with MHC specificity HLA-A3 [50]. It was later found that the causative gene for hemochromatosis is *HFE* [51], and that the main susceptibility allele of this gene is on the subset of a CEH ([HLA-B7, SC31, DR2]) with HLA-A3 at high frequency [50]. About 37% of the CEH [HLA-B7, SC31, DR2] bears HLA-A3 [22]. What is remarkable is that the marker (*HLA-A*) and the susceptibility locus are over 4.5 Mb apart, illustrating how long-range haplotype structure can yield distant associations that can be misleading if ignored.

Although causal gene localization is made more difficult by the existence of CEHs, the knowledge that CEHs carry most, if not all, of the MHC and MHC-linked susceptibility alleles for some diseases suggests additional ways to explore their genetics. For example, CEH analysis of GSE [52, 53] and dermatitis herpetiformis (DH) [53] patients and their family members provides evidence for recessive inheritance of the MHC-associated gene(s) for those diseases. The former study [52] confirmed earlier family association and linkage evidence [54] for the recessive nature of the GSE MHC-associated gene(s). Not all genetic diseases mapping to the MHC are recessive: direct observation of inheritance patterns of single MHC alleles/haplotypes in families with ankylosing spondylitis [55] and PV [56] is consistent with dominant inheritance. In the case of ankylosing spondylitis, the disease was associated with specific HLA-B27 allelic variants [57], but not specific haplotype(s). In PV patient families, the presence of antibody to the epidermal antigen cadherin is completely correlated with one of the two known CEHs associated with PV or an *HLA-DRB1*, *-DQB1* fragment of them [56]. PV only occurred in a small percentage of the antibody-positive individuals, suggesting a possible causal role for other genes [57].

Another method to determine the mode of inheritance of (a) regional gene(s) in a complex disease or trait involves *prospective* analysis of CEH carriers. If a specific CEH has an increased frequency in patients suffering from a complex disorder, one can determine the frequency of that disorder among homozygotes and heterozygotes for that CEH selected for other reasons [58]. For example, it was known that the CEH [HLA-B8, SC01, DR3] is at increased frequency among patients with IgA deficiency (IgAd) [59]. Individuals who are homozygotes, heterozygotes, and noncarriers of this CEH were tested for serum concentrations of IgA, IgD, IgE, IgM, and IgG subclasses [60]. No instances of deficiency

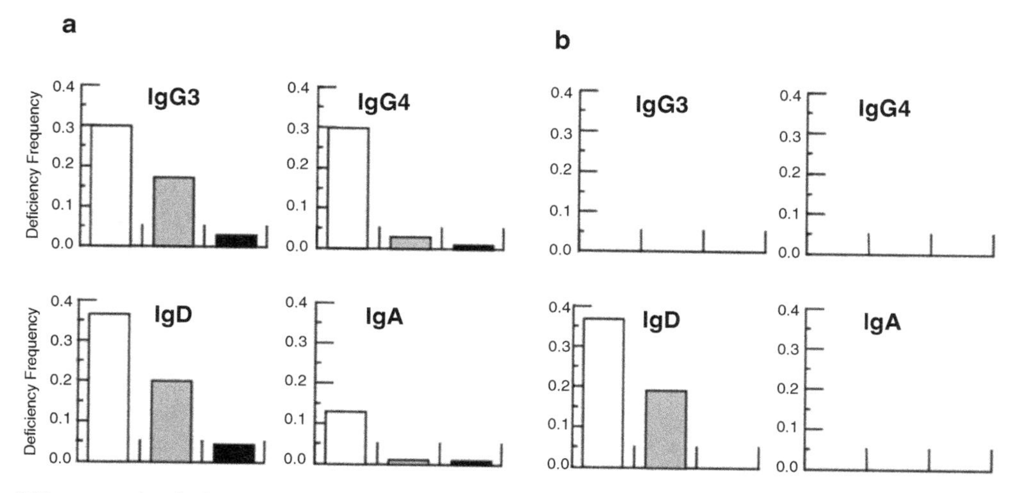

Fig. 2 Homozygotes, heterozygotes, and non-carriers of the MHC CEHs (**a**) [HLA-B8, SC01, DR3] and (**b**) [HLA-B18, F1C30, DR3] were studied for serum immunoglobulin (Ig) levels. The prevalence of Ig deficiency in haplotype homozygotes (*open bars*), heterozygotes (*gray bars*), and non-carriers (*black bars*) is shown. The height of each bar represents the relative frequency of individuals deficient in that Ig class or subclass as a fraction of the number of subjects in that category tested. (Reprinted with permission from ref. 49, Elsevier, Ltd., Amsterdam, The Netherlands, and modified (**a**) with permission from ref. 60, John Wiley & Sons, Hoboken, NJ and (**b**) with permission from ref. 62, Springer, New York, NY.)

of IgM, IgE, IgG1, or IgG2 occurred among these individuals. But IgAd was detected among 13.3% of the homozygotes, compared to only 1.7% of heterozygotes and 1.6% of non-carriers (Fig. 2). These findings suggested that the MHC gene or linked genes for IgAd are recessive and that some (but not all) of the susceptibility genes for IgAd are on the [HLA-B8, SC01, DR3] CEH. A similar pattern was observed for IgG4 deficiency (IgG4d) in 30% of homozygotes compared to 1.7% in heterozygotes and 3.4% of non-carriers of the CEH, consistent with IgG4d also showing recessive inheritance of (an) MHC gene(s).

Finally, MHC CEH fragment analysis may be helpful in localizing specific susceptibility genes in CEH-associated diseases. The specific CEHs carrying susceptibility genes to GSE and DH among Europeans are the same: [HLA-B8, SC01, DR3] and [HLA-B44, FC31, DR7] [52, 53]. However, many patients carried fragments of these CEHs rather than the complete haplotypes. In GSE, these fragments were more commonly centromeric, whereas they were more commonly telomeric in DH [53]. This suggests that different susceptibility genes for the two diseases are on, or linked to, the same CEHs.

3.2 MHC CEHs Elucidate Intrinsic Penetrance of MHC-Linked Disease Genes

Incomplete penetrance is the failure of genetically fully susceptible individuals to exhibit a trait or disease. It is most clearly illustrated by the failure of some identical (monozygotic) twins (MZTs) of a proband with a trait or disease to be similarly affected (incomplete penetrance of susceptibility). Broadly, incomplete penetrance could be due either to variation in an extrinsic process (i.e., environmental trigger) or an intrinsic, gene-associated process [58]. We studied subjects prospectively who carried particular MHC CEHs to determine whether we could distinguish these two possibilities for immunoglobulin deficiencies (Igds) with a dominant mode of inheritance of MHC or MHC-linked genes [58, 60–62]. In contrast to the findings for IgAd and IgG4d, significant numbers of heterozygotes in addition to homozygotes for the [HLA-B8, SC01, DR3] CEH [60] had IgD (IgDd) or IgG3 (IgG3d) deficiency (Fig. 2), suggesting a dominant mode of inheritance. The ratio of homozygotes to heterozygotes affected by each of these diseases was roughly 2:1, suggesting that penetrance of these two dominant Igds affects the expression of genes on each chromosome independently. If an environmental trigger had been responsible for expression differences of the dominant Igds, one would expect the effect to be on the whole person, with homozygotes and heterozygotes having the same frequencies of the traits. Similar studies of Igds were conducted among homozygotes, heterozygotes, and non-carriers for two other CEHs, [HLA-B18, S042, DR2] [61] (the C2-deficiency CEH) and a second HLA-DR3-containing CEH [HLA-B18, F1C30, DR3] [62] (Fig. 2). Remarkably, only one Igd was increased in frequency in homozygotes and heterozygotes of each CEH, that of IgDd. The ratio of IgDd among homozygotes to heterozygotes was roughly 2:1 for both CEHs, as previously observed for carriers of the CEH [HLA-B8, SC01, DR3].

These findings led to the concept of a random process (e.g., epigenetic change in expression of a susceptibility gene) being the basis of penetrance in autoimmune diseases like T1D [49, 58]. Further insight into the nature of incomplete penetrance in complex disease was obtained in the prospective study of Igds among MZTs discordant for T1D [63]. As discussed, several CEHs carry susceptibility for T1D as well as IgAd, IgG4d, and/or IgDd. Therefore, as expected, there were elevated incidences of IgAd (6%), IgDd (33%), and IgG4d (12%), but, surprisingly, not of IgG3d, among MZTs discordant for T1D [63]. Concordance among MZT pairs was 50% for IgAd, 57% for IgDd, and 50% for IgG4d [63], remarkably similar to concordance for T1D in MZTs [64, 65]. Whether one (or more, should they exist) of the MHC-linked susceptibility genes for T1D is/are shared by the MHC-determined Igds remains to be explored. However, there was no association between the presence of T1D and the presence of any Igd among the individual twins. Thus, the specific random (possibly epigenetic) processes responsible for penetrance in the deficiencies of IgA, IgD, and IgG4 are likely different and operate independently of those operative in T1D.

3.3 Non-MHC Susceptibility Genes for Complex Diseases

About 6–8 % of the siblings of T1D patients also have T1D, whereas 16–18 % of MHC-identical siblings are affected [65]. The difference between 16–18 % concordance for MHC-identical siblings and the much higher 40–50 % probandwise concordance for MZTs is consistent with non-MHC susceptibility gene alleles also being required for disease susceptibility. This suggests the possibility of there being at least one [66–68] or two [47, 48, 67–69] required non-MHC T1D susceptibility genes interacting with susceptibility alleles at the MHC locus to provide full genetic susceptibility to T1D.

Still another method [69] for determining the mode of inheritance and estimating the aggregate general population frequency of the MHC susceptibility haplotypes was developed based on the possible mating pairs that could produce MHC-susceptible offspring. Among siblings affected by T1D, 55 % share two MHC susceptibility alleles, 38 % share one, and 7 % share none [68]. Based on this pattern, it was concluded that the MHC T1D susceptibility gene is recessive, has a population aggregate disease allele frequency of 0.525 [69] (Fig. 3), and that the MHC susceptibility haplotypes do not show a dominant inheritance pattern. For T1D, if there are two additional non-MHC loci with aggregate susceptibility alleles of equal frequency, the latter would occur at frequencies of approximately 0.38 each; if there are three loci, their aggregate susceptibility allele frequencies would each be 0.53, similar to the aggregate frequency of MHC susceptibility alleles. For other diseases, the mode of inheritance is more difficult to

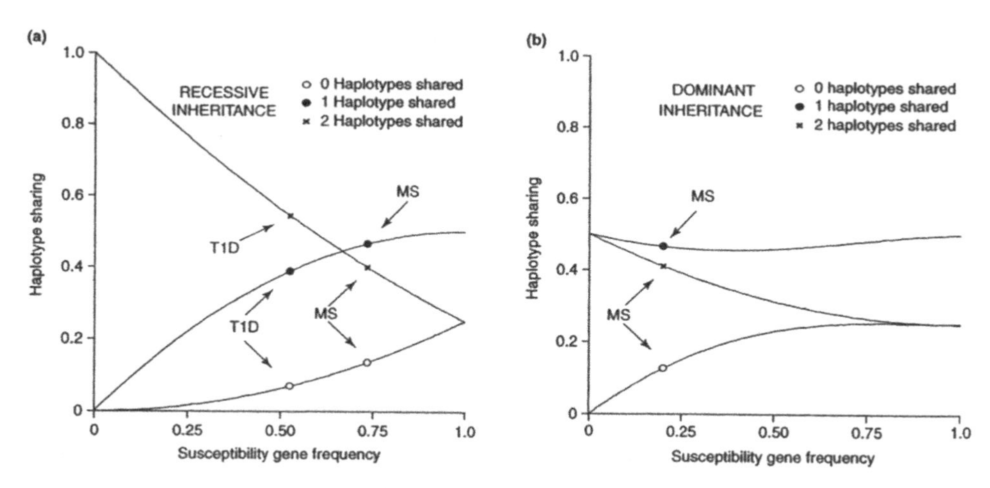

Fig. 3 Predicted frequencies of 2, 1, and 0 haplotypes shared for recessive (**a**) and dominant (**b**) inheritance of an MHC disease susceptibility gene *D* at various frequencies of *D*. Observed haplotype distribution frequencies for multiple sclerosis (MS) and type 1 diabetes (T1D) are marked. Note that for MS, the distribution fits either form of inheritance. There is no dominant solution for T1D. (Reprinted with permission from ref. 49, Elsevier, Ltd., Amsterdam, The Netherlands and modified with permission from ref. 69, Karger Publishing, Basel, Switzerland.)

determine. For MS, based on ASP data of 40% sharing two haplotypes, 47% sharing one, and 13% sharing none [69], either recessive or dominant inheritance was possible (Fig. 3).

4 Conceptual Problems for Complex Disease Genetics When Ignoring Haplotypes

There are several conceptual problems with the current, widely held view of the genetics of T1D and other complex diseases. Too often, complex genetic analysis is based on GWAS and ASP analysis of individual SNPs (rather than pedigree-analyzed haplotypes) in patients compared with controls. In that view, the MHC-determined susceptibility for T1D is "strong" and resides primarily in *HLA-DRB1, -DQB1* [70–72]. The evidence that MHC susceptibility is Mendelian recessive and carried by a gene or genes on specific CEHs is usually ignored or forgotten. Although haplotype analysis is often noted, genetic analysis of T1D (and of other complex diseases) too often is based on statistical analysis of susceptibility "risk," although some analyses are more nuanced than others. Non-MHC additive genes are discovered by GWAS and determined by individual SNPs, but often show little to no genetic linkage to the disease in linkage studies. Because LOD scores on linkage analysis of these non-MHC genes are lower than the 3.6 required for significance [18], the genes are considered to be of small individual risk-elevating effect [69–72].

5 Summary

Causal genes are the keys to understanding genetic diseases [9, 10, 67]. In our view, any causal gene will significantly increase allele/haplotype sharing by ASPs (i.e., show genetic linkage). As pointed out above, almost none of the genes of small effect in T1D meet these requirements as judged by LOD scores or by chi-squared analysis of ASP allele/haplotype sharing in terms of deviation from the random sharing of two, one, or no haplotypes of 25%, 50%, and 25%, respectively (unpublished observations).

The great challenge in modern human genetics is unraveling the causal genes of complex disease. We have focused our discussion of haplotypes on the MHC and T1D, but the concerns and considerations relate to complex diseases in general. It is highly likely that CEHs occur throughout the genome. If so, and if a causal gene occurs on a CEH, the gene's flagging is facilitated, but its actual identification is made more difficult. Defining non-MHC CEHs requires both direct observation in pedigrees and the consideration of long-range sequence conservation ("fixity") within populations. Whole genome sequencing will soon allow direct determination of full haplotype sequences if analyzed appropriately.

This requires either sequencing individual chromosomes after physical isolation or sequencing moderate to large pedigrees to phase pedigree data directly [2, 40, 41]. The latter allows both sequence integrity crosschecking and directly observed recombination.

References

1. Ceppellini R, Curtoni E, Mattiuz P, Miggiano V, Seudelder G, Serra A et al (1967) Genetics of leukocyte antigens: a family study of segregation and linkage. In: Curtoni E, Mattiuz P, Tosi R (eds) Histocompatibility testing. Munksgaard, Copenhagen, pp 149–185
2. Glusman G, Cox HC, Roach JC (2014) Whole-genome haplotyping approaches and genomic medicine. Genome Med 6:73–88
3. Schulte PA (1987) Simultaneous assessment of genetic and occupational risk factors. J Occup Med 29:884–891
4. Svejgaard A, Platz P, Ryder LP (1981) Insulin dependent diabetes mellitus. In: Terasaki PI (ed) Histocompatibility testing 1980. UCLA Tissue Typing Laboratory, Los Angeles, pp 638–656
5. Ludwig H, Polymenidis Z, Granditsch Z, Wick G (1973) [Association of HL-A1 and HL-A8 with childhood celiac disease]. Z Immunitätsforsch Exp Klin Immunol 146:158–167
6. McDevitt HO, Bodmer WF (1974) HL-A, immune-response genes, and disease. Lancet 1(7869):1269–1275
7. Park MS, Terasaki PI, Ahmed AR, Tiwari JL (1979) HLA-DRw4 in 91% of Jewish pemphigus vulgaris patients. Lancet 2(8140):441–442
8. Hanis CL, Boerwinkle E, Chakraborty R, Ellsworth DL, Concannon P, Stirling B et al (1996) A genome-wide search for human non-insulin-dependent (type 2) diabetes genes reveals a major susceptibility locus on chromosome 2. Nat Genet 13:161–166
9. Spence MA, Greenberg DA, Hodge SE, Vieland VJ (2003) The emperor's new methods. Am J Hum Genet 72:1084–1087
10. Madsen AM, Hodge SE, Ottman R (2011) Causal models for investigating complex disease: I. A primer. Hum Hered 72:54–62
11. Rubinstein P, Walker M, Carpenter C, Carrier C, Krassner J, Falk C et al (1981) Genetics of HLA disease associations: the use of the haplotype relative risk (HRR) and the "haplo-delta" (Dh) estimates in juvenile diabetes from three racial groups. Hum Immunol 3:384 (abstr)
12. Raum D, Awdeh Z, Yunis EJ, Alper CA, Gabbay KH (1984) Extended major histocompatibility complex haplotypes in type I diabetes mellitus. J Clin Invest 74:449–454
13. Onengut-Gumuscu S, Concannon P (2006) A haplotype-based analysis of the *PTPN22* locus in type 1 diabetes. Diabetes 55:2883–2889
14. Morton NE (1955) Sequential tests for the detection of linkage. Am J Hum Genet 7:277–318
15. Ott J (1999) Analysis of human genetic linkage. Johns Hopkins University Press, Baltimore
16. Lander E, Kruglyak L (1995) Genetic dissection of complex traits: guidelines for interpreting and reporting results. Nat Genet 11:241–247
17. Concannon P, Chen WM, Julier C, Morahan G, Akolkar B, Erlich HA et al (2009) Genome-wide scan for linkage to type 1 diabetes in 2,496 families from the Type 1 Diabetes Genetics Consortium. Diabetes 58:1018–1022
18. Morahan G, Mehta M, James I, Wei-Min C, Akolkar B, Erlich HA et al (2011) Tests for genetic interactions in type 1 diabetes. Linkage and stratification analyses of 4,422 affected sib-pairs. Diabetes 60:1030–1040
19. Lewontin RC (1964) The interaction of selection and linkage. I. General considerations; heterotic models. Genetics 49:49–67
20. Alper CA, Larsen CE, Dubey DP, Awdeh ZL, Fici DA, Yunis EJ (2006) The haplotype structure of the human major histocompatibility complex. Hum Immunol 67:73–84
21. Gabriel SB, Schaffner SF, Nguyen H, Moore JM, Roy J, Blumenstiel B et al (2002) The structure of haplotype blocks in the human genome. Science 296:2225–2229
22. Awdeh ZL, Raum D, Yunis EJ, Alper CA (1983) Extended HLA/complement allele haplotypes: evidence for T/t-like complex in man. Proc Natl Acad Sci U S A 80:259–263
23. Degli-Esposti MA, Leaver AL, Christiansen FT, Witt CS, Abraham LJ, Dawkins RL (1992) Ancestral haplotypes: conserved population MHC haplotypes. Hum Immunol 34:242–252
24. Smith WP, Vu Q, Li SS, Hansen JA, Zhao LP, Geraghty DE (2006) Toward understanding MHC disease associations: partial resequencing of 46 distinct HLA haplotypes. Genomics 87:561–571
25. Bilbao JR, Calvo B, Costaño L, Aransay AM, Martin-Pagola A, Perez de Nanclares G et al (2006) Conserved extended haplotypes discriminate HLA-DR3-homozygous Basque

patients with type 1 diabetes mellitus and celiac disease. Genes Immun 7:550–554

26. Aly T, Eller E, Ede A, Gowan K, Babu SR, Erlich HA et al (2006) Multi-SNP analysis of MHC region: remarkable conservation of HLA-A1-B8-DR3 haplotype. Diabetes 55: 1265–1269
27. Szilágyi A, Bánlaki Z, Pozsonyi E, Yunis EJ, Awdeh ZL, Hossó A et al (2010) Frequent occurrence of conserved extended haplotypes (CEHs) in two Caucasian populations. Mol Immunol 47:1899–1904
28. Larsen CE, Alford DR, Trautwein MR, Jalloh YK, Tarnacki JL, Kunnenkeri SK et al (2014) Dominant sequences of human major histocompatibility complex conserved extended haplotypes from *HLA-DQA2* to *DAXX*. PLoS Genet 10:e1004637
29. Alper CA, Raum D, Karp S, Awdeh ZL, Yunis EJ (1983) Serum complement 'supergenes' of the major histocompatibility complex in man (complotypes). Vox Sang 45:62–67
30. Horton R, Gibson R, Coggill P, Miretti M, Allcock RJ, Almeida J et al (2008) Variation analysis and gene annotation of eight MHC haplotypes: the MHC haplotype project. Immunogenetics 60:1–18
31. (1968) Nomenclature for factors of the HL-A system. Bull World Health Organ 39: 483–486
32. Robinson J, Halliwell JA, Hayhurst JD, Flicek P, Parham P, Marsh SGE (2015) The IPD and IMGT/HLA database: allele variant databases. Nucleic Acids Res 43(D1):D423–D431
33. Walsh EC, Mather KA, Schaffner SF, Farwell L, Daly MJ, Patterson N et al (2003) An integrated map of the human major histocompatibility complex. Am J Hum Genet 73:580–590
34. Yunis EJ, Larsen CE, Fernandez-Vina M, Awdeh ZL, Romero T et al (2003) Inheritable variable sizes of DNA stretches in the human MHC: conserved extended haplotypes and their fragments or blocks. Tissue Antigens 62:1–20
35. Williamson JF, McLure CA, Baird PN, Male D, Millman J, Lawley B et al (2008) Normal sequence elements define ancestral haplotypes of the region encompassing complement factor H. Hum Immunol 69:207–219
36. Allcock RJ, Atrazhev AM, Beck S, de Jong PJ, Elliott JF, Forbes S et al (2002) The MHC haplotyping project: a resource for HLA-linked association studies. Tissue Antigens 5:520–521
37. Stewart CA, Horton R, Allcock RJ, Ashurst JL, Atrazhev AM, Coggill P et al (2004) Complete MHC haplotype sequencing for common disease gene mapping. Genome Res 14:1176–1187
38. Traherne JA, Horton R, Roberts AN, Miretti MM, Hurles ME, Stewart CA et al (2006) Genetic analysis of completely sequenced disease-associated MHC haplotypes identifies shuffling of segments in recent human history. PLoS Genet 2:e9
39. Kagale S, Koh C, Clarke WE, Bollina V, Parkin IA, Sharpe AG (2016) Analysis of genotyping-by-sequencing (GBS) data. Methods Mol Biol 1374:269–284
40. Amini S, Pushkarev D, Christiansen L, Kostem E, Royce T, Turk C et al (2014) Haplotype-resolved whole-genome sequencing by contiguity-preserving transposition and combinatorial indexing. Nat Genet 46:1343–1349
41. Snyder MW, Adey A, Kitzman JO, Shendure J (2015) Haplotype-resolved genome sequencing: experimental methods and applications. Nat Rev Genet 16:344–358
42. Rubinstein P, Suciu-Foca N, Nicholson JF (1976) Genetics of juvenile diabetes mellitus. A recessive gene closely linked to HLA-D and with 50 per cent penetrance. N Engl J Med 297:1036–1040
43. Raum D, Awdeh Z, Alper CA (1981) BF types and the mode of inheritance of insulin-dependent diabetes mellitus (IDDM). Immunogenetics 12:59–74
44. Rubinstein P, Walker M, Mollen N, Carpenter C, Beckerman S, Suciu-Foca N et al (1990) No excess of DR*3/4 in Ashkenazi Jewish or Hispanic IDDM patients. Diabetes 39: 1138–1143
45. Boehm BO, Schifferdecker E, Rosak C, Kuehnl P, Driesel AJ, Schöffling K (1990) The HLA-DR4-associated DQw8 allele is confined to HLA-DR3/DR4 heterozygous type 1 (insulin-dependent) diabetics. Tissue Antigens 36:81–82
46. Rotter JI, Anderson CE, Rubin R, Congleton JE, Terasaki PI, Rimoin DL (1983) HLA genotypic study of insulin-dependent diabetes: the excess of DR3/DR4 heterozygotes allows rejection of the recessive hypothesis. Diabetes 32:169–174
47. Awdeh ZL, Alper CA (2005) Mendelian inheritance of polygenic diseases: a hypothetical basis for increasing incidence. Med Hypotheses 64:495–498
48. Awdeh ZL, Yunis EJ, Audeh MJ, Fici D, Pugliese A, Larsen CE et al (2006) A genetic explanation for the rising incidence of type 1 diabetes, a polygenic disease. J Autoimmun 27:174–181

49. Larsen CE, Alper CA (2004) The genetics of HLA-associated disease. Curr Opin Immunol 16:660–667
50. Simon M, Bourel M, Fauchet R, Genetet B (1976) Association of HLA-A3 and B14 antigens with idiopathic haemochromatosis. Gut 17:332–334
51. Feder JN, Gnirke A, Thomas W, Tsuchihashi Z, Ruddy DA, Basava A (1996) A novel MHC class I-like gene is mutated in patients with hereditary haemochromatosis. Nat Genet 13:399–408
52. Alper CA, Fleischnick E, Awdeh Z, Katz AJ, Yunis EJ (1987) Extended major histocompatibility complex haplotypes in patients with gluten-sensitive enteropathy. J Clin Invest 79:251–256
53. Ahmed AR, Yunis JJ, Marcus-Bagley D, Yunis EJ, Salazar M, Katz AJ et al (1993) Major histocompatibility complex susceptibility genes for dermatitis herpetiformis compared with those for gluten-sensitive enteropathy. J Exp Med 178:2067–2075
54. Greenberg DA, Hodge SE, Rotter JI (1982) Evidence for recessive and against dominant inheritance at the HLA-"linked" locus in coeliac disease. Am J Hum Genet 34:263–277
55. Thomson G, Bodmer W (1977) The genetic analysis of HLA and disease associations. In: Dausset J, Svejgaard A (eds) HLA and disease. Munksgaard, Copenhagen, pp 84–93
56. Ahmed AR, Mohimen A, Yunis EJ, Mirza NM, Kumar V, Beutner EH et al (1993) Linkage of pemphigus vulgaris antibody to the major histocompatibility complex in healthy relatives of patients. J Exp Med 177:419–424
57. Reveille JD (2006) Major histocompatibility genes and ankylosing spondylitis. Best Pract Res Clin Rheumatol 20:601–609
58. Alper CA, Awdeh Z (2000) Incomplete penetrance of MHC susceptibility genes: prospective analysis of polygenic MHC-determined traits. Tissue Antigens 56:199–206
59. Van Thiel DH, Smith WI Jr, Rabin BS, Fisher SE, Lester R (1977) A syndrome of immunoglobulin A deficiency, diabetes mellitus, malabsorption, a common HLA haplotype. Immunologic and genetic studies of forty-three family members. Ann Intern Med 86:10–19
60. Alper CA, Marcus-Bagley D, Awdeh Z, Kruskall MS, Eisenbarth GS, Brink SJ et al (2000) Prospective analysis suggests susceptibility genes for deficiencies of IgA and several other immunoglobulins on the [HLA-B8, SC01, DR3] conserved extended haplotype. Tissue Antigens 56:207–216
61. Alper CA, Xu J, Cosmopoulos K, Dolinski B, Stein R, Uko G et al (2003) Immunoglobulin deficiencies and susceptibility to infection among homozygotes and heterozygotes for C2 deficiency. J Clin Immunol 23:297–305
62. Calvo B, Castaño L, Marcus-Bagley D, Fici DA, Awdeh Z, Alper CA (2000) The [HLA-B18, F1C30, DR3] conserved extended haplotype carries a susceptibility gene for IgD deficiency. J Clin Immunol 20:216–220
63. Alper CA, Husain Z, Larsen CE, Dubey DP, Stein R, Day C et al (2006) Incomplete penetrance of susceptibility genes for MHC-determined immunoglobulin deficiencies in monozygotic twins discordant for type 1 diabetes. J Autoimmun 27:89–95
64. Redondo MJ, Yu L, Hawa M, Mackenzie T, Pyke DA, Eisenbarth GS et al (2001) Heterogeneity of type 1 diabetes: analysis of monozygotic twins in Great Britain and the United States. Diabetologia 44:354–362
65. Raffel LJ, Goodarzi MO (2013) Diabetes mellitus. In: Rimoin DL, Connor JM, Pyeritz RE, Korf BR (eds) Emery and Rimoin's principles and practice of medical genetics, 6th edn. Churchill-Livingstone, New York, pp 1–58, Chapter 86
66. Risch N (1990) Linkage strategies for genetically complex traits. I. Multilocus models. Am J Hum Genet 46:222–228
67. Madsen AM, Ottman R, Hodge SE (2011) Causal models for investigating complex disease: II. What causal models can tell us about penetrance for additive, heterogeneity, and multiplicative two-locus models. Hum Hered 72:68–72
68. Thomson G (1980) A two locus model for juvenile diabetes. Ann Hum Genet 43: 383–398
69. Alper CA, Dubey DP, Yunis EJ, Awdeh Z (2000) A simple estimate of the general population frequency of the MHC susceptibility gene for autoimmune polygenic disease. Exp Clin Immunogenet 17:138–147
70. Florez JC, Hirschhorn J, Altshuler D (2003) The inherited basis of diabetes mellitus: implications for the genetic analysis of complex traits. Annu Rev Genomics Hum Genet 4:257–291
71. Pociot F, Akolkar B, Concannon P, Erlich HA, Julier C et al (2010) Genetics of type 1 diabetes: what's next? Diabetes 59:1561–1571
72. Polychronakos C, Li Q (2011) Understanding type 1 diabetes through genetics: advances and prospects. Nat Rev Genet 12:781–792

Part III

Haploid-cell Typing

Chapter 7

Haplotyping a Non-meiotic Diploid Fungal Pathogen Using Induced Aneuploidies and SNP/CGH Microarray Analysis

Judith Berman and Anja Forche

Abstract

The generation of haplotype information has recently become very attractive due to its utility for identifying mutations associated with human disease and for the development of personalized medicine. Haplotype information also is crucial for studying recombination mechanisms and genetic diversity, and for analyzing allele-specific gene expression. Classic haplotyping methods require the analysis of hundreds of meiotic progeny. To facilitate haplotyping in the non-meiotic human fungal pathogen *Candida albicans*, we exploited trisomic heterozygous chromosomes generated via the *UAU1* selection strategy. Using this system, we obtained phasing information from allelic biases, detected by SNP/CGH microarray analysis. This strategy has the potential to be applicable to other diploid, asexual *Candida* species that are important causes of human disease.

Key words Haplotyping, Phased genome, *Candida albicans*, Non-meiotic organism, Aneuploidy, *UAU1* cassette, Heterozygosity, SNP/CGH microarrays, Ymap

1 Introduction

Haplotyping, in general terms, is the process by which the phased haploid sequence is determined from unordered (unphased) genotype data. Knowing the specific haplotype carried by a given individual is essential for understanding the phenotypic consequences of gene regulation, genetic diversity, recombination mechanisms, and for identifying mutations associated with monogenic and multigenic human diseases [1]. To give two examples, haplotype information facilitates the detection of loss of heterozygosity (LOH) in cancer cells, and enables identification of the alleles carried at the major histocompatibility complex locus, important for treating autoimmune disorders and organ transplantation [2]. The classical genetic approach to produce haplotype maps, or "hapmaps,"

Electronic supplementary material: The online version of this chapter (doi:10.1007/978-1-4939-6750-6_7) contains supplementary material, which is available to authorized users.

Irene Tiemann-Boege and Andrea Betancourt (eds.), *Haplotyping: Methods and Protocols*, Methods in Molecular Biology, vol. 1551, DOI 10.1007/978-1-4939-6750-6_7, © Springer Science+Business Media LLC 2017

involves genotyping hundreds of meiotic progeny, and comparing recombinant and paternal genotypes. Current technologies, such as whole genome sequencing, rarely resolve diploid sequence data into their respective haplotypes, and haplotypes are most often inferred using statistical approaches [1].

Non-meiotic organisms pose a special challenge for haplotyping methods, as they do not normally produce recombinant offspring and can have highly divergent homologs. *C. albicans*, an opportunistic fungal pathogen of humans and the major cause of fungal infections in the clinic, is such a highly heterozygous ameiotic diploid [3]. Despite no evidence of meiosis, the *C. albicans* genome rapidly adapts to in vitro and in vivo stresses, including antifungal drugs [4–8]. To better understand this rapid adaptation, we constructed a hapmap for the standard laboratory strain SC5314 [9], which facilitated both linkage mapping and the analysis of recombination mechanisms. To generate this hapmap, we exploited the ameiotic nature of *C. albicans* and the ability to generate trisomic individuals with two copies of one homologous chromosome and a single copy of the other homolog. Thus, by determining allelic ratios along complete chromosomes, we were able to assign SNPs to differentiate homologs of the homozygous and/or aneuploid chromosome.

Initially, we used allelic ratios of a series of isolates from two different strain backgrounds (laboratory strain SC5314 and clinical isolate T118) that were either aneuploid or homozygous for whole chromosomes [allelic ratios of SNPs (1:2 or 2:1)] to assign alleles to haplotypes [5, 10]. This allowed us to generate hapmaps for a limited set of SNP markers (150) [9, 11]. Next, we used tiling SNP/CGH arrays to analyze ~39,000 additional SNPs for the SC5314 strain background [12]. Muzzey et al. followed up with massively parallel sequencing of the same set of strains derived from SC5314 to construct whole chromosome hapmaps for all the SNPs in this particular strain background [13]. Markers on Chr3R are homozygous in the original laboratory strain and were designated "allele a" in earlier studies. Unfortunately, Muzzey et al. analysis renamed the alleles on Chr3R such that all are named "allele b," which then became the allele assignments in the CGD database [14].

Since only a limited number of strains homozygous or triploid for whole chromosomes were available, we tested a *UAU1* transformation strategy to artificially induce whole chromosome trisomy of specific chromosomes. We investigated the *UAU1* approach, originally developed to delete both copies of a desired gene during a single round of transformation, as it was already known to result in trisomic alleles when targeting essential genes [15]. To produce gene knockouts, a *UAU1* cassette consisting of a central functional gene (*ARG4*) flanked by two partial and overlapping fragments of the *URA3* gene is used; homologous recombination of the

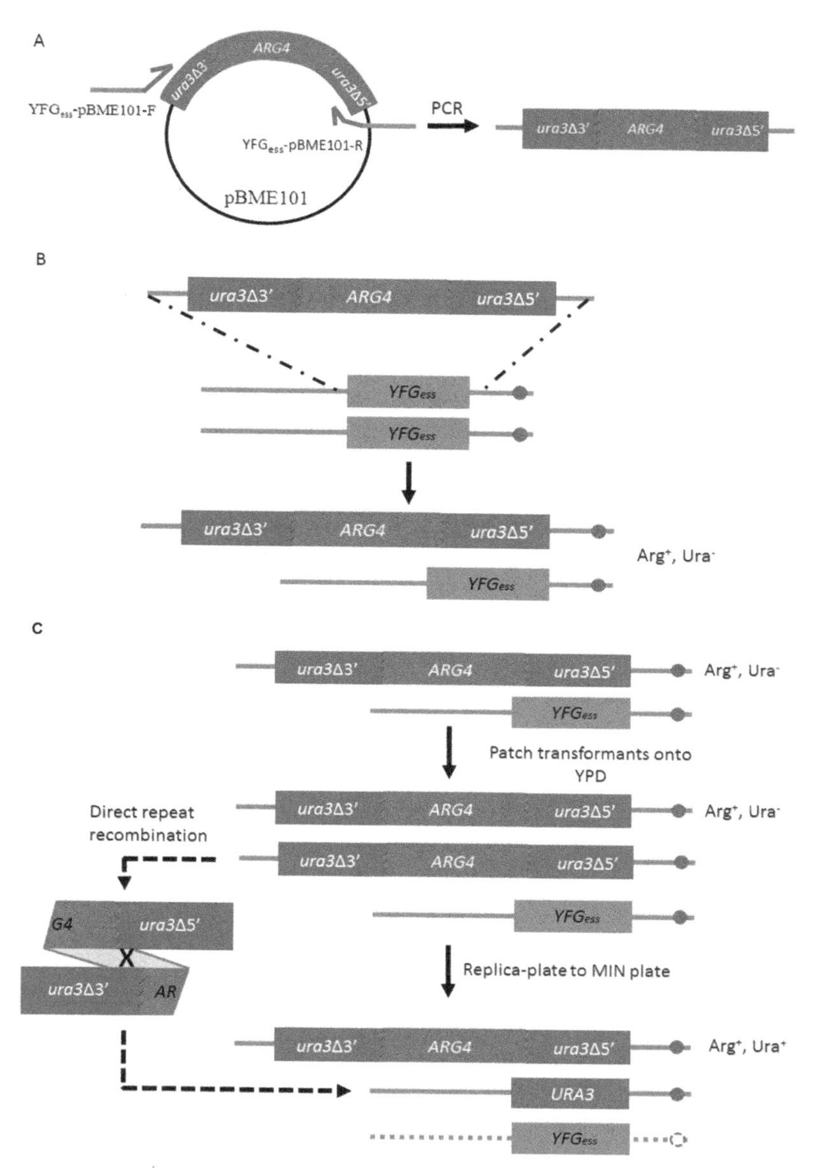

Fig. 1 (**a**). Amplification of the *UAU1* cassette from plasmid pBME101, (**b**). Transformation of *UAU1* cassette into the *Candida albicans* genome. Adapted from [15]. (**c**). Two consecutive recombination events result in transformants that are Arg$^+$ and Ura$^+$ with the wild type allele deleted or, especially in the case of essential genes, with the wild type allele present on a third copy

overlapping regions brings together the *URA3* fragments, resulting in a complete and functional copy of *URA3* [15, 16] (Fig. 1a, b). To delete both copies of a given gene (Your Favorite Gene 1, *YFG1*), a *C. albicans* strain lacking *ARG4* and *URA3* is transformed with a custom *UAU1* construct carrying sequences homologous to those flanking the open reading frame (ORF) of *YFG1*.

Homologous recombination between these sequences on the construct and those on the chromosome will yield transformants where the *YFG1* coding sequence has been replaced with the *UAU1* construct and can be detected because they are prototrophic for arginine (Arg⁺) (Fig. 1b). To obtain knockouts of both copies of *YFG1*, Arg⁺ transformants are grown on nonselective media to allow for two sequential rounds of recombination (Fig. 1c). The first recombination event results in the deletion of the second remaining *YFG1* allele by homologous recombination, resulting in loss of heterozygosity of the locus such that both copies contain the *UAU1* cassette. This recombination event can occur via a double crossover that directly replaces one allele for the other, but more frequently involves a single crossover between the centromere and the *YFG1* locus followed by the segregation of the two *UAU1* alleles into the same daughter cell. The second recombination event results in a functional *URA3* allele by direct repeat recombination between the two *URA3* fragments. These cells will be both Ura⁺ and Arg⁺, as retention of the original *UAU1* copy on the other allele maintains the Arg⁺ phenotype, and can be easily identified by selection on medium lacking arginine and uridine (Fig. 1c).

Interestingly, the results for essential genes differed: repeated efforts to delete these genes did not eliminate the wild-type copy of *YFG1*, despite the presence of both the complete *UAU1* and the recombined *URA3* cassettes. Diagnostic PCR of these strains always amplified all three possible alleles for the *YFG1* locus, the wild-type allele, the *UAU1* cassette, and the functional *URA3* gene. In many of these cases, the transformants were trisomic for the chromosome carrying the essential *YFG1* (*YFG1ess*) locus, with all known SNPs on the chromosome having a 1:2 allele ratio bias. Accordingly, by targeting the essential gene *ORC1* on chromosome 1 with *UAU1*, we were able to confirm whole chromosome trisomy in 43% of the Arg⁺ Ura⁺ transformants (four of seven transformants analyzed by SNP microarray and aCGH) [9].

The following protocols will enable the reader to target a specific chromosome of interest in *C. albicans* by using the *UAU1* strategy to replace an essential gene on this chromosome. Once transformants tri-allelic for the targeted chromosome are obtained, the phased genotype can be determined by inferring the ratio of DNA amounts for each allele, using commercially available custom SNP/CGH arrays. Alternatively, the genotypes can be analyzed by deep sequencing, or by cost-effective reduced representation sequencing methods such as double-digest restriction site associated DNA sequencing (ddRADseq), and allele ratios for SNPs can be determined from the sequence data [12, 17–21].

One limitation of this method is that it is only applicable to strains that are auxotrophic for arginine and uridine. This limitation can be potentially circumvented by modifying the transformation protocol to use two drug resistance markers, such as norceothricin

(*NAT1*) and hygromycin B (*HYG*) (e.g., *NA-HYG-AT*). This strategy should then be adaptable not only to other strains of *C. albicans*, but also to other diploid *Candida* species such as *C. dubliniensis* and *C. parapsilosis*, thereby facilitating the analysis of genome structure and recombination mechanism in these species.

2 Materials

2.1 Custom-Ordered: Agilent 8× 60 K SNP/CGH Arrays for *C. albicans* (Amadid 039364)

Probes were 19–33 nt long with melting points (Tm) near 55 °C [22] and the SNP at the middle position. Initially, probes were designed for 43,658 SNP loci and 28,563 non-SNP loci (including regions that do not have high frequencies of SNPs). Probes were eliminated if they had more than one significant alignment (*e*-value, 0.001) with the reference genome using BLAST analysis [23]. The acceptable Tm range was reduced to optimize the number of probes to the Agilent 8× 60 K microarray format (41,616 SNP allele probes (two probes per SNP locus, representing the two SNP alleles) plus 20,363 non-SNP (CGH) probes (one per locus). Quality control of the final array design resulted in a usable SNP allele probe count of 39,222. A total of 19,016 SNP loci (97% of usable SNP loci) were informative and were mapped to one of the two homologs (*see* Table S1 for SNP probes) [12].

2.2 Equipment

1. Shaking incubator.
2. Eppendorf centrifuge 5430R (or similar).
3. Table top centrifuge (for 50 mL conical tubes).
4. Thermocycler machine for PCR reactions.
5. Digital water bath.
6. Gel electrophoresis setup.
7. Microscope (×40 magnification).
8. Plastic nitrogen purge desiccator cabinet for microarray slide storage.
9. Agilent hybridization oven with rotator rack and rotator rack conversion rod.
10. Agilent hybridization chamber.
11. Slide staining dish (×4).
12. Magnetic stir plate (×3).
13. Agilent microarray scanner.

2.3 Growth Media

1. Yeast extract Peptone Dextrose medium (YPD) for strain recovery from stock: 1% yeast extract, 2% bacto peptone, 2% dextrose, 1.5% agar (omit for broth), 5 mL uridine (20 mg/L), 995 mL deionized water, autoclave for 30 min.

2. Minimal medium (MIN) (*see* **Note 1**): 0.67% yeast nitrogen base without amino acid (with ammonium sulfate), 2% dextrose, 1.5% agar (omit for broth), 1 L deionized water, autoclave for 20 min.
3. MIN + Uridine (MIN + Uri): 0.67% yeast nitrogen base without amino acid (with ammonium sulfate), 2% dextrose, 1.5% agar (omit for broth), 5 mL uridine (20 mg/L), 995 mL deionized water, autoclave for 20 min.

2.4 Buffers

1. TENTS buffer: 100 mM NaCl, 10 mM Tris–HCl, pH 8.0, 1 mM EDTA pH 8.0, 2% triton X, 1% SDS.
2. 10× TE buffer: 5 mL 1 M Tris, pH 7.5, 1 mL 0.5 M EDTA pH 8.0, 44 mL deionized water, filter sterilize.
3. TELiAc buffer (*see* **Note 2**): 8 vol deionized water, 1 vol 10× TE buffer, 1 vol 1 M LiAc.
4. 50% Polyethylenglycol (PEG) (*see* **Note 3**): to a 150 mL glass beaker add a magnetic stir bar, 50 g polyethylene glycol MW 3350, 40 mL deionized water, seal with parafilm, let stir overnight, raise volume to 100 mL, filter sterilize.
5. PLATE mix: 8 vol 50% PEG, 1 vol 1 M LiAc, 1 vol 10× TE buffer.
6. Agilent 10× aCGH blocking agent.
7. Agilent 2× hybridization buffer.
8. Agilent OligoCGH wash buffer 2.
9. Agilent stabilization and drying solution.
10. 10× TBE gel electrophoresis buffer (1 L): 108 g Tris base, 55 g Boric acid, 40 mL 0.5 M EDTA pH 8.0, raise volume to 1 L with deionized water.

2.5 Chemicals

1. Phenol/Chloroform/Isoamylalcohol (25:24:1).
2. Isopropanol.
3. 3 M Na Acetate, pH 5.5.
4. Ethanol.
5. Ethidium bromide solution.
6. Acetonitrile (caution: extremely flammable).
7. Agarose (Seakem).

2.6 Molecular Biology

1. dNTP mix (for PCR).
2. Phusion Taq Polymerase.
3. Oligonucleotides YFG_{ess}-pBME101-F (70 nt of 5′ YFG_{ess}-GTTTTCCCAGTCACGACGTT) and YFG_{ess}-pBME101-R (70 nt of 3′ YFP_{ess}-TGTGGAATTGTGAGCGGATA) [15] for amplification of *UAU1* transformation cassette (Fig. 1a).

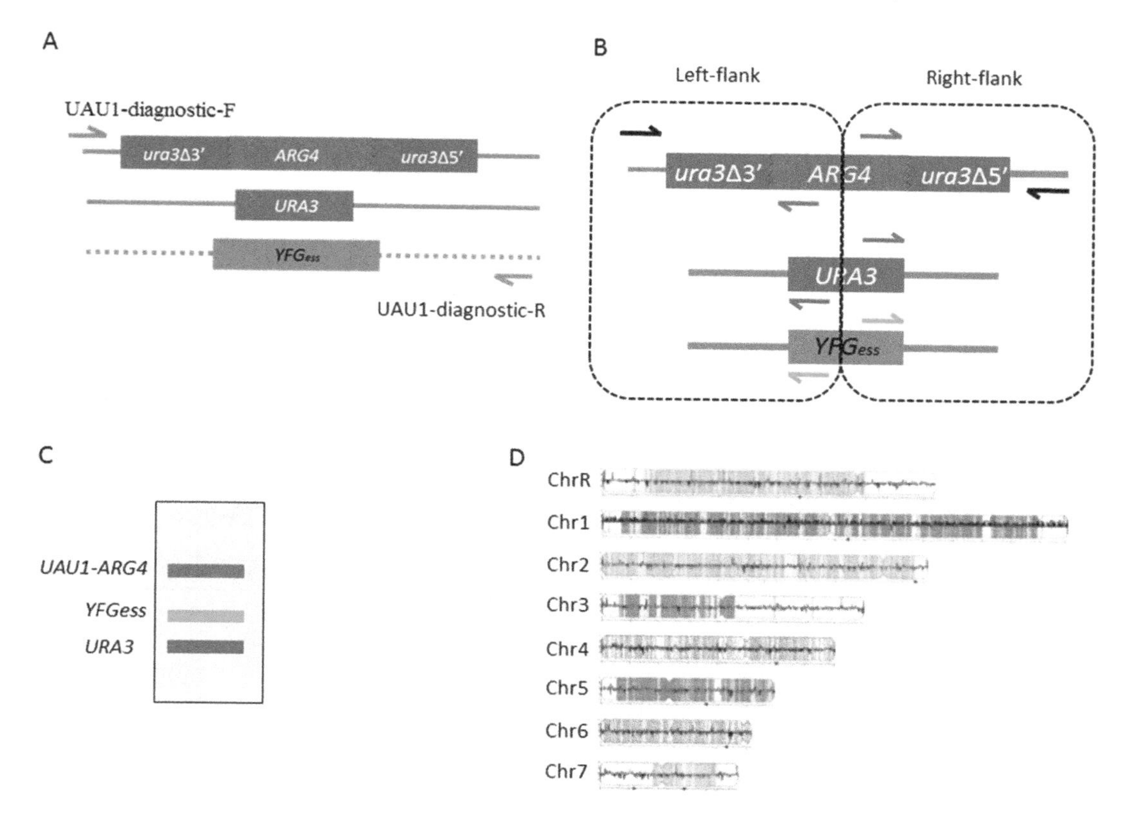

Fig. 2 Options for diagnostic PCR include (**a**). One-step PCR when *URA3* and *YFGess* are of sufficiently different sizes to be distinguished by gel electrophoresis. (**b**). Left-flank and Right-flank multiplex PCRs when *URA3* and *YFGess* are indistinguishable by gel electrophoresis. (**c**). Expected result of diagnostic PCR for both options (**a** and **b**) for a trisomic locus. (**d**). Illustrative image generated with Y_{MAP} for a strain exhibiting Chr1 trisomy and major regions of homozygosity on Chrs R, 2, 3, 5 and 7. White = homozygous in parental strain SC5314; grey = heterozygous; cyan = homozygous for 'a' homolog; magenta = homozygous for 'b' homolog; purple = bba allelic ratio (for trisomic chromosome 1 in this diagram)

4. Oligonucleotides for diagnostic PCR (Fig. 2a, b) (*see* **Note 4**).
5. Plasmid carrying *UAU1* cassette (pBME101) [15].
6. 10 mg/mL sheared salmon sperm DNA.
7. *Hae*III restriction enzyme and buffer.
8. Invitrogen Bioprime kit.
9. dNTP mix (1.2 mM dATP, dGTP, dCTP, 0.9 mM dTTP) (for SNP/CGH array).
10. Cy3 and Cy5 Fluorolink monofunctional dyes.

2.6.1 Kits

1. Qiagen MinElute PCR purification kit.
2. Agilent Oligo aCGH/CHiP-Chip hybridization kit.

2.6.2 Disposables

1. Sterile, powder-free nitrile gloves.
2. Glass beads (for gDNA extraction).
3. Sterile 1.5 mL microfuge tubes.
4. Sterile 1.5 mL Eppendorf Safe-lock tubes.
5. Sterile 0.2 mL PCR tubes.
6. Sterile pipet tips.
7. Microcon 30 KDa centrifugal filters (Millipore).
8. Array gasket backings for Agilent slides.

2.6.3 Other

1. *C. albicans* strains (stocks in 50% glycerol stored at −80 °C).
2. Glass beads for plating.
3. Magnetic stir bars (4×).
4. BlueGnome microarray analysis software (Cambridge).

3 Methods

3.1 Amplification and Purification of UAU1 Transformation Cassette

1. For each transformation, prepare eight 50 μL PCR reactions as follows (*see* **Note 5**):

   32 μL deionized water.

   10 μL GC buffer.

   2 μL dNTPs (2.5 mM each).

   2 μL YFG_{ess}-pBME101-F (10 μM).

   2 μL YFG_{ess}-pBME101-R (10 μM).

   4.5 μL of 25 mM $MgCl_2$.

   0.5 μL Phusion Taq Polymerase (two units).

   2 μL of plasmid pBME101 (1:100 dilution of a quick plasmid prep).
2. Run PCR under the following conditions (*see* **Note 6**):

   98 °C—30 s

   34 cycles of:

   98 °C—10 s

   55 °C—30 s

   72—2 min

   72—10 min

   15-hold
3. Run 5 μL of each PCR reaction on a 1% agarose gel to check for amplification of *UAU1* cassette (4.15 kb).
4. Pool all eight PCR reactions into a 1.5 mL Eppendorf tube, add 1 mL cold 95% ethanol and 50 μL 3.5 M Na Acetate pH 5.5, precipitate DNA at −20 °C for at least 20 min.

5. Centrifuge sample for 20 min at 13,000 rpm in Eppendorf centrifuge (M5430R), remove ethanol, air dry pellet, and resuspend DNA in 30 μL deionized water. Store at −20 °C until use.

3.2 Lithium Acetate Transformation

1. Streak out *C. albicans* strains to be transformed onto YPD medium and incubate at 30 °C for 2 days.
2. Inoculate a single colony into 2 mL YPD and grow overnight at 30 °C with shaking.
3. Add 600 μL overnight culture to 50 mL YPD (Erlenmeyer flask).
4. Incubate 4–5 h (OD_{600} should be between 0.5 and 0.8) at 30 °C with shaking.
5. Check cells for active budding by microscopy. Transformation efficiency of actively budding cells is better because the cell walls of growing buds are not as rigid as stationary phase (non-budding) cells and budding cells are also more metabolically active.
6. Transfer entire culture to 50 mL conical tubes, pellet cells at 1200 rpm for 5 min, and discard medium.
7. Resuspend pellet in 5 mL deionized water, spin for 5 min at 1200 rpm, and discard water.
8. Resuspend cells in 500 μL TELiAc, centrifuge for 2 min at 1200 rpm in microfuge.
9. Remove TELiAc with pipette and resuspend cells in 250 μL TELiAc (total volume including pellet ~300 μL).
10. To one 1.5 microfuge tube, add in this order: 5 μL sheared salmon sperm DNA and 150 μL cells (negative control).
11. To another microfuge tube add 5 μL carrier DNA, entire 30 μL of *UAU1* transformation cassette and 150 μL cells, incubate for 30 min at room temperature (RT).
12. Add 700 μL PLATE mix (*see* **Note 7**), invert to mix, and incubate overnight at RT.
13. Heat shock transformation reactions at 42 °C for 1 h in water bath.
14. Centrifuge cells for 3 min at 3000 rpm in microfuge.
15. Remove supernatant with pipette and resuspend in 150 μL deionized water.
16. Plate entire mixture on MIN + Uri and incubate for up to 3 days at 30 °C.
17. Patch-streak Arg^+ transformants onto YPD and grow at 30 °C for 2 days.
18. Replica-plate onto MIN plates to select transformants that underwent recombination (Arg^+Ura^+) (*see* Fig. 1c).

3.3 gDNA Extraction for Arg^+Ura^+ Transformants

1. Streak out Arg^+Ura^+ transformants for single colonies onto MIN plates (to maintain selection) and incubate plates at 30 °C for 2 days.
2. Transfer a single colony to 2 mL MIN broth and incubate overnight at 30 °C with shaking.
3. Transfer 1.0 mL *C. albicans* culture to 1.5 mL Eppendorf Safelock tube (*see* **Note 8**), spin down at max speed in Eppendorf centrifuge for 6 s, and discard liquid.
4. Add 1 mL deionized water, vortex tube until pellet is resuspended, spin down at max speed for 6 s, and discard water.
5. Add 500 μL TENTS buffer (*see* Subheading 2), 200 μL glass beads (use 0.2 mL scoop for measuring), and 500 μL Phenol/Chloroform/Isoamylalcohol (25:24:1) (UNDER THE HOOD).
6. Close tubes WELL, turn tubes upside down (very carefully) to check for leaking.
7. Put tubes in tube adaptor of vortex shaker and shake for 20 min.
8. Spin tubes at max speed for 10 min, remove upper phase, and transfer to new, labeled 1.5 mL tube.
9. Add 1 mL ice cold isopropanol and 50 μL 3.5 M Na acetate, mix gently by inversion.
10. Put tubes in a freezer for at least 20 min to allow for maximal DNA precipitation.
11. Spin tubes for 20 min at max speed, discard liquid (careful not to lose pellet).
12. Add 500 μL ice cold 70% ethanol to remove any residual salt (make sure you rinse the walls of tubes); remove ethanol with pipette.
13. Either air dry or speed vac dry the pellets.
14. Add 60 μL TE buffer (with 20 μg/mL RNase), vortex well, and incubate for 45 min at 37 °C.

3.4 Diagnostic PCR to Test for Presence of Potential Trisomy

To confirm the presence of three alleles at the *YFGess* locus in Arg^+Ura^+ transformants:

1. *URA3* and *YFGess* are of similar length—One-step diagnostic PCR (Fig. 2a, c):

   32 μL deionized water.

   10 μL GC buffer.

   2 μL dNTPs (2.5 mM each).

   2 μL YFGess-pBME101-F (10 μM).

   2 μL YFGess-pBME101-R (10 μM).

4.5 μL of 25 mM $MgCl_2$.

0.5 μL Phusion Taq Polymerase (two units).

2 μL gDNA from transformant (30 ng/μL).

Run PCR under the following conditions (*see* **Note 6**):

98 °C—30 s

34 cycles of:

98 °C—10 s

55 °C—30 s

72—2 min

72—10 min

15-hold

Run out 10 μL of each PCR reaction on a 1% agarose gel to check for amplification of the correct-size products.

2. *URA3* and *YFGess* are of different length—Left-flank and Right-flank multiplex diagnostic PCRs (Fig. 2b, c):

Left-flank

32 μL deionized water.

10 μL GC buffer.

2 μL dNTPs (2.5 mM each).

1.5 μL UAU1-diagnostic-F (10 μM).

0.5 μL diagnostic-URA3-R (10 μM).

0.5 μL diagnostic-ARG4-R (10 μM).

0.5 μL diagnostic-YFGess-R (10 μM).

4.5 μL of 25 mM $MgCl_2$.

0.5 μL Phusion Taq Polymerase (two units).

2 μL gDNA from transformant (30 ng/μL).

Right-flank

32 μL deionized water.

10 μL GC buffer.

2 μL dNTPs (2.5 mM each).

1.5 μL UAU1-diagnostic-R (10 μM).

0.5 μL diagnostic-URA3-F (10 μM).

0.5 μL diagnostic-ARG4-F (10 μM).

0.5 μL diagnostic-YFGess-F (10 μM).

4.5 μL of 25 mM $MgCl_2$.

0.5 μL Phusion Taq Polymerase (two units).

2 μL gDNA from transformant (30 ng/μL).

Run PCRs under the following conditions (*see* **Note 6**):

98 °C—30 s

34 cycles of:

98 °C—10 s

55 °C—30 s

72—2 min

72—10 min

15-hold

3.5 SNP/CGH Arrays (See Note 9)

3.5.1 gDNA Restriction Enzyme Digests

1. Per sample (pipette into 1.5 mL Eppendorf tubes):

   3 μg gDNA.

   3 μL *Hae*III restriction enzyme.

   6 μL 10× M restriction enzyme buffer
2. Add deionized water to total volume of 60 μL.
3. Vortex well, spin down tubes briefly, and incubate at 37 °C for at least 4 h and up to overnight (~16 h).

3.5.2 Purification of Restriction Digests

1. Add 60 μL DNA to a Qiagen minElute spin column. Add 120 μL NT1 (Binding) buffer, spin at max speed for 1 min, and discard flow-through.
2. Add 700 μL NT3 (Wash) buffer, spin at max speed for 1 min, and discard flow-through; repeat.
3. Spin one last time at max speed for 1 min and transfer column to a new 1.5 mL tube.
4. Add 10 μL AE (Elution) buffer, incubate for 1 min at RT, and spin at max speed for 1 min; repeat.
5. Transfer flow through to the top of the column (~20 μL), incubate for 1 min at RT, and spin at max speed for 1 min.

3.5.3 Labeling Reaction

1. Per sample, add 20 μL purified DNA and 20 μL random primer.
2. Incubate sample at 95 °C for 5 min and keep on ice for 5 min.
3. Per sample, add 5 μL dNTP mix (1.2 mM dATP, dGTP, dCTP, 0.9 mM dTTP), 1.5 μL Cy-dye (Cy3 for sample, Cy5 for diploid control strain), and 1 μL Klenow enzyme.
4. Incubate at 37 °C for 2 h.

3.5.4 Wash and Concentrate Probes

1. Combine corresponding Cy3- and Cy5-labeled samples into a Microcon 30 KDa centrifugal filter and add 450 μL deionized water.
2. Spin at 12,000 rpm in microfuge for 8 min. There should be 20–50 μL left in the filter. Discard flow through and add 450 μL deionized water.

3. Spin at 12,000 rpm in microfuge for 5 min in batches until there is approximately 20 μL left in the filter. The concentrate should be purple in color (mix of Cy3 and Cy5 dyes).
4. Discard flow through. Invert Microcon 30 KDa centrifugal filter into new 1.5 mL tubes and spin at max speed for 2 min to collect the concentrate.

3.5.5 Hybridization

1. To a 0.2 mL PCR tube, add 20 μL labeled concentrated DNA, 5 μL 10× blocking buffer, and 25 μL 2× hybridization buffer.
2. Incubate at 95 °C for 3 min followed by 30 min at 65 °C in PCR machine.
3. Set the hybridization oven to 65 °C. Place the gasket slide into the hybridization chamber, gasket side up.
4. Add 45 μL of each sample to CENTER of each gasket. LABEL SAMPLE LOCATIONS ON SLIDE and SLIDE ID#.
5. Place array slide onto gasket slide with AGILENT labeled slide facing gasket slide.
6. Assemble the rest of the hybridization chamber (Fig. 3).
7. Check to see that any bubbles in gaskets are able to move.
8. Secure hybridization chambers into hybridization oven and start rotation, incubate at 65 °C for 24 h.

3.5.6 Slide Washing

1. Prepare five wash dishes:
   (a) Wash Buffer 1, RT and glass container—to pry gasket slide from array slide.
   (b) Wash Buffer 1, RT, glass container with stir bar.
   (c) Wash Buffer 2, 37 °C, glass container with stir bar.
   (d) Acetonitrile, RT, glass container with stir bar—FUME HOOD.
   (e) Stabilization and drying solution, RT, glass container with stir bar—FUME HOOD.

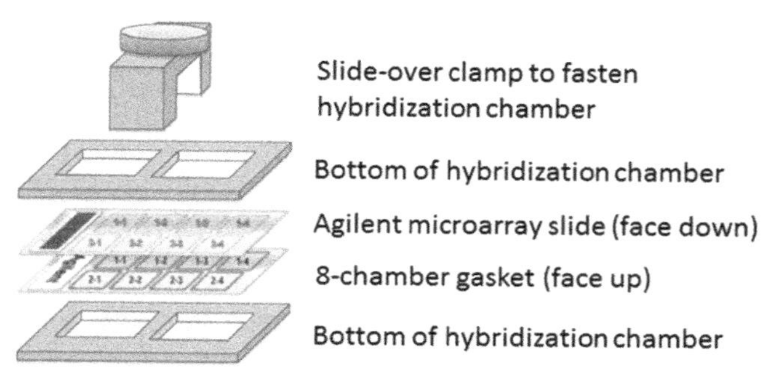

Fig. 3 Assembly of hybridization chamber for SNP/CGH arrays. Adapted from [25, 26]

2. Remove array slide from oven and disassemble hybridization chamber.
3. Place joined slides into (a) Wash Buffer 1, using a razor blade, pry gasket slide from array slide at the barcode labeled end. Place array in slide rack.
4. Wash in (b) Wash Buffer 1 for 5 min, with stirring.
5. Wash in (c) Wash Buffer 2 for 1 min, with stirring.
6. Wash in (d) Acetonitrile for 1 min, with stirring in FUME HOOD.
7. Wash in (e) Stabilization and drying solution for 30 s, with stirring.
8. SLOWLY (5–10 s) remove slide rack.
9. Store slides in dry and dark slide box under nitrogen until scanned.

3.5.7 Scanning, Data Analysis, and Data Visualization

1. Place slide with AGILENT side up in scanner chamber.

   Parameters: Profile: AgilentHD_CGH (61 × 21.6 mm).

   Resolution: 2 μm TIFF 16-bit.

   RPMT/GPMT: 100 %.

   XDR: None.
2. Analyze images using "BlueFuse for Microarrays" (BlueGnome, Cambridge).
3. Normalize data using Block Lowess method (analyze at two levels of stringency [one with no data excluded; the second, data excluded when "Quality less than 1; Confidence less than 0.4; PON Ch1 or Ch2 less than 0.6"]) (*see* **Note 10**).
4. Create Ymap account at http://lovelace.cs.umn.edu/Ymap/ (*see* **Note 11**).
5. Click on "Manage Dataset" tab.
6. Click on "Install New Dataset."
7. Fill in "Dataset Name," "Ploidy of experiment" (default is 2.0), and "Baseline Ploidy" (default is 2.0).
8. Under "Data type" select "SnpCgh microarray."
9. Click on "Create New Dataset."
10. Click on "Reload" button.
11. Click on "Add: SnpCgh array data" to upload your data file.
12. Data download and processing will start automatically.
13. When processing is done, click on "Visualize Datasets" tab to obtain graphic representation of your data (for example, *see* Fig. 2d).
14. Data can be saved in multiple formats (e.g., jpg, png, tif).

4 Notes

1. Instead of MIN medium, a semidefined dropout medium (SDC) can be used [24] (for selection of Arg^+, Ura^- transformants plate on SDC-Arg, when screening for Arg^+Ura^+ transformants plate on SDC-Arg-Uri).
2. TELiAc buffer can be prepared as stock, and filter-sterilized. It can be stored for several months.
3. Following the protocol for making PEG is very important, fresh stock should be prepared every month. Using older PEG will decrease transformation efficiency.
4. Primers for diagnostic PCR need to be designed for each transformation. Primer 3 software is freely available online (http://bioinfo.ut.ee/primer3/).
5. Eight 50 μL PCR reactions are needed for high transformation efficiency.
6. All PCRs in this chapter may need some optimization based on brand of polymerase, brand of PCR thermocycler machine, and primers used.
7. PLATE mix should be made fresh every time; TELiAc can be made as a stock and used for a couple of months.
8. To avoid leaking of tubes use Safe-lock Eppendorf tubes for gDNA extractions.
9. Each SNP/CGH hybridization requires the preparation of the experimental strain and of the diploid control strain. The control strain is typically labeled with Cy5 fluorolink dye and the experimental strain with Cy3 fluorolink dye.
10. For more detailed description of how to process scanned microarray images, contact the authors.
11. Ymap has been optimized to work on Macintosh computers using Safari browser.

Acknowledgments

We like to thank Mathura A. Thevandavakkam for critical reading of the manuscript. A.F. is supported by a grant from the NIAID 2 R15 AI090633. J.B. is supported by the People Programme (Marie Curie Actions) of the European Union's Seventh Framework Programme (FP7/2007-2013) REA grant agreement number 303635; by an European Research Council Advanced Award, number 340087, RAPLODAPT, by grants from the Israel Science foundation (340/13), and by the National Institute of Allergy and Infectious Disease (R01AI075096 and R01AI0624273).

References

1. Xu J (2006) Extracting haplotypes from diploid organisms. Curr Issues Mol Biol 8:113–122
2. Glusman G, Cox H, Roach J (2014) Whole-genome haplotyping approaches and genomic medicine. Genome Med 6(9):73
3. Bennett RJ, Forche A, Berman J (2015) Rapid mechanisms for generating genome diversity: whole ploidy shifts, aneuploidy, and loss of heterozygosity. Cold Spring Harb Perspect Med 4(10), pii:a019604
4. Forche A, Abbey D, Pisithkul T, Weinzierl MA, Ringstrom T et al (2011) Stress alters rates and types of loss of heterozygosity in *Candida albicans*. mBio 2, e00129-11
5. Forche A, Magee PT, Selmecki A, Berman J, May G (2009) Evolution in *Candida albicans* populations during single passage through a mouse host. Genetics 182:799–811
6. Bouchonville K, Forche A, Tang KE, Selmecki A, Berman J (2009) Aneuploid chromosomes are highly unstable during DNA transformation of *Candida albicans*. Eukaryot Cell 8:1554–1566
7. Selmecki AM, Gerami-Nejad M, Paulsen C, Forche A, Berman J (2008) An isochromosome confers drug resistance in vivo by amplification of two genes, *ERG11* and *TAC1*. Mol Microbiol 68:624–641
8. Gerstein AC, Berman J (2015) Shift and adapt: the costs and benefits of karyotype variations. Curr Opin Microbiol 26:130–136
9. Legrand M, Forche A, Selmecki A, Chan C, Kirkpatrick D, Berman J (2008) Haplotype mapping of a diploid non-meiotic organism using existing and induced aneuploidies. PLoS Genet 4:e1
10. Forche A, Alby K, Schaefer D, Johnson A, Berman J, Bennett R (2008) The parasexual cycle in *Candida albicans* provides an alternative pathway to meiosis for the formation of recombinant strains. PLoS Biol 6:e110
11. Cowen LE, Sanglard D, Calabrese D, Sirjusingh C, Anderson JB, Kohn LM (2000) Evolution of drug resistance in experimental populations of *Candida albicans*. J Bacteriol 182:1515–1522
12. Abbey D, Hickman M, Gresham D, Berman J (2011) High-resolution SNP/CGH microarrays reveal the accumulation of loss of heterozygosity in commonly used *Candida albicans* strains. G3 (Bethesda) 1:523–530
13. Muzzey D, Schwartz K, Weissman J, Sherlock G (2013) Assembly of a phased diploid *Candida albicans* genome facilitates allele-specific measurements and provides a simple model for repeat and indel structure. Genome Biol 14:R97
14. Arnaud M, Costanzo M, Skrzypek M, Binkley G, Lane C et al (2005) The Candida Genome Database (CGD), a community resource for *Candida albicans* gene and protein information. Nucleic Acids Res 33:D358–D363
15. Enloe B, Diamond A, Mitchell AP (2000) A single-transformation gene function test in diploid *Candida albicans*. J Bacteriol 182:5730–5736
16. Nobile C, Mitchell AP (2009) Large-scale gene disruption using the *UAU1* cassette. Methods Mol Biol 499:175–194
17. Abbey D, Funt J, Lurie-Weinberger M, Thompson D, Regev A et al (2014) YMAP: a pipeline for visualization of copy number variation and loss of heterozygosity in eukaryotic pathogens. Genome Med 6:100
18. Sirr A, Cromie GA, Jeffery EW, Gilbert TL, Ludlow CL et al (2015) Allelic variation, aneuploidy, and nongenetic mechanisms suppress a monogenic trait in yeast. Genetics 199:247–262
19. Tan Z, Hays M, Cromie GA, Jeffery EW, Scott AC et al (2013) Aneuploidy underlies a multicellular phenotypic switch. Proc Natl Acad Sci U S A 110:12367–12372
20. Ludlow CL, Scott AC, Cromie GA, Jeffery EW, Sirr A et al (2013) High-throughput tetrad analysis. Nat Methods 10(7):671–675
21. Hickman MA, Paulson C, Dudley A, Berman J (2015) Parasexual ploidy reduction drives population heterogeneity through random and transient aneuploidy in *Candida albicans*. Genetics 200:781–794
22. Gresham D, Curry B, Ward A, Gordon DB, Brizuela L et al (2010) Optimized detection of sequence variation in heterozygous genomes using DNA microarrays with isothermal-melting probes. Proc Natl Acad Sci U S A 107:1482–1487
23. Altschul SF, Madden TL, Schaffer AA, Zhang J, Zhang Z et al (1997) Gapped BLAST and PSI-BLAST: a new generation of protein database search programs. Nucleic Acids Res 25:3389–3402
24. Rose MD (1987) Isolation of genes by complementation in yeast. Methods Enzymol 152:481–504
25. Berger MF, Bulyk ML (2009) Universal protein-binding microarrays for the comprehensive characterization of the DNA-binding specificities of transcription factors. Nat Protoc 4:393–411
26. Painter HJ, Altenhofen LM, Kafsack BF, Llinas M (2013) Whole-genome analysis of *Plasmodium spp.* Utilizing a new agilent technologies DNA microarray platform. Methods Mol Biol 923:213–219

Chapter 8

Whole-Genome Haplotyping of Single Sperm of *Daphnia pulex* (Crustacea, Anomopoda)

Sen Xu and Kim Young

Abstract

Sequencing the entire genome of single sperm cells can provide valuable information of the distribution of meiotic recombination events in eukaryotic genomes. Here, we provide a description of the experimental work flow for isolating single sperm cells from the microcrustacean *Daphnia pulex* using fluorescence-activated cell sorting. Moreover, we describe the application of a whole-genome amplification technique (i.e., Multiple Annealing and Looping Based Amplification Cycles method) to single sperm of *Daphnia* to generate enough DNA for library preparation of next-generation sequencing.

Key words Single sperm, Whole-genome amplification, Meiosis, Recombination, Males

1 Introduction

Direct genomic sequencing of haploid gametes such as sperm provides an important approach for examining meiotic recombination events [1–4]. However, there are substantial technical challenges associated with rapidly isolating a large number of single sperm and sequencing the extremely small amount of DNA (e.g., ~3 pg DNA in a human sperm) using standard polymerase chain reaction (PCR). In the 1980s, the combination of manual isolation of single sperm and highly optimized PCR generated the first sperm haplotyping data [5], yielding insight into human meiotic recombination. More recently, the development of new sequencing technologies has rendered feasible the sequencing of large numbers of single sperm [4, 6]. Furthermore, the development of high-throughput cell sorting technology has facilitated the application of whole-genome sequencing of single sperm to expand into non-human organisms such as the microcrustacean *Daphnia pulex* [7]. Compared to the classic method of crossing experiments, genomic sequencing of sperm saves substantial amount of time in creating a marker-dense genetic map and allows pinpointing the genomic locations of recombination events [7]. However, because of the

Irene Tiemann-Boege and Andrea Betancourt (eds.), *Haplotyping: Methods and Protocols*, Methods in Molecular Biology, vol. 1551, DOI 10.1007/978-1-4939-6750-6_8, © Springer Science+Business Media LLC 2017

extremely small amount of DNA contained in a single sperm, a whole-genome amplification (WGA) procedure has to be performed to generate enough DNA for genome sequencing. Compared to existing WGA techniques, the newly developed Multiple Annealing and Looping Based Amplification Cycles (MALBAC) method [8] allows an unbiased recovery of the entire genome by employing quasi-linear amplification in the first few amplification cycles (i.e., initially amplified fragments do not get amplified in subsequent amplification cycles). Because of the great potential utility of single-genome sequencing for research on other organisms, this chapter summarizes the technical procedures for isolating single sperm using fluorescence-activated cell sorting (FACS) and the WGA method of MALBAC to prepare single sperm samples that can be subject to whole-genome sequencing and/or regular gene-specific PCR amplifications.

2 Materials

We prepared all solutions and culture media using ultrapure MilliQ water unless noted otherwise.

2.1 Methyl Farnesoate-COMBO Media for Male Induction

Sex is environmentally determined in *Daphnia* (see Subheading 3.1). To induce the production of males, we use the chemical methyl farnesoate (MF), a juvenile hormone that determines the sex of *Daphnia* offspring to be male [9, 10]. To prepare the stock solution of MF, add 0.100152 mg MF per 30 μl EtOH in a glass vial (*see* **Note 1**). Then, 30 μl of the MF stock solution is added to1 L of COMBO media at room temperature. To prevent the precipitation of MF in COMBO media, the COMBO media is vigorously stirred on a stir plate for at least 30 min.

2.2 Needle Pipette

To make needle pipette used for collecting sperm sample, follow the steps below.

1. Capillary glass (Sutter Instrument Co., catalogue no. BF150-86-15) is soaked in SigmaCote (Sigma-Aldrich, catalogue no. 190895). Air dry the capillary glass. All operations involving Sigmacote should be performed in a fume hood.
2. To pull a needle pipette using the capillary glass, a P-87 Flaming/Brown micro pipette puller (Sutter Instrument Co.) is used. The setting for the machine is: Heat: 888, Pull: 080, Velocity: 010, Time: 250, Pressure: 600.

2.3 Cover Slips, Microscope Slides, and Dissection Microscope

1. Cover slips and microscope slides need to be soaked in SigmaCote under the fume hood.
2. Dry the cover slips and microscope slides with paper towel and then let them air dry in fume hood.

3. We use a dissection microscope for manipulating *Daphnia* individuals.

2.4 Aspirator Tube Assembly

An aspirator tube assembly (VWR catalog number 53507-278) is used to collect sperm from the microscope slide to buffer. Load the capillary needle constructed above into the needle fitting of an aspirator tube assembly (*see* **Note 2**).

2.5 PBS Solution

The extracted fluid containing *Daphnia* sperm needs to be stored in 1× PBS solution (pH 7.4). To prepare PBS solution, start with 800 mL H_2O, measure and add the component chemicals according to Table 1. Adjust pH to 7.4 using HCl. Fill up to 1 L with H_2O.

2.6 Hoechst Dye

Prepare 100 μg/μl Hoechst 33528 solution (Sigma-Aldrich). Store the solution at 4 °C in the dark.

2.7 AriaII Flow Cytometry

We use FACS Aria II SORP Flow Cytometer (BD Biosciences) to isolate single sperm. Lasers used are a 488 nm 100 mW for light scatter detection and a 355 nm 20 mW for Hoechst detection. A FSC-PMT was used for optimal small particle discrimination. A 70 μm nozzle was used at 45 psi. Sperm cells were dispensed into regular 96-well PCR plates.

2.8 Sperm Cell Lysis Buffer

Prepare sperm cell lysis buffer that contains 30 mM Tris–Cl PH 7.8, 2 mM EDTA, 20 mM KCl, 0.2% Triton X-100, 50 mM DTT, and 500 μg/ml Qiagen Protease.

2.9 Reagents for Whole-Genome Amplification

For performing whole-genome amplification using the Multiple Annealing and Looping Based Amplification Cycles (MALBAC) technique [6], we need the following reagents 10× ThermoPol Buffer (New England Biolabs, catalogue no. B9004S), dNTP (10 mM, Bioline), ultra-pure H_2O, DeepVentR exo- (New England Biolabs, catalogue no. M0259L), Bst large fragment (New England Biolabs, catalogue no. M0275L), and Pyrophage 3173 exo- (Lucigen).

Table 1
Recipe for 1× PBS solution (pH 7.4)

Chemical	Gram per liter
Na_2HPO_4 (0.01 M)	1.42
KH_2PO_4 (0.0018 M)	0.24
NaCl (0.137 M)	8.0
KCl (0.0027 M)	0.2

Table 2
First batch of stock solutions for preparing COMBO artificial medium. All the solutions are prepared in 125 ml volume

Chemical	Gram per 125 mL H_20
LiCl (3.7 M)	19.68
RbCl (1.3 M)	19.90
$SrCl_2 \cdot 6H2O$ (0.56 M)	18.75
KI (0.02 M)	0.4125
NaBr (0.16 M)	2.00
Na_2SeO_3 (6.2×10^{-3} M)	2.70×10^{-4}

In the first few cycles of amplification (*see* Subheading 3), the MALBAC procedure requires the randomly annealing primers (10 μM) consisting of a 27mer (5′-GTGAGTGATGGTTGA GGTAGTGTGGAG-3′) and 8 degenerate nucleotides at the 3′ end. At the completion of the first stage of amplification, all the newly formed amplicons carry the 27mer and its complementary sequence at their ends. In the second stage of amplification for MALBAC, the 27mer 5′-GTGAGTGATGGTTGAGGTAGTG TGGAG-3′ (10 μM) is used for further amplifying all the amplicons accumulated in the first stage.

2.10 Combo Medium

Stock solutions of Na_2SeO_3, LiCl, RbCl, $SrCl_2 \cdot 6H2O$, KI, NaBr $MgSO_4 \cdot 7H_2O$, $NaHCO_3$, $Na_2SiO_3 \cdot 9H_2O$, H_3BO_3, KCl, $CaCl_2 \cdot 2H_2O$, Na_2SeO_3, following Table 2.

3 Methods

3.1 Daphnia Culture Media

Because *Daphnia pulex* reproduces clonally under favorable conditions, we can maintain a single clone in mass culture in laboratory conditions essentially for indefinite period of time. We use COMBO artificial lake water medium [11] to maintain mass culture of *Daphnia*. The procedure for preparing 1 l of COMBO medium is below.

1. Prepare a batch of 125-mL stock solutions of Na_2SeO_3 and trace elements LiCl, RbCl, $SrCl_2 \cdot 6H_2O$, KI, NaBr following Table 2.
2. Prepare a batch of 1-L stock solutions of $MgSO_4 \cdot 7H_2O$, $NaHCO_3$, $Na_2SiO_3 \cdot 9H_2O$, H_3BO_3, KCl, $CaCl_2 \cdot 2H_2O$, Na_2SeO_3, and a 1 L stock of combined trace elements following Table 3.
3. For each liter of the COMBO medium, add 1 mL of each of the eight stock solutions from **step 2** and fill up with H_2O.

Table 3
Second batch of stock solutions for preparing COMBO artificial medium. All the solutions are prepared in 1 L volume

Chemical	Gram per liter
$MgSO_4 \cdot 7H_2O$ (0.15 M)	36.97
$NaHCO_3$ (0.15 M)	12.60
$Na_2SiO_3 \cdot 9H_2O$ (0.1 M)	28.42
H_3BO_3 (0.39 M)	24.00
KCl (0.10 M)	7.45
$CaCl_2 \cdot 2H_20$ (0.25 M)	36.71
Na_2SeO_3 (6.2×10^{-6})	1 ml of Na_2SeO_3 stock solution from Table 2
Trace elements	1 ml of LiCl, RbCl, $SrCl_2 \cdot 6H_2O$, KI, NaBr stock solution from Table 2

4. Maintain the COMBO medium under constant aeration, and allow 24 h to pass before using to allow the pH to stabilize and the oxygen level in the media to reach saturation.

3.2 Inducing the Production of Male

The sex of offspring produced by female *Daphnia* through parthenogenesis is determined by the environment [9, 10]. Under favorable conditions, all the offspring are females. Environmental conditions such as high population density and food shortage can induce the production of males through parthenogenesis. Furthermore, male production is a polymorphic trait for *Daphnia*, i.e., some *Daphnia* females are not capable of producing male offspring [10]. It is possible that the *Daphnia* clone of interest is a non-male producer.

1. Start a culture of *Daphnia pulex* with a single female or several females of the same clone. Keep the culture at 20 °C, in a 12 h light and 12 h dark cycle.
2. The green algae *Scenedesmus obliquus* is fed ten times per week at a rate of approximately 5×10^6 cells/ml of media.
3. We place four adult mature female *Daphnia* (carrying parthenogenetic eggs, if possible) in 100 mL MF-COMBO media in a glass beaker. Feed the animals with algae as described above. After 48 h, we transfer the females to fresh MF-COMBO media. After another 48 h, we start to check the gender of the offspring (if any present in the beaker). The major morphological characteristics of male *Daphnia* can be found in Fig. 1. If none of the offspring appears to be male, check the offspring again after another 2 days. If there are no males at this point, this clone is most likely a non-male producer.

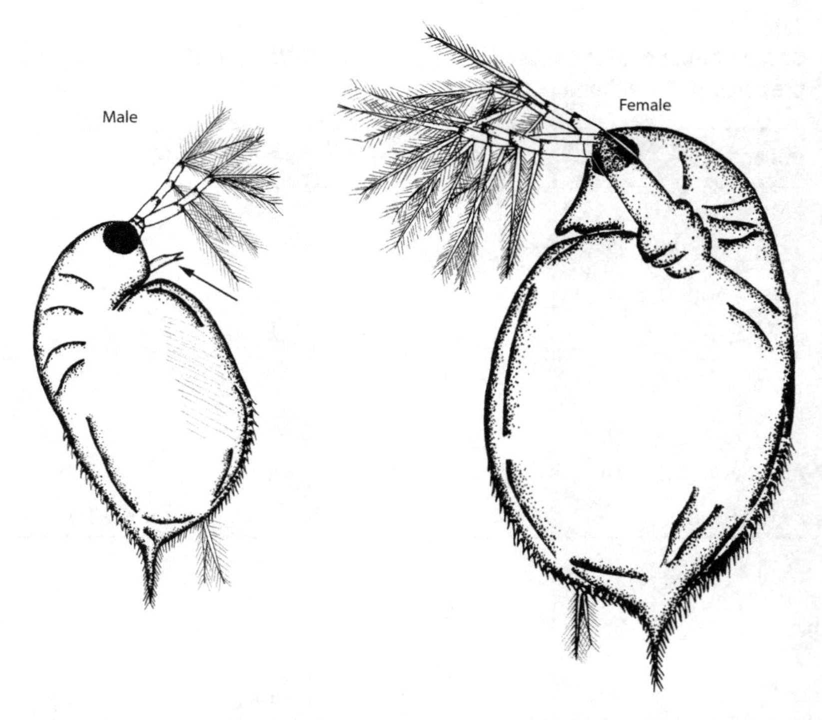

Fig. 1 Drawing of *Daphnia pulex* male and female. Diagnostic features of male *Daphnia* include the presence of a reduced second antennae (arrow) and smaller body sizes compared to females. Note that this graph is only meant to provide a quick guide for identifying males. More details about the morphological sexual differentiation should be sought elsewhere (e.g., [13])

4. After collecting the male offspring produced by the females of the same clone, separate each male into COMBO media in a 50 mL centrifuge tube. Feed them with algae as described above. Once they become mature (usually after 7–10 days), we can collect sperm from them.

3.3 Collecting Sperm

1. Prepare 50 μl PBS buffer in a 1.5 ml Eppendorf tube.
2. Pipette 1 μl H_2O onto the center of a microscope slide (*see* Subheading 2.4) and place a male in the water (*see* **Note 3**). Observe the animal using a dissection microscope. Make sure the abdomen is extended beyond the carapace. Then, use a dissection needle to gently touch the cover slip ten times and do not attempt to crush the animals. Remove the cover slip and use mouth pipette to collect all the liquid on the cover slip and microscope slide. Transfer the collected liquid into the same tube containing PBS buffer. Repeat this step until finishing extraction from all males.
3. It is optional, but recommended, to confirm presence of sperm using a light microscope at 100× (Fig. 2). This only needs to be done for a few individuals.

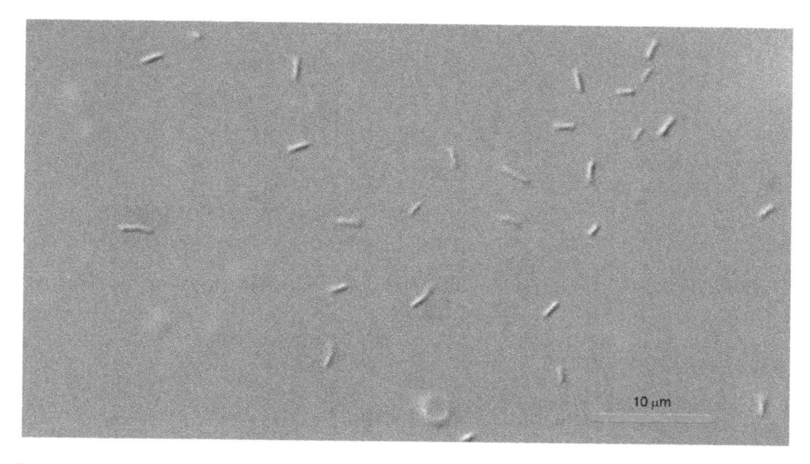

Fig. 2 Sperm of *Daphnia pulex*. Sperm is rod-shaped and ~2–3 μm long

4. After collecting all the sperm sample, add 5 μl 100 μg/μl Hoechst 33528 into the PBS solution containing sperm sample. Incubate at 20 °C for 2–3 h.

3.4 Isolating Single Sperm Using Flow Cytometry

1. Place 5 μl sperm cell lysis buffer into each well of a 96-well PCR plate.
2. Set up the AriaII flow cytometry as specified in Subheading 2.8 and begin single cell isolation (*see* **Note 4**).
3. Once the isolation is done, lyse the sperm in the PCR plate by incubating at 50 °C for 3 h, 75 °C for 20 min, 80 °C for 5 min.

3.5 Whole-Genome Amplification

The Multiple Annealing and Looping Based Amplification Cycles (MALBAC) technique [6] is used for whole genome amplification of single sperm. For each sample, add 3.0 μl ThermoPol Buffer, 1.0 μl dNTP (10 mM), 26 μl H_2O, and 0.75 μl randomly annealing primer (10 μM). Then,

1. Denature samples at 94 °C for 3 min.
2. Quench the samples on ice immediately following denaturation, which allows the annealing of random primers.
3. For each sample, add a mixture of 0.375 μl *Bst* DNA Polymerase (Large Fragment) and 0.225 μl Pyrophage 3173 exo-. Then, load the samples on a PCR thermocycler and perform the following cycling regime that consists of 45 s at 10 °C, 45 s at 15 °C, 45 s at 20 °C, 45 s at 30 °C, 45 s at 40 °C, 45 s at 50 °C, 2 min at 65 °C, and 20 s at 95 °C.
4. Quench the samples on ice immediately. For each sample, add a mixture of 0.375 μl *Bst* DNA Polymerase (Large Fragment) and 0.225 μl Pyrophage 3173 exo-. Load the samples on a PCR thermocycler and perform the following cycling regime that consists of 45 s at 10 °C, 45 s at 15 °C, 45 s at 20 °C, 45 s

at 30 °C, 45 s at 40 °C, 45 s at 50 °C, 2 min at 65 °C, 20 s at 95 °C, and 30 s at 58 °C. Note that, compared to the cycling in **step 3**, there is an extra step of 58° at the end of this cycling program. The optimal temperature for this polymerase is 65 °C.

5. Quench samples on ice and add enzyme again.
6. Repeat five times the **steps 4** and **5**, which enables five rounds of linear amplification of DNA in sperm sample.
7. Normal PCR amplification is performed on samples. For each sample, perform normal PCR using the primer (5′-GT GAG TGA TGG TTG AGG TAG TGT GGA G-3′) and the program consisting of 22 cycles of 20 s at 94 °C, 20 s at 59 °C, 1 min at 65 °C, 2 min at 72 °C, which is followed by 5 min at 72 °C and a hold at 4 °C. Normally, the amount of DNA after PCR is between 1 and 2 μg.
8. To verify whether the whole-genome amplification is successful, we perform regular PCR for multiple gene fragments using the diluted (10–20×) whole-genome amplified product (*see* **Note 5**).

3.6 Sequencing Library Preparation and Sequencing

After the whole amplification, most of the amplified DNA fragments are 1.5 kb long. The DNA samples are ready for library preparation for Illumina sequencing platforms. Typically, we can run 96 sperm samples in plate format for library preparation using the entire whole-genome amplified product. There is currently a large selection of library preparatory kits available in the market. The standard procedures recommended by the manufacturers should be followed to produce quality sequencing libraries. After pooling equal molar of each of the 96 libraries to ensure minimal variation of sequence coverage among samples, whole-genome sequencing of 96 samples can be done on any compatible Illumina next-generation sequencer.

3.7 Recombination Event Detection

The primer sequences used for the single sperm whole-genome amplification reactions were computationally removed from the ends of raw reads using the software CLC Genomics Workbench (version 7, CLC Bio). Then the raw reads for each sample were mapped to the *Daphnia* reference assembly using the short-read mapper Novoalign 3.2 (www.novocraft.com) with default settings. However, reads mapped to multiple locations were removed from further analysis. The haplotype for each nucleotide position of each sample was determined based on a consensus approach, where a call is made with the support of >80 % of the reads. To avoid sequencing errors, PCR artifacts, and potential mapping errors, at least two forward and two reverse reads are necessary to validate the consensus call. Only sites that are heterozygous (i.e., sites where two different nucleotides were found across the entire set of sperm samples) were used for downstream analyses because only heterozygous loci are informative for revealing recombination events.

Given this set of sperm sequences is derived from the recombination of two parental haplotypes, the haplotype of each sperm consists of a continuous series of 0 or 1(0 and 1 designate the paternal and maternal haplotypes, respectively) if there are no recombination events. However, crossover and gene conversion events can cause different patterns of switching of haplotype, with a double switch over a genomic interval suggesting gene conversion events (e.g., 00000011000000) and a single switch suggesting crossover events (e.g., 00000001111111). According to models of DNA double strand break repair, crossovers are accompanied by gene conversion event near the DNA double strand breakpoint. A haplotype configuration of triple switch (e.g., 00000001110011111) represents such gene conversion associated with crossover.

We developed an algorithm to simultaneously determine the phase for each informative site and detect crossover and gene conversion events that occurred in each sperm [7]. In brief, this algorithm first aims to establish haplotype blocks where there is no evidence of recombination using four gamete test [12]. We assign either 00 or 11 (0 and 1 representing the two parental haplotypes) to the two-locus haplotype for the first pair of sites on each scaffold. Given we can safely ignore de novo mutations due to their low rate (5×10^{-9} per site per generation) and a single generation time, every pair of sites across the samples have a maximum of four haplotypes when a crossover or gene conversion event happens (00, 01, 10, 11), 3 haplotypes for gene conversion events (00, 01 or 10, 11), and two haplotypes for no recombination (00, 11). We then consecutively examine each pair of sites and extend the haplotype assignment. A haplotype block is terminated once the extension cannot continue due to the presence of three or four haplotypes. The haplotype phase of two adjacent haplotypes is computed such that maximal phase consistency between blocks is achieved and the minimum number of recombination events needs to be invoked. Furthermore, if the number of recombination events between two adjacent blocks exceeds 15, the upstream haplotype block is labeled with a flag and discarded for further analysis as this pattern may indicate assembly problems. Instead, the second nearest upstream block is used for computing haplotype phase. Once the haplotype phase assignment is done, crossover and gene conversion events are called using the criteria mentioned above. This algorithm has been implemented in a python script, which can be downloaded from https://github.com/matthew-ackerman/PHASESPERM/.

4 Notes

1. Methyl farnesoate binds strongly to plastic. Therefore, it cannot be stored in a plastic container or applied to *Daphnia* in a falcon tube. However, transferring methyl farnesoate with plastic tips for micropipettes should not be an issue.

2. To make sure that it does not take extra efforts to aspirate liquid, it is helpful to make sure that the needle capillary has a length around 4 cm. If it is hard to aspirate liquid when mouth pipetting, remove a bit of the tip of the capillary needle.
3. When collecting *Daphnia* sperm, it is necessary to take appropriate measures to prevent any potential non-*Daphnia* DNA contamination because any contaminant DNA will be amplified in the whole-genome amplification process. It is therefore necessary to wear gloves and perform the experiment with extra care.
4. The AriaII flow cytometer is a sophisticated cell sorter that requires extensive training before one can use it effectively. We recommend readers to consult with the manufacturer for training or work with a trained technician for using the equipment. The sperm quantity for *Daphnia* is a lot lower than mammals, so sperm dilution may not work as well as in other organisms.
5. Because the whole-genome amplified product mainly consists of DNA fragments ranging up to 1.5 kb, the PCR verification should aim at DNA fragments between 300 and 800 bp. Fragments longer than the specified length may not be amplified from the whole-genome amplified product because the fragment may not be entirely contained in any whole-genome amplified DNA fragments. We also recommend performing PCR verification on multiple loci because some genomic region may be amplified better than other regions. Users can determine their own criteria for selecting high quality amplified samples for downstream experiment. For example, one can decide that only samples with successful PCR verification for more than seven out of ten loci are considered suitable for whole-genome sequencing preparation.

Acknowledgments

We would like to thank Drs. L. J. Bright, T. Doak, K. Spitze, M. Ackerman, and M. Lynch for their technical help and stimulating discussion. Dr. Hongan Long prepared the drawing of Fig. 1.

Reference List

1. Li HH, Gyllensten UB, Cui XF, Saiki RK, Erlich HA, Arnheim N (1988) Amplification and analysis of DNA sequences in single human sperm and diploid cells. Nature 335(6189):414–417. doi:10.1038/335414a0
2. Cui XF, Li HH, Goradia TM, Lange K, Kazazian HH, Galas D, Arnheim N (1989) Single sperm typing - determination of genetic distance between the G-gamma-globin and parathyroid hormone loci by using the polymerase chain reaction and allele specific oligomers. Proc Natl Acad Sci U S A 86(23):9389–9393. doi:10.1073/pnas.86.23.9389
3. Arbeithuber B, Betancourt AJ, Ebner T, Tiemann-Boege I (2015) Crossovers are associated with mutation and biased gene conversion at recombination hotspots. Proc Natl Acad Sci U S A 112(7):2109–2114. doi:10.1073/pnas.1416622112

4. Wang JB, Fan HC, Behr B, Quake SR (2012) Genome-wide single-cell analysis of recombination activity and de novo mutation rates in human sperm. Cell 150(2):402–412. doi:10.1016/j.cell.2012.06.030
5. Cui XF, Li HH, Goradia TM, Lange K, Kazazian HH, Galas D, Arnheim N (1989) Single-sperm typing: determination of genetic distance between the G-gamma globin and parathyroid hormone loci by using the polymerase chain reaction and allele-specific oligomers. Proc Natl Acad Sci U S A 86(23):9389–9393. doi:10.1073/pnas.86.23.9389
6. Lu S, Zong C, Fan W, Yang M, Li J, Chapman AR, Zhu P, Hu X, Xu L, Yan L, Bai F, Qiao J, Tang F, Li R, Xie XS (2012) Probing meiotic recombination and aneuploidy of single sperm cells by whole-genome sequencing. Science 338(6114):1627–1630. doi:10.1126/science.1229112
7. Xu S, Ackerman MS, Long H, Bright L, Spitze K, Ramsdell JS, Thomas WK, Lynch M (2015) A male-specific genetic map of the microcrustacean *Daphnia pulex* based on single-sperm whole-genome sequencing. Genetics 201(1):31–38. doi:10.1534/genetics.115.179028
8. Zong C, Lu S, Chapman AR, Xie XS (2012) Genome-wide detection of single-nucleotide and copy-number variations of a single human cell. Science 338(6114):1622–1626. doi:10.1126/science.1229164
9. Olmstead AW, Leblanc GA (2002) Juvenoid hormone methyl farnesoate is a sex determinant in the crustacean *Daphnia magna*. J Exp Zool 293(7):736–739. doi:10.1002/jez.10162
10. Toyota K, Miyakawa H, Hiruta C, Furuta K, Ogino Y, Shinoda T, Tatarazako N, Miyagawa S, Shaw JR, Iguchi T (2015) Methyl farnesoate synthesis is necessary for the environmental sex determination in the water flea *Daphnia pulex*. J Insect Physiol 80:22–30. doi:10.1016/j.jinsphys.2015.02.002
11. Kilham SS, Kreeger DA, Lynn SG, Goulden CE, Herrera L (1998) COMBO: a defined freshwater culture medium for algae and zooplankton. Hydrobiologia 377:147–159. doi:10.1023/A:1003231628456
12. Hudson RR, Kaplan NL (1985) Statistical properties of the number of recombination events in the history of a sample of DNA sequences. Genetics 111(1):147–164
13. Benzie JAH (2005) Cladocera: the genus *Daphnia* (including *Daphniopsis*) (Anomopoda: Daphniidae). Guides to the identification of the microinvertebrates of the continental waters of the world, vol no 21. Backhuys, Leiden

Part IV

Haplotyping Single Chromosomes

Chapter 9

Chromosome-Range Whole-Genome High-Throughput Experimental Haplotyping by Single-Chromosome Microdissection

Li Ma, Wenzhi Li, and Qing Song

Abstract

Haplotype is fundamental genetic information; it provides essential information for deciphering the functional and etiological roles of genetic variants. As haplotype information is closely related to the functional and etiological impact of genetic variants, it is widely anticipated that haplotype information will be extremely valuable in a wide spectra of applications, including academic research, clinical diagnosis of genetic disease and in the pharmaceutical industry. Haplotyping is essential for LD (linkage disequilibrium) mapping, functional studies on cis-interactions, big data imputation, association studies, population studies, and evolutionary studies. Unfortunately, current sequencing technologies and genotyping arrays do not routinely deliver this information for each individual, but yield only unphased genotypes. Here, we describe a high-throughput and cost-effective experimental protocol to obtain high-resolution chromosomal haplotypes of each individual diploid (including human) genome by the single-chromosome microdissection and sequencing approach.

Key words Experimental, Haplotype, Chromosome-length, Whole-genome, High-throughput

1 Introduction

A "haplotype" refers to a group of alleles inherited on a single chromosome (Fig. 1) [1, 2]. A large number of statistical and computational methods have been developed to reconstruct haplotypes from conventional unphased genotype data [3–7]. These methods suffer from short-phasing distance, switch errors, and ambiguities [4, 8]. These uncertainties or ambiguities on haplotype configuration create complications in genetic analysis [9]. A number of experimental approaches have also been developed to determine haplotypes in recent years [10–22].

Chromosomal haplotypes will be essential for functional interpretation of genomes, especially for studying the impact of cis-interactions on gene expression (Fig. 2). Recent ENCODE data shows that only 5 % of DNase I hypersensitive sites (DHSs) lie

Irene Tiemann-Boege and Andrea Betancourt (eds.), *Haplotyping: Methods and Protocols*, Methods in Molecular Biology, vol. 1551,
DOI 10.1007/978-1-4939-6750-6_9,

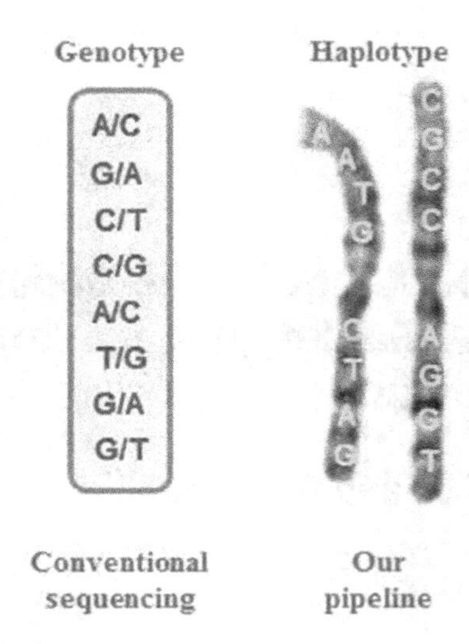

Fig. 1 Difference between genotypes and haplotypes

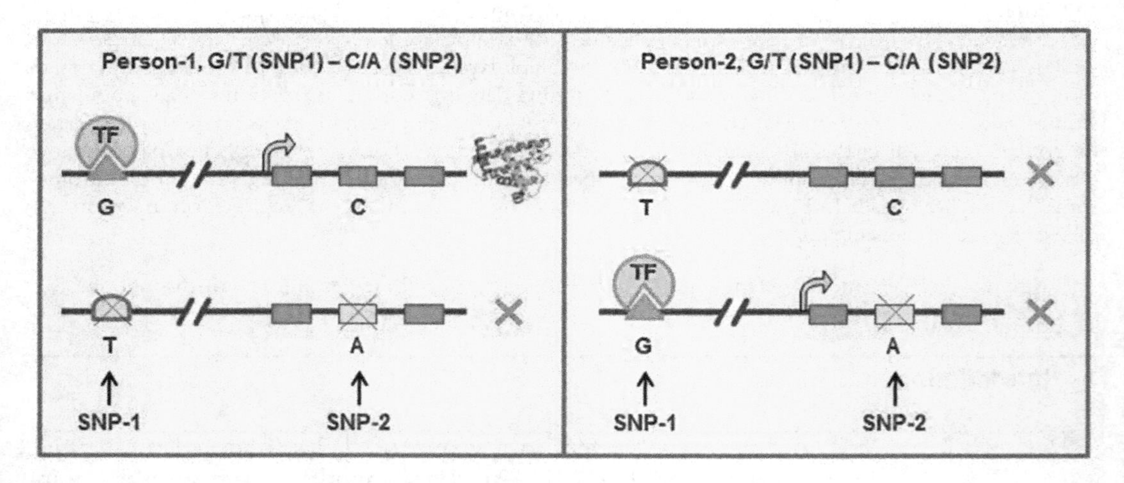

Fig. 2 An illustration that haplotypes rather than genotypes are functionally relevant. In this special case, two different individuals have the same genotypes (G/T and C/A) but different haplotypes at two SNP loci. SNP-1 (G/T) occurred in an enhancer (G/T) that is essential for the expression of this gene, in which Allele-T is a null allele. SNP-2 (C/A) resides in an exon of this gene, in which Allele-A will disrupt the translation with an early stop codon. The production of this protein requires a cis-relationship between the enhancer and exon. Thus, Person-1 can produce this protein because one of his gene copies contains both functional alleles on the same chromosome (cis); Person-2 cannot produce this protein because neither of his gene copies contains two functional alleles. Please note that the enhancers may be close to a gene, or as far as 1 million base away from its regulatory target

within 2.5 kb of transcriptional start sites (TSSs); the remaining 95% of DHSs are positioned distally [23]. In fact, proximity in sequence data may be a poor predictor of interactions generally. Chromatin is highly packed and organized in the nucleus [24], and regulatory elements and genes that are far apart in terms of genomic

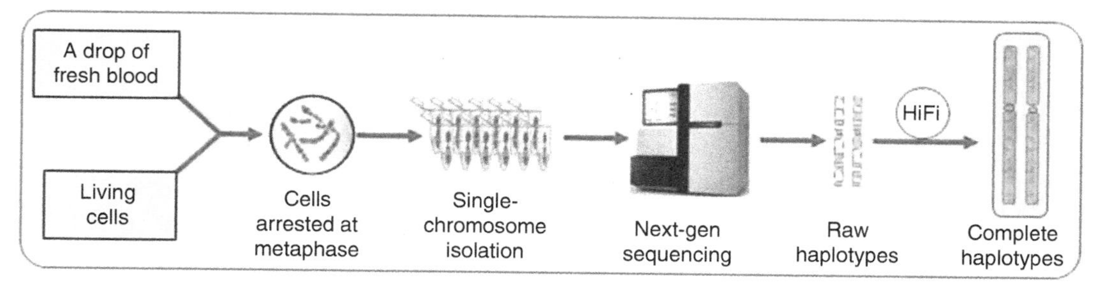

Fig. 3 The pipeline of haplotype determination described in this chapter. In this pipeline, we will first culture the cells to metaphase and then isolate single-chromosomes; the isolated single-chromosomes will be then subjected to whole-genome amplification (WGA) followed by high-throughput sequencing; the sequencing data will be computationally analyzed (removing the noises, increasing the resolution and output the haplotypes)

sequence can be brought together in the nucleus via the formation of three-dimensional loops [25–28]. Only ~7% of these looping interactions are with the nearest gene. And *cis*-interactions appear to occur more often than expected by chance even at distances greater than 200 Mb (that is, intra-chromosomal contact probabilities at this distance are much greater than the average contact probability between different chromosomes [29]). Thus, chromosomal haplotypes will be necessary for identifying the causal genetic variants by genetic studies.

In this chapter, we will describe an experimental pipeline to obtain chromosomal haplotypes in a high-throughput manner (Fig. 3). Besides the features on high-accuracy, high-throughput, low-cost, and applicability to all types of genetic markers [13, 17, 30–32], a unique feature of this technology is the extremely long length (or full chromosome-length) of the recovered haplotype.

2 Materials

1. A drop (100 μL) of fresh whole blood or living cells of human or any multiploid organisms (*see* **Note 1**).
2. Gibco® PB-MAX™ Karyotyping Medium, stored at 4 °C.
3. L-Glutamine, stored at room temperature.
4. Actinomycin D (20 μg/mL). Working solution: add 0.5 mL of acetone to the 10 mg vial containing Actinomycin-D powder. Mix with 100 mL of RPMI 1640 in limited light conditions. Distribute into labeled and dated polystyrene tubes in 3 mL aliquots under sterile conditions. Immediately place into −10 °C to −20 °C, store aliquots at −20 °C.
5. Ethidium Bromide (EB) (1 mg/mL). Working solution: add 1 mL of stock solution (10 mg/mL) to 9 mL dH_2O in a light proof bottle with parafilm around the lid (EB is light sensitive). This working solution will expire in 6 months or at expiration date of stock solution, whichever is sooner.

6. Phytohemagglutinin-PHA, stored at 4 °C.
7. Gentamicin, stored at 4 °C.
8. Colcemid™ Solution in PBS (10 μg/mL), stored at 4 °C.
9. 0.075 M KCl hypotonic solution, shelf life 2 weeks at room temperature.
10. Carnoys Fixative (Methanol:Acetic Acid, 3:1). This is to be made fresh immediately before needed and is to be tightly capped when not in use.
11. UV-light sliceable foiled slide (Vashaw Scientific).
12. Giemsa staining solution.
13. 0.2-mL Leica collecting tube for microdissection (Leica Microsystems).
14. A whole-genome amplification (WGA) kit, such as Sigma GenomePlex WGA4 kit.
15. QIAquick PCR purification kit.
16. A high-throughput sequencer or a high-throughput array genotyping platform.
17. A regular desktop or laptop computer with CPU of 3.0 GHz and 8 GB RAM on a 32 or 64 Window XP/7/8 or an Ubuntu 12 LTS system.

3 Methods

3.1 Single-Chromosome Isolation

1. Collect about 100 μL of whole blood (anticoagulated by sodium heparin) into a conical 15-mL centrifuge tube containing 5 mL of prewarmed complete PB Max Karyotyping medium with fetal bovine serum (FBS), L-glutamine, phytohemagglutinin (PHA), and gentamicin. Incubate at 37 °C for about 48 h (*see* **Note 2**).
2. Thaw actinomycin-D in 37 °C water bath. Add 200 μL of Act-D solution and 100 μL of EB working solution to each blood culture tube under sterile conditions in a hood. Mix gently by inverting tubes. Incubate cultures for 30 min at 37 °C.
3. Add 50 μL of colcemid to a final concentration of 0.083 μg/mL. Mix by inverting again gently and incubate at 37 °C for 30 min.
4. While tubes are incubating, make fresh fixative (3:1 methanol:glacial acetic acid) and place in the freezer to chill. Also at this time, put 75 mM KCl in the water bath to warm to 37 °C.
5. After incubation with EB and colcemid, centrifuge at 1000 rpm (135 × *g*) for 10 min.

6. Aspirate all but 0.3 mL of supernatant, gently resuspend cell pellet. Add 37 °C prewarmed 75 mM KCl, vortex gently to assure KCl is mixed well with the pellet. Incubate at room temperature (25 °C) for 15 min.
7. Add four to five drops cold fixative, gently mix by inverting tubes, and centrifuge 1000 rpm for 10 min. Return fix to freezer. Repeat this step three times. Finally, resuspend cell pellet in 5 mL of cold fixative.
8. Drip the cells onto a UV-light sliceable foiled slide to spread chromosomes.
9. Briefly stain the chromosomes with Giemsa (1:20) for 10 min, dip the slide into dH_2O, and air dry.
10. Put the UV-light sliceable foiled slide under a laser microdissection microscope (ASLMD; Leica) (*see* **Note 3**). Use a computer mouse to draw a circle on the computer monitor, and then the computer will direct a laser beam to cut a small foil containing target single-chromosomes (Fig. 4). The foil will be collected into a Leica collecting tube with 9 μL of dH_2O in it (*see* **Note 4**).

3.2 Whole-Genome Amplification (WGA)

11. The collected foil will be directly used in subsequent experiments without DNA extraction.
12. Amplify the single chromosomes with a whole-genome amplification kit following the manufacturer's protocol. The following steps are described for the WGA with the Sigma GenomePlex WGA4 kit.
13. Add 1 μL of the Lysis and Fragment buffer, and then incubate at 50 °C for 1 h.

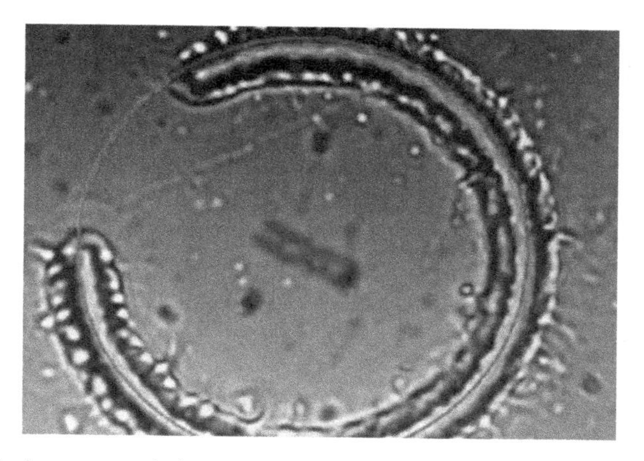

Fig. 4 A chromosome is being microdissected by a laser beam directed by a computer

14. Heat the sample to 99 °C for EXACTLY 4 min and chill on ice.
15. Add 2 μL of the Single Cell Library Preparation buffer and 1 μL of Library Stabilization solution. Incubate at 95 °C for 2 min.
16. Cool the sample on ice, consolidate the sample by centrifugation, and place on ice.
17. Add 1 μL of Library Preparation Enzyme, mix thoroughly, and centrifuge briefly.
18. Incubate with the following cycles: 16 °C for 20 min, 24 °C for 20 min, 37 °C for 20 min, and 75 °C for 5 min.
19. Incubate the samples at 95 °C for 3 min followed by 35 cycles of 94 °C for 30 s and 65 °C for 5 min.
20. Purify the amplified product with the QIAquick PCR purification kit.

3.3 Haplotyping by Single-Chromosome Sequencing

21. The amplified DNA will be ready for a whole-genome genotyping or next-generation sequencing following the manufacturer's User Guide.
22. If using next-generation sequencing, users need to map the reads and call SNP alleles, producing a VCF formatted file. If using genotyping arrays, SNP alleles will be directly output (*see* **Note 5**).
23. Read out the single-chromosomal sequence directly from the next-gen sequence data. The SNP alleles from the single-chromosome sequencing or genotyping arrays will be chromosomal haplotypes.
24. Input the experimental haplotype data into HiFi software to obtain high-resolution whole-genome haplotypes (*see* **Note 6**).

4 Notes

1. The blood cells can be directly subjected to cell culture. If a tissue specimen is used, please follow the experimental procedure for the primary cell culture. The living cells released from tissue specimens (such as by trypsin digestion) will be grown in the media and arrested at metaphase following the same protocol described above.
2. Either fresh blood or lymphoblastoid cells or any living cells can be used. Correspondingly, the medium should be switched to the growth medium for the corresponding cell type. For example, when lymphoblastoid cells are used, cells will be grown in RPMI 1640 containing 15% FBS and a mitogen (such as PHA) for 45 h, followed by the same experimental

steps from colcemid treatment to chromosome microdissection as described above.

3. Isolation of single chromosomes can be done with chromosome sorting or microfluidics or any other new device.
4. The higher volume of dH_2O in the collection tube, the higher chance to successfully collect the dropped foil, and also a higher usage and cost to use all subsequent reagents. So the users need to adjust the volume to ensure a successful collection of the microdissected foil. Direct visualization of the foil in the tube after each microdissection under the microscope is usually recommended.
5. When choosing the Illumina genotyping arrays, and the GenomeStudio software (Illumina part # 11207066) to call alleles, please choose the Forward model rather than Top/Bottom calling.
6. HiFi software needs three input files, the low-resolution experimental haplotypes obtained in the procedure described above, an unphased genotype dataset, and a reference panel. The reference panels can be downloaded from the International HapMap Project database (phase 2 public release 22, phase 3 public draft release 1, and phase 2+3 February 2009 release 27), the 1000 Genomes Project database, and any other databases that contain the haplotype data for specific populations. For Windows user, HiFi requires three input files, haplotype.txt, genotype.txt, and refHaplotype.txt. The input files are named following the above examples. The names should match with them exactly. Run the software with double click the software. For Linux user, HiFi requires three input files, haplotype.txt genotype.txt, and refHaplotype.txt. If the input files are named following the above examples. HiFi can be run as /HiFi; otherwise, the file names can be named by the user and provided in the haplotype, genotype, refHaplotype order. Run the software with the following Command example: HiFi haplotype_fileName genotype_fileName refHaplotype_fileName. HiFi takes a fourth parameter called MAFSTEP, which is the changing step of minor allele frequency. Default value for MAFSTEP is 0.1. Its value can be set between 0 and 0.5. Command example: HiFi haplotype.txt genotype.txt refHaplotype.txt 0.01

Acknowledgment

This work was supported by US National Institutes of Health grants (R21HG006173, RC4MD005964, HL003676, RR014758, RR003034, GM74913, G12MD007602, U54MD007588), an American Heart Association grant (09GRNT2300003).

References

1. Slatkin M (2008) Linkage disequilibrium—understanding the evolutionary past and mapping the medical future. Nat Rev Genet 9:477–485
2. Liu N, Zhang K, Zhao H (2008) Haplotype-association analysis. Adv Genet 60:335–405
3. Browning SR, Browning BL (2007) Rapid and accurate haplotype phasing and missing-data inference for whole-genome association studies by use of localized haplotype clustering. Am J Hum Genet 81:1084–1097
4. Browning SR, Browning BL (2011) Haplotype phasing: existing methods and new developments. Nat Rev Genet 12:703–714
5. Howie BN, Donnelly P, Marchini J (2009) A flexible and accurate genotype imputation method for the next generation of genome-wide association studies. PLoS Genet 5, e1000529
6. Li Y, Willer CJ, Ding J, Scheet P, Abecasis GR (2010) Mach: using sequence and genotype data to estimate haplotypes and unobserved genotypes. Genet Epidemiol 34:816–834
7. Delaneau O, Marchini J, Zagury JF (2012) A linear complexity phasing method for thousands of genomes. Nat Methods 9:179–181
8. Kukita Y, Miyatake K, Stokowski R, Hinds D, Higasa K, Wake N, Hirakawa T, Kato H, Matsuda T, Pant K, Cox D, Tahira T, Hayashi K (2005) Genome-wide definitive haplotypes determined using a collection of complete hydatidiform moles. Genome Res 15: 1511–1518
9. Kong A, Masson G, Frigge ML, Gylfason A, Zusmanovich P, Thorleifsson G, Olason PI, Ingason A, Steinberg S, Rafnar T, Sulem P, Mouy M, Jonsson F, Thorsteinsdottir U, Gudbjartsson DF, Stefansson H, Stefansson K (2008) Detection of sharing by descent, long-range phasing and haplotype imputation. Nat Genet 40(9):1068–75
10. Fan HC, Wang J, Potanina A, Quake SR (2011) Whole-genome molecular haplotyping of single cells. Nat Biotechnol 29:51–57
11. Kirkness EF, Grindberg RV, Yee-Greenbaum J, Marshall CR, Scherer SW, Lasken RS, Venter JC (2013) Sequencing of isolated sperm cells for direct haplotyping of a human genome. Genome Res 23:826–832
12. Kitzman JO, Mackenzie AP, Adey A, Hiatt JB, Patwardhan RP, Sudmant PH, Ng SB, Alkan C, Qiu R, Eichler EE, Shendure J (2011) Haplotype-resolved genome sequencing of a Gujarati Indian individual. Nat Biotechnol 29:59–63
13. Rao W, Ma Y, Ma L, Zhao J, Li Q, Gu W, Zhang K, Bond VC, Song Q (2013) High-resolution whole-genome haplotyping using limited seed data. Nat Methods 10:6–7
14. Kuleshov V, Xie D, Chen R, Pushkarev D, Ma Z, Blauwkamp T, Kertesz M, Snyder M (2014) Whole-genome haplotyping using long reads and statistical methods. Nat Biotechnol 32:261–266
15. Selvaraj S, Dixson JR, Bansal V, Ren B (2013) Whole-genome haplotype reconstruction using proximity-ligation and shotgun sequencing. Nat Biotechnol 31:1111–1118
16. Suk EK, McEwen GK, Duitama J, Nowick K, Schulz S, Palczewski S, Schreiber S, Holloway DT, McLaughlin S, Peckham H, Lee C, Huebsch T, Hoehe MR (2011) A comprehensively molecular haplotype-resolved genome of a European individual. Genome Res 21: 1672–1685
17. Ma L, Xiao Y, Huang H, Wang Q, Rao W, Feng Y, Zhang K, Song Q (2010) Direct determination of molecular haplotypes by chromosome microdissection. Nat Methods 7:299–301
18. Yang H, Chen X, Wong WH (2011) Completely phased genome sequencing through chromosome sorting. Proc Natl Acad Sci U S A 108:12–17
19. Zhang K, Zhu J, Shendure J, Porreca GJ, Aach JD, Mitra RD, Church GM (2006) Long-range polony haplotyping of individual human chromosome molecules. Nat Genet 38:382–387
20. Ding C, Cantor CR (2003) Direct molecular haplotyping of long-range genomic DNA with M1-PCR. Proc Natl Acad Sci U S A 100:7449–7453
21. Peters BA, Kermani BG, Sparks AB, Alferov O, Hong P, Alexeev A, Jiang Y, Dahl F, Tang YT, Haas J, Robasky K, Zaranek AW, Lee JH, Ball MP, Peterson JE, Perazich H, Yeung G, Liu J, Chen L, Kennemer MI, Pothuraju K, Konvicka K, Tsoupko-Sitnikov M, Pant KP, Ebert JC, Nilsen GB, Baccash J, Halpern AL, Church GM, Drmanac R (2012) Accurate whole-genome sequencing and haplotyping from 10 to 20 human cells. Nature 487:190–195
22. Xiao M, Wan E, Chu C, Hsueh WC, Cao Y, Kwok PY (2009) Direct determination of haplotypes from single DNA molecules. Nat Methods 6:199–201
23. Thurman RE, Rynes E, Humbert R, Vierstra J, Maurano MT, Haugen E, Sheffield NC, Stergachis AB, Wang H, Vernot B, Garg K, John S, Sandstrom R, Bates D, Boatman L,

Canfield TK, Diegel M, Dunn D, Ebersol AK, Frum T, Giste E, Johnson AK, Johnson EM, Kutyavin T, Lajoie B, Lee BK, Lee K, London D, Lotakis D, Neph S, Neri F, Nguyen ED, Qu H, Reynolds AP, Roach V, Safi A, Sanchez ME, Sanyal A, Shafer A, Simon JM, Song L, Vong S, Weaver M, Yan Y, Zhang Z, Lenhard B, Tewari M, Dorschner MO, Hansen RS, Navas PA, Stamatoyannopoulos G, Iyer VR, Lieb JD, Sunyaev SR, Akey JM, Sabo PJ, Kaul R, Furey TS, Dekker J, Crawford GE, Stamatoyannopoulos JA. The accessible chromatin landscape of the human genome. Nature. 2012;489:75–82.

24. Dekker J (2006) The three 'c' s of chromosome conformation capture: controls, controls, controls. Nat Methods 3:17–21
25. Cremer T, Cremer C (2001) Chromosome territories, nuclear architecture and gene regulation in mammalian cells. Nat Rev Genet 2:292–301
26. Dekker J (2008) Gene regulation in the third dimension. Science 319:1793–1794
27. Miele A, Dekker J (2008) Long-range chromosomal interactions and gene regulation. Mol BioSyst 4:1046–1057
28. Phillips JE, Corces VG (2009) Ctcf: master weaver of the genome. Cell 137:1194–1211
29. Lieberman-Aiden E, van Berkum NL, Williams L, Imakaev M, Ragoczy T, Telling A, Amit I, Lajoie BR, Sabo PJ, Dorschner MO, Sandstrom R, Bernstein B, Bender MA, Groudine M, Gnirke A, Stamatoyannopoulos J, Mirny LA, Lander ES, Dekker J (2009) Comprehensive mapping of long-range interactions reveals folding principles of the human genome. Science 326:289–293
30. Li W, Fu G, Rao W, Xu W, Ma L, Guo S, Song Q (2015) Genomelaser: fast and accurate haplotyping from pedigree genotypes. Bioinformatics 31:3984–3987
31. Li W, Xu W, Fu G, Ma L, Richards J, Rao W, Bythwood T, Guo S, Song Q (2015) High-accuracy haplotype imputation using unphased genotype data as the references. Gene 572:279–284
32. Ma Y, Zhao J, Wong JS, Ma L, Li W, Fu G, Xu W, Zhang K, Kittles RA, Li Y, Song Q (2014) Accurate inference of local phased ancestry of modern admixed populations. Sci Rep 4:5800

Chapter 10

Phased Genome Sequencing Through Chromosome Sorting

Xi Chen, Hong Yang, and Wing Hung Wong

Abstract

Phase information of an individual genome provides fundamentally useful genetic information for the understanding of genome function, phenotype, and disease. With the development of new sequencing technology, much interest has been focused on the challenges in obtaining long-range phase information. Here, we present the detailed protocol for a method capable of generating genomic sequences completely phased across the entire chromosome through FACS-mediated chromosome sorting and next generation sequencing, known as Phase-seq.

Key words Haplotype, Phased sequencing, Single chromosome sequencing, Chromosome sorting, SNP

1 Introduction

The human genome contains two homologous sets of chromosomes, one from each parent, with the particular combination of alleles on the chromosome termed a "haplotype." Haplotypes consist of combinations of genetic variants, including single nucleotide polymorphisms (SNPs), copy number variations (CNVs), insertions/deletions (Indels), etc. Since the genetic makeup is one of the most important factors for the susceptibility, progression, and prevention of many heritable diseases, knowledge of complete haplotype structure of an individual is therefore essential for the study of disease and the advancement of personalized medicine [1, 2].

With the development of microarrays and massively parallel sequencing technologies, several approaches have been proposed to solve the whole-genome haplotyping (or phasing) challenge, including population-based inference, genetic analysis, and molecular haplotyping [3, 4]. In the population-based inference, haplotypes are often inferred from distantly related individuals in an isolated community using statistical methods [5]. This approach, however, only works well for markers that are very close to each other [6]. In genetic analysis, haplotypes are instead constructed

Irene Tiemann-Boege and Andrea Betancourt (eds.), *Haplotyping: Methods and Protocols*, Methods in Molecular Biology, vol. 1551, DOI 10.1007/978-1-4939-6750-6_10, © Springer Science+Business Media LLC 2017

from family trio or quartets using high-throughput paired-end short reads [7, 8]. The limitation of this approach is that it is not always feasible to recruit participants required for family-based studies. Furthermore, positions in which all family members are heterozygous cannot be phased using this method. Recently, molecular haplotyping has become attractive in which the direct observation of alleles on a single molecule can be achieved through various means of separating homologous chromosomes, including microdissection [9], microfluidics [10], and fluorescence-activated cell sorting [11]. Once separated, the chromosomes are either individually tagged or first combined into pools that tend to contain no more than one copy of each homologous chromosome prior to genotyping and sequencing. Despite the fact that molecular haplotyping can be labor intensive, time-consuming, and expensive, this approach is still advantageous since long-range haplotype spanning as far as whole chromosomes can be directly observed from sequence data and it is highly accurate without switch errors observed in molecular haplotyping methods via synthetic long reads [12, 13]. Switch errors denote the haplotyping events where consecutive variants are assigned incorrectly to opposite parental haplotypes [12].

In this chapter, we will describe in detail the phased genome sequencing method through FACS-mediated chromosome sorting. Several major procedures are involved, including cell culture, chromosome preparation, chromosome sorting, single chromosome amplification, and chromosome verification by Real-time PCR, followed by library construction, multiplex Illumina sequencing, and data analysis.

2 Materials

2.1 Cell Culture

1. 10 mL EDTA-coated blood collection tubes.
2. Phosphate-buffered saline (PBS), pH 7.4.
3. 50 mL conical tubes.
4. Ficoll-Paque™ PLUS solution (GE Healthcare Bio-sciences AB, Uppsala, Sweden).
5. Centrifuge with swinging bucket rotor.
6. 250 and 500 mL Nalgene rapid-flow filters.
7. PBS supplemented with 2% fetal bovine serum (FBS), freshly made, filtered.
8. *Complete medium*. RPMI 1640 medium with 10% FBS, 2 mM L-glutamine, 100 U/mL Penicillin-Streptomycin (Thermo fisher scientific, Waltham, MA). Filtered.
9. Ultrapure™ DNase/RNase-free distilled water (Life technologies, Carlsbad, CA) (*see* **Note 1**).

10. Phytohemagglutinin-M (PHA-M, Roche applied science, Mannheim, Germany) stock solution: 10 mg/mL in Ultrapure™ distilled water. Filter, aliquot, and store at −20 °C.
11. *Complete medium with PHA-M*: *Complete medium* supplemented with 10 μg/mL PHA-M.
12. Hemocytometer.
13. 75 cm^2 cell culture flasks.
14. Cell culture incubator.
15. Demecolcine solution (Sigma-Aldrich, St. Louis, MO), store at 4 °C.

2.2 Chromosome Preparation

1. 15 and 50 mL conical tubes.
2. 0.25% Trypsin-EDTA.
3. Benchtop centrifuge.
4. Kimwipes.
5. 50 mL tube top filters.
6. Spermidine trihydrochloride/spermine tetrahydrochloride stock solution: 0.26 g spermidine (Sigma-Aldrich) and 0.14 g spermine (Sigma-Aldrich) in 1 mL Ultrapure™ distilled water. Filter, aliquot, and store at −20 °C.
7. *Hypotonic solution* (30 mL): 5.6 mg/mL KCl and 15 μL spermidine/spermine stock solution. Freshly made. Filter before use.
8. *10× chromosome isolation buffer 1 (10× CIB1)*: 200 mM NaCl, 800 mM KCl, 150 mM Tris–HCl, 2 mM spermine tetrahydrochloride, 5 mM spermidine trihydrochloride in Ultrapure™ distilled water. Adjust the pH to 7.2. This buffer can be stored at 4 °C for 1 month.
9. Digitonin (Sigma-Aldrich), store at 4 °C.
10. 5 and 10 mL Pasteur pipets.
11. Propidium iodide (PI) stock solution: 50 μg/mL in PBS.
12. Fluorescence microscope.
13. 12×75 mm polystyrene round-bottom tubes for the flow cytometer (Corning life sciences, Tewksbury, MA).
14. 12×75 mm polystyrene round-bottom tube with cell-strainer cap (Corning life sciences).
15. 12×75 mm polypropylene round-bottom tubes (Corning life sciences).
16. Hoechst 33258 (Sigma-Aldrich) stock solution: 1 mg/mL in Ultrapure™ distilled water. Heat to 65 °C to dissolve. Filter, aliquot, and store in amber tubes at −20 °C.
17. Chromomycin-A3 (Sigma-Aldrich) stock solution: 2 mg/mL in 100% ethanol (200 proof). Aliquot and store in amber tubes at −20 °C.

18. $MgCl_2$ stock solution: 15 mM in Ultrapure™ distilled water. Filter and store at room temperature.
19. Sodium citrate/sodium sulfite stock solution: 0.29 g sodium citrate tribasic dehydrate (Sigma-Aldrich), 0.32 g sodium sulfite (Sigma-Aldrich) in 10 mL Ultrapure™ distilled water. Filter and store at room temperature.

2.3 Chromosome Sorting

1. 1 M Tris sterile solution, pH 7.5 (Amresco, Solon, OH).
2. 25 mL disposable presterile polystyrene reservoir (VWR international, Radnor, PA).
3. 10 and 100 μL multichannel pipettor.
4. ABgene SuperPlate skirted PCR plates (Thermo fisher scientific).
5. Adhesive PCR foil seals (Thermo fisher scientific).
6. Ice bucket.
7. BD Influx cell sorter.
8. 1.5 mL DNase/RNase-free DNA LoBind tubes (Eppendorf, Hamburg, Germany).

2.4 Single Chromosome Amplification

1. Tissue culture hood.
2. 70% ethanol.
3. Eppendorf® PCR cooler, for 96-well plates (Sigma-Aldrich).
4. Microarray Microplate Centrifuge (Arrayit Corporation, Sunnyvale, CA).
5. 10, 20, 200, and 1000 μL filter pipet tips.
6. 10 mM Tris sterile solution: 1:100 dilution of 1 M Tris sterile solution (pH 7.5) in Nuclease-free water.
7. PicoPLEX™ WGA kit (Rubicon genomics, Ann Arbor, MI) (*see* **Note 2**).
8. *Extraction cocktail.* 4.8 μL Extraction enzyme dilution buffer and 0.2 μL Cell extraction enzyme (provided by PicoPLEX™ WGA kit).
9. PCR thermal cycler.
10. *Pre-Amp cocktail.* 4.8 μL Pre-Amp buffer and 0.2 μL Pre-Amp enzyme (provided by PicoPLEX™ WGA kit).
11. *Amplification cocktail.* 25 μL Amplification buffer, 0.8 μL Amplification enzyme, and 34.2 μL Nuclease-free water (provided by PicoPLEX™ WGA kit).
12. Qubit dsDNA HS assay kit (Life technologies).
13. Qubit 0.5 mL assay tubes (Life technologies).
14. Qubit 2.0 Fluorometer (Life technologies).

2.5 Chromosome Verification by Real-Time PCR

1. Low-profile unskirted 96-well PCR plates (Bio-rad, Hercules, CA), specific for Bio-rad CFX real-time PCR thermocycler.
2. iTaq universal SYBR® Green supermix (2×) (Bio-rad).
3. DNA primers for each chromosome: stock concentration 200 μM (Table 1, *see* **Note 3**).

Table 1
Chromosome-specific primers

Chr.	Forward	Reverse
1-L[a]	CCCAAATGGATCTTTGCTTG	GAATCTCTTTTCCCTGGCTTG
1-R[b]	GTCTGTGTGGTGTGGGTGTC	TGCTGCTTCATGCTGTCTCT
2-L	CAACATGGACAAATGCCAAA	GCCCCAATGAAAGACTCAAA
2-R	GCAGGATACGAACTGCTTCC	TGTTTGCATAGCTCACATCAGTC
3-L	CTCATCTGGAACCCCACACT	CACATTCAGGACTGCGAAAA
3-R	TCAACTGAGCAGGGAGAGGT	TCATGTCAGCCCTCAAAAAG
4-R-1	AAACAGGCACAGAGAAGTCCA	TCCTTGGGAAACTCCACAGA
4-R-2	TATGCGTGTCCTTGTTGCTG	TCAGTCCCAGTCAGTTCATCC
5-R-1	AAAAGGCCCATCCATTCATT	TCAAAGAGCACCAGAGCAGA
5-R-2	TCTTGCTTTCCACCCTGACT	AGCTACCCATCCCTCATCCT
6-L	GAACCACCCCTAGAGCCTTC	AAGGAATGCACCAAATCGTC
6-R	TAGCCATCAGGGACCACTTC	GCAAGCCAGACATTTCCTTT
7-L	TTTCAGCCTCCCAAGAGTGT	CGTGGTGGCCTTGACTATTT
7-R	TATTCGGAGGCAAGTTCACC	CTCCAGTTGATCCCACGTCT
8-R-1	CCCAAATGGTAAGTGGTCCT	TGCTAGTCAGATCCCTGGAAA
8-R-2	TCACACAGGGTCCACTCTCA	TGTAGAGCCCACTGCAAAATAA
9-R-1	CATCAAAGGTACAGGCAGCA	TGACACCCAGCCTCATTTTT
9-R-2	GGCCAGGTGGAGATAACTGA	GCTTATGCAGGCAGAGGAAC
10-R-1	GATGGGCTTGGCTTCACTAA	TGTTCTGTTGGTCCGTGTGT
10-R-2	TGCACTTGTCTGGCTCTTTG	TGGCCTGTAGCCTTCTGAGT
11-R-1	TGGAGAAGAACCGAGACACC	CTTCCCGAACAACAGAAACC
11-R-2	CTTTTCCATCCCTGTCTCCA	CTAACCATGCCAGCAACAGA
12-R-1	CAGTAGGTAAATCACACTGGAGGA	CCCACCTGTCTCTCAAAGGA
12-R-2	TCTTGAAAGGTGGCTTTGGT	CTCTTTGCCGTGCTGTGTT

(continued)

Table 1
(continued)

Chr.	Forward	Reverse
13-R-1	GCAATCTCGGTCCATTCATT	GGGAAAGTGTTGGAAGTTTGA
13-R-2	AATATCGCTGGCTCACATCC	AGGACCATCTTTCCTGCTGA
14-R-1	GAAGCCAATTTCTGGATGGA	GGGAATGTTTTGGGACTTTG
14-R-2	TCATACCACCAAAGGCACAA	TGAACATGGCGAGAAGTCAG
15-R-1	ACTGCCCCAATTCATGTTTC	CCTCACCTTTCCTGTTTCCA
15-R-2	GAGGTGGTCGTTTTTGAGGA	ATGGAGGACAGCGAGAAGAA
16-L	AAAAGCCAAACCTGATGGTG	ACATCCTGCCCCTTCTATCC
16-R	TGTGGTCCCTAGCACTTTCC	AGCAAGCAGATTCCAGCATT
17-L	AACTCAATCTTGGGGCACAC	TGGTACTCGGCACACAGAAG
17-R	CACCCTGACACTTCTCCACA	TTGCACACCACAGGACAGTT
18-L	GTGACCTTTGGGCAAGTCAT	TCAATGACAAGCAGCCTTTG
18-R	CATCTAGCCCAGTGCCTAGC	GTCTGCCTTGAAGCAAGTCC
19-L	AACTGACCACGTTCCTCACC	GTAGCGTCTGTTCCCTCTGC
19-R	CACAGAGCCACAGGACTTCA	GGCTCACTGAAGCCTCAAAC
20-L-1	CTATGTGACTTGGGGCGTGT	GCTAAACCTGACTGGGTGGA
20-L-2	GAAAACCTCTCTGCCCTTCC	TTTACCCTCCCCTCACACAC
21-R-1	CCAGGCTTGACCCCTAGTAA	TGTTGTGACCCATGCAAGTC
21-R-2	TCCCAATTTCCTTGTGGTTT	AATTCCTGGCTGCTTGAATG
22-R-1	GGTAATCAGCCTCCCCTTGT	TTCTTCCCCTTTCCCAAATC
22-R-2	TTTATTCCATGTGGGGCTGT	TGGGGGCTAAACAAATCAAA
X-L	AGGCCACCCTTCACCTATTC	CCTGCCTGTCACCTGTTTTT
X-R	TTGGAGCTTTTCCGTCTCTC	ATGCTTTTTGGTCCTGCTGT
Y-L	GGTTCAGTCCCATCTCCTCA	TCAGGCTACGGCTTTGTTTT
Y-R	CCTGGACACACTCCACAAAG	AGGCAGAAAACGACATCACC

[a]L: Left arm of chromosome
[b]R: Right arm of chromosome

4. *PCR master mix*: 15 μL iTaq SYBR® green supermix (2×), 1.125 μL Forward primer of chromosome of interest, 1.125 μL Reverse primer of chromosome of interest, and 10.25 μL Nuclease-free water.
5. Human male genomic DNA (Promega, Madison, WI): adjust stock concentration to 20 ng/μL.

6. Microseal "B" adhesive seals (Bio-rad).
7. CFX96 real-time PCR thermocycler (Bio-rad).
8. QIAquick PCR purification kit (Qiagen, Hilden, Germany).

2.6 Library Construction and Multiplex Illumina Sequencing

1. NEBNext dsDNA fragmentase (New England Biolabs, Ipswich, MA).
2. *Fragmentation mixture*. 1 μL NEBNext dsDNA fragmentase, 1 μL 10× Fragmentation buffer, and 1 μL 10× BSA (provided by NEBNext dsDNA fragmentase kit).
3. 30 °C and 37 °C water bath.
4. 0.5 M EDTA.
5. Elution buffer: 10 mM Tris-Acetate, pH 8.0, reagent grade water.
6. *Stop solution*: 2 μL 0.5 M EDTA and 8 μL of Elution buffer.
7. Agencourt® AMPure XP magnetic beads (Beckman coulter, Brea, CA).
8. 80% ethanol, freshly made.
9. Kapa library preparation kit with real-time PCR library amplification for Illumina (Kapa Biosystems, Wilmington, MA).
10. *End-repair mixture*. 10 μL of 10× End-repair buffer, 5 μL of End-repair enzyme mix, 0.5 μL of *E. coli* DNA ligase, and 34.5 μL of Nuclease-free water (provided by Kapa library preparation kit).
11. *A-tailing mixture*. 5 μL of 10× A-tailing buffer, 3 μL of A-tailing enzyme, and 12 μL of Nuclease-free water (provided by Kapa library preparation kit).
12. Liglinker_truS_a: TACACTCTTTCCCTACACGACGCTCT TCCGATC*T (stock concentration: 60 μM).
13. Liglinker_truS_b: /5Phos/GATCGGAAGAGCACACGTCT GAACTCCAGTCA*C (stock concentration: 60 μM).
14. 30 μM Annealed ligation linkers (*see* **Note 4**).
15. *Adaptor ligation mixture*: 10 μL of 5× Ligation buffer, 5 μL of DNA ligase (provided by Kapa library preparation kit), and 5 μL of 30 μM Annealed ligation linkers.
16. 0.2 mL PCR eight-tube strip with attached cap.
17. Adaptor_truS_A (Universal primer, 100 μM). Dilute to 2 μM in elution buffer. AATGATACGGCGACCACCGAGA TCTACACTCTTTCCCTACACGAC.
18. Adaptor_truS_B (Index primer, total 96 indexes, 100 μM). Dilute to 1 μM in elution buffer. 96-plex Barcode sequences are listed in Kozarewa and Turner (2011) [15].

    5′-CAAGCAGAAGACGGCATACGAGAT-[Barcode]-GTGACTGGAGTTCAGACGTGT-3′

19. Truseq PCR primer 1 (100 μM): AATGATACGGCGA CCACCGA. Dilute to 50 μM in elution buffer.
20. Truseq PCR primer 2 (100 μM): CAAGCAGAA GACGGCATACGA. Dilute to 50 μM in elution buffer.
21. *Library amplification mixture*: 0.5 μL of Adaptor_truS_A, 0.25 μL of Truseq PCR primer 1, 0.25 μL of Truseq PCR primer 2, and 25 μL of 2× Kapa HiFi HotStart ready mix (provided by Kapa library preparation kit).
22. E-Gel® iBase™ power system (Life technologies).
23. E-Gel® SizeSelect 2 % Agarose gel (Life technologies).
24. 50 or 100 bps DNA ladder.
25. Agilent 2100 Bioanalyzer (Agilent technologies, Santa Clara, CA).
26. *Second PCR mixture*: 1 μL of Truseq PCR primer 1 (25 μM), 1 μL of Truseq PCR primer 2 (25 μM), and 25 μL of 2× Kapa HiFi HotStart ready mix (provided by Kapa library preparation kit).
27. Illumina HiSeq sequencing system.

3 Methods

3.1 Cell Culture

1. Collect 40 mL of Human whole blood in the EDTA-coated blood collection tubes, stored at 4 °C (*see* **Note 5**).
2. Dilute the whole blood 1:1 with PBS (pH 7.4) in 50 mL conical tubes, mix gently.
3. Pipette 15 mL of well-mixed Ficoll-Paque solution into a new 50 mL conical tube.
4. Carefully layer 20 mL of diluted blood over the Ficoll-Paque solution.
5. Centrifuge at 400 × *g* for 30 min in a swinging bucket rotor *at room temperature without break.*
6. Carefully transfer the interphase cells to a new 50 mL conical tube.
7. Add at least three volumes of PBS with 2 % FBS, mix, and centrifuge at 60–100 × *g* for 10 min. Discard the supernatant completely.
8. Combine the lymphocytes into one 50 mL conical tube.
9. Resuspend the lymphocytes in 15 mL of PBS with 2 % FBS, mix gently.
10. Spin again at 60–100 × *g* for 10 min. Discard the supernatant completely.
11. Resuspend the lymphocytes in 20 mL *complete medium with PHA-M.*

12. Count the cells with the hemocytometer. Dilute the cells at 0.5×10^6 cells/mL in *complete medium with PHA-M*.
13. Add 30 mL of culture in each 75 cm^2 cell culture flask. Incubate the cells at 37 °C incubator with 5% CO_2. Let flask flat even.
14. After 48 h incubation, demecolcine is added to the medium at a final concentration of 0.1 μg/mL to arrest cells in metaphase.
15. Harvest the cells 14 h later.

3.2 Chromosome Preparation

1. Shake off the cells by giving the flask a sharp rap and transfer the cells to one 50 mL conical tube per flask. If necessary, trypsinize the remaining cells with 8 mL 0.25% Trypsin-EDTA.
2. Centrifuge at $100 \times g$ for 10 min, discard the supernatant, and resuspend the cells in 5 mL fresh *complete medium*.
3. Combine the cells from two 50 mL conical tubes together, centrifuge again at $100 \times g$ for 10 min.
4. Discard the supernatant by inverting the tube. Remove the last few drops from inside the tube with Kimwipes. Gently flick the tube to disaggregate the cell pellet.
5. Add 30 mL freshly made *hypotonic solution*, mix gently, and leave at room temperature for 20 min.
6. While the cells are in the *hypotonic solution*, dissolve 24 mg of digitonin in 10 mL Ultrapure™ distilled water by heating on a hot plate for 5 min. Allow the digitonin solution to cool down before adding 2 mL of *10× CIB1* and make the volume up to 20 mL with Ultrapure™ distilled water. Adjust to pH 7.2 if necessary. Filter and place on ice.
7. Centrifuge the cells at $100 \times g$ for 10 min, remove the supernatant carefully with a 10 mL Pasteur pipet.
8. Add 3 mL cold digitonin in *1× CIB1* solution as prepared in **step 6** (approximately ten times the volume of the cell pellet).
9. Aspirate gently a few times with a 5 mL Pasteur pipet. Avoid too much force, as it can result in chromosome damage.
10. To check if the chromosomes are released from the cell membrane, mix equal volumes of chromosome preparation with Propidium iodide. View with a fluorescence microscope. If the chromosomes are still maintained within the cell, mix the preparation gently a few more times with 5 mL Pasteur pipet.
11. The chromosome suspension can be stored at 4 °C for several weeks with little deterioration of the flow karyotype.
12. To stain, transfer 1.1 mL of the well-mixed chromosome suspension into a 12×75 mm polystyrene tube. Centrifuge at $200 \times g$ for 1 min.

13. Take out 1 mL of chromosome containing supernatant carefully without disturbing the pellet, and filter through a new 12×75 mm polystyrene tube with a cell-strainer cap.
14. As a medium only control, transfer 1.1 mL of digitonin in *1× CIB1* solution into another 12×75 mm polystyrene tube. Centrifuge and filter as indicated in **steps 12** and **13** (*see* **Note 6**).
15. Before staining, transfer 500 μL of filtered chromosome suspension or digitonin in *1× CIB1* solution into a 12×75 mm polypropylene tube, separately (*see* **Note 7**).
16. To stain, add 2 μL of Hoechst 33258, mix immediately. Then, add 33.3 μL of Chromomycin A3 and 26.7 μL of 15 mM $MgCl_2$. Mix and leave the sample at 4 °C for overnight in the dark.
17. To improve the chromosome profile, add 100 μL of sodium citrate/sodium sulfite to the chromosome suspension at least 15 min prior to running on flow cytometer. Keep both the chromosome suspension and the medium only control on ice before sorting.

3.3 Chromosome Sorting

1. Dilute 1 M Tris sterile solution (pH 7.5) 1:100 with Ultrapure™ distilled water in a 25 mL presterile polystyrene reservoir.
2. Add 2.5 μL of 10 mM Tris in a well of an ABgene SuperPlate skirted PCR plate (*see* **Note 8**). Total 24 wells (well: B3-B10, D3-D10, and F3-F10). Prepare at least six plates per chromosome.
3. Cover each PCR plate with adhesive PCR foil seal immediately to prevent contamination. Keep the plates on the ice bucket (*see* **Note 9**).
4. Chromosome sorting is performed exclusively by the flow cytometry specialist on a BD Influx cell sorter, and the Influx is set up per manufacturer's instructions with the following specificity.
5. Set up the sorter with a constant 40 psi air supply with a 0.2 μM filter in line.
6. Chill the sample port and sort reservoirs to 4 °C.
7. Fit the sorter with a 70 μM diameter nozzle tip to get the resolution required for the display of the flow karyotype.
8. Warm up the lasers for at least 30 min prior to alignment.
9. Excitation of Hoechst and Chromomycin is done with a solid state 100 mW, 355 nm laser and a 200 mW, 457 nm laser, respectively.
10. Emission of the Hoechst fluorescence is collected with a 460/50 bandpass filter, and the Chromomycin fluorescence is collected with a 550/50 bandpass filter.

11. The UV laser is timed as the first laser, and detection is triggered by Hoechst fluorescence.
12. Select the designated wells for single chromosome sorting using computer software.
13. Always run the medium only control first to saturate the sample line with fluorescence dye on the flow cytometer before sorting. Once the expected flow karyotype is identified, set the gates for sorting a specific chromosome from peaks (*see* **Note 10**).
14. To collect chromosomes, first remove the adhesive PCR foil seal from the PCR plate and place the plate on the sorter.
15. Once the chromosomes are collected, the plate is sealed immediately with new adhesive PCR foil seal and stored on ice. For long-term storage, always keep the plates in a −80 °C freezer.
16. Collect the stream control in a 1.5 mL DNA LoBind tube.

3.4 Single Chromosome Amplification

1. Disinfect the tissue culture hood with 70% ethanol. Turn on the UV light for 30 min.
2. Disinfect the pipettors, PCR cooler, and Microarray Microplate Centrifuge with 70% ethanol (*see* **Note 11**).
3. Thaw one plate containing sorted chromosomes on ice.
4. Briefly spin down plate in a Microplate Centrifuge.
5. Disinfect the plate with 70% ethanol before putting on a PCR cooler.
6. Single chromosomes are amplified using PicoPLEX™ WGA kit. To prevent DNA contamination, the following procedures are performed in the tissue culture hood unless otherwise specified.
7. First, remove the adhesive PCR foil seal from the plate, add 2.5 μL of 10 mM Tris sterile solution to each well containing sorted chromosome. As a control, add 2.5 μL of 10 mM Tris sterile solution to an empty well containing 2.5 μL of stream control. The total volume in each well is 5 μL.
8. Add 5 μL of *extraction cocktail* to each well.
9. Cover the plate with new adhesive PCR foil seal. Briefly spin the plate in a Microplate Centrifuge.
10. Incubate the plate in a PCR thermal cycler as follows: 75 °C for 10 min, 95 °C for 4 min. Hold at room temperature.
11. Briefly spin the plate in a Microplate Centrifuge.
12. Disinfect the plate with 70% ethanol before putting on the PCR cooler.
13. Add 5 μL of *pre-Amp cocktail* to each well. Cover the plate with new adhesive PCR foil seal.
14. Briefly spin the plate in a Microplate Centrifuge.

15. Incubate the plate in the PCR thermal cycler as follows: 95 °C for 2 min, 12 cycles of 95 °C for 15 s, 15 °C for 50 s, 25 °C for 40 s, 35 °C for 30 s, 65 °C for 40 s, and 75 °C for 40 s. Hold at 4 °C.
16. Briefly spin the plate in a Microplate Centrifuge.
17. Disinfect the plate with 70% ethanol before putting on the PCR cooler.
18. Add 60 μL of *amplification cocktail* to each well. Mix. Cover the plate with new adhesive PCR foil seal.
19. Briefly spin the plate in a Microplate Centrifuge.
20. Incubate the plate in the PCR thermal cycler as follows: 95 °C for 2 min, 17 cycles of 95 °C for 15 s, 65 °C for 1 min, and 75 °C for 1 min. Hold at 4 °C.
21. Briefly spin the plate in a Microplate Centrifuge.
22. Disinfect the plate with 70% ethanol before putting on the PCR cooler.
23. Immediately transfer PCR products from each well to the 1.5 mL DNA LoBind tube individually. Rinse the well with 20 μL of Nuclease-free water.
24. Measure DNA concentration using Qubit dsDNA HS assay kit (*see* **Note 12**).
25. Store the tube at −20 °C.

3.5 Chromosome Verification by Real-time PCR

1. Thaw amplified DNA on ice.
2. Add 27.5 μL of *PCR master mix* in a well of a low-profile 96-well unskirted PCR plate, including the control well.
3. Add 2.5 μL of amplified DNA (<500 ng) to each well, separately.
4. Add 2.5 μL of Human male genomic DNA to the control well.
5. Cover the plate with Microseal "B" adhesive seal.
6. Spin the plate in a benchtop centrifuge at 300 × *g* for 1 min.
7. Real-time PCR is performed as follows: 95 °C for 3 min, 45 cycles of 95 °C for 10 s, 55 °C for 30 s, and a final melting curve from 65 °C to 95 °C with a 0.5 °C stepwise increment to determine the specificity of the amplification. All data are analyzed using the BioRad CFX manager software.
8. To confirm the specificity of the amplification, the melting dissociation curve for each DNA sample is compared to the control that used Human male genomic DNA as DNA template. A similar (or matching) melting dissociation curve indicates that the amplified DNA is likely from the chromosome of interest.

9. DNA from specific chromosome is then purified with QIAquick PCR purification kit and elute in 50 μL EB.
10. Determine DNA concentration using Qubit dsDNA HS assay kit.

3.6 Library Construction and Multiplex Illumina Sequencing

1. For each chromosome, DNA from at least 4–6 verified individual chromosomes are chosen for library preparation.
2. DNA fragmentation: 160 ng of chromosomal DNA is individually fragmented with 3 μL of *fragmentation mixture* at 37 °C water bath for 20 min (*see* **Note 13**). Keep on ice.
3. Stop the reaction by adding 10 μL of *stop solution*. Keep on ice.
4. Purify fragmented DNA using 30 μL AMPure XP magnetic beads (*see* **Note 14**).
5. Elute DNA in 52.5 μL of elution buffer, recover DNA in 50 μL of supernatant, store at −20 °C if not proceeding immediately.
6. Libraries are constructed using Kapa library preparation kit as follows. All reactions must be set up on ice.
7. Add 50 μL *end-repair mixture* to the sample, incubate for 30 min at 20 °C.
8. Purify DNA using 160 μL of AMPure XP magnetic beads.
9. Elute DNA in 32.5 μL of elution buffer, recover DNA in 30 μL of supernatant, store at −20 °C if not proceeding immediately.
10. Add 20 μL *A-tailing mixture* to the sample, incubate for 30 min at 30 °C water bath.
11. Purify DNA using 90 μL of AMPure XP magnetic beads.
12. Elute DNA in 32.5 μL of elution buffer, recover DNA in 30 μL of supernatant, store at −20 °C if not proceeding immediately.
13. Add 20 μL of *adaptor ligation mixture* to the sample, incubate for 15 min at 20 °C.
14. Purify DNA using 50 μL of AMPure XP magnetic beads **TWICE**.
15. Elute DNA in 25 μL of elution buffer, recover DNA in 23 μL of supernatant, store at −20 °C if not proceeding immediately.
16. Transfer DNA sample to 0.2 mL PCR eight-tube strip with attached cap. It is better to have an empty well between two samples to prevent cross contamination.
17. Add 26 μL of *library amplification mixture* to the sample.
18. Add 1 μL of Adaptor_truS_B (indexed) to the sample (*see* **Note 15**).

19. Each library is amplified as follows: 72 °C for 3 min, 98 °C for 45 s, nine cycles of 98 °C for 10 s, 63 °C for 30 s, and 72 °C for 3 min. Final extension at 72 °C for 8 min. Hold at 4 °C.
20. Immediately transfer PCR products from each well to the 1.5 mL DNA LoBind tube individually. Rinse the well with 10 μL of Nuclease-free water.
21. Purify DNA using 60 μL of AMPure XP magnetic beads.
22. Elute DNA in 30 μL of elution buffer, recover DNA in 28 μL of supernatant, store at −20 °C if not proceeding immediately.
23. Measure DNA concentration using Qubit dsDNA HS assay Kit.
24. Mix equal amount DNA from each library together.
25. Size-select using 2 % E-Gel® agarose gel on the E-Gel® iBase™ power system. The total sample volume should be between 20 and 25 μL for each well and the total amount of DNA should not exceed 700 ng per well.
26. Select the fragment size at 350, 400, 450, 500, 550, 600 bps using 50 or 100 bps DNA ladder as control (*see* **Note 16**).
27. Purify DNA fragments using equal volume of AMPure XP magnetic beads.
28. Elute DNA in 28 μL of elution buffer, and recover DNA in 26 μL of supernatant.
29. Measure DNA concentration using Qubit dsDNA HS assay kit.
30. Determine the fragment size distribution using 1 μL of DNA by Agilent 2100 Bioanalyzer.
31. Based on the Bioanalyzer data, choose the sample containing the correct fragment size for second PCR in order to obtain enough DNA for sequencing.
32. Transfer 23 μL of DNA sample (~50 ng) to 0.2 mL PCR eight-tube strip with attached cap.
33. Add 27 μL of *second PCR mixture* to the sample, and perform PCR as follows: 72 °C for 3 min, 98 °C for 45 s, five cycles of 98 °C for 10 s, 63 °C for 30 s, and 72 °C for 3 min. Final extension at 72 °C for 8 min. Hold at 4 °C (*see* **Note 17**).
34. Purify PCR products using 50 μL of AMPure XP magnetic beads.
35. Elute DNA in 30 μL of elution buffer, and recover DNA in 28 μL of supernatant.
36. Measure DNA concentration using Qubit dsDNA HS assay kit.
37. Analyze the DNA fragment size again by Agilent 2100 Bioanalyzer before submitting the sample for sequencing.

38. The library is sequenced with 150 base paired-end reads on an Illumina HiSeq sequencing system (*see* **Note 18**). Sequencing read primers are:

    Truseq read 1 primer: TACACTCTTTCCCTACACGACG CTCTTCCGATC*T. Truseq read 2 primer: GTGACT GGAGTTCAGACGTGTGCTCTTCCGATC. Truseq index read primer: GATCGGAAGAGCACACGTCTGAACTCCA GTCAC.

3.7 Sequence Data Analysis

Sequence data analysis is modified from Yang et al. [11]. Base-calling and variant calling can be done using standard procedures, e.g., via BWA-mem and GATK. Note that it may be helpful to remove the beginning (up to 29) bases, as the 18-base linker sequence plus 11-base semi-random sequence introduced in the PicoPLEX™ WGA procedure frequently occupy the beginning of the reads. FastQC can be used to help gauge the linker effect. When variant calling is done, a list of variant sites (of those differ from the reference base identity) for each chromosome is compiled from the reads. Then, the consensus of each individual chromosomes of one specific chromosome type at each variant site (from the variant sites list) is compared pairwise. The single chromosomes from the same parental origin will have close to a perfect match at those variant sites (i.e., having a variant consistency ratio close to 100% (e.g., 91–100%, Yang et al. [11]), whereas the single chromosomes from different parental origins will have a variant consistency ratio significantly less than 1 (e.g., 22–61%, Yang et al. [11]).

4 Notes

1. In order to obtain high quality chromosomes and prevent chromosome degradation, it is better to use DNase/RNase-free water. Therefore, all solutions are made using Ultrapure™ DNase/RNase-free distilled water, manufactured by Life technologies.
2. Alternative to the PicoPLEX™ WGA procedure, a similar result can be achieved by using the MDA-based (such as REPLI-g) or MALBAC amplification procedures [14].
3. To verify chromosomes, two sequence-specific PCR primers corresponding to the left and the right arm of each chromosome are designed using Primer3. In some chromosomes, two PCR primers might be designed from the same arm if the other arm is too short.
4. To make 30 μM Annealed ligation linkers, mix equal amounts of primer Liglinker_truS_a (60 μM) and Liglinker_truS_b (60 μM) in a 1.5 mL DNA LoBind tube. Incubate the tube in a beaker with approximately 400 mL boiling water on a hot

plate for 5 min. Remove the beaker from the hot plate and let the water cool down slowly to room temperature. DNA adaptor can be stored at 4 °C for short-term storage or −20 °C for long-term storage.

5. Prior to the implementation of this protocol, all personnel involved in the processing of human blood samples should complete safety training for the hazards of blood-borne pathogens. Always handle blood samples in a biological safety cabinet with appropriate personal protective equipment, including laboratory coats, gloves, masks, and eye goggles. For better protection, it might be helpful to request an infectious disease test result on the blood sample. Disinfect any blood spill with 10% bleach. Once the blood is drawn, it is good practice to keep it at 4 °C and process within 5–6 h.
6. This medium only control is useful when sorting the chromosomes on the flow cytometer. It is used to saturate the sample line with fluorescence dye on the flow cytometer before actually running the chromosome samples, to prevent any dye efflux.
7. Polypropylene tubes are recommended for the flow cytometer. Polystyrene tubes are only used during centrifugation to prevent uptake of loose pellets containing the nuclei.
8. Selection of appropriate 96-well plates depends on the type of PCR machine you used in the following single chromosome amplification step. Note that stackable plates save space when stored at −80 °C freezer.
9. Keeping both the stained chromosome suspension and the sorted chromosomes cold provides better sorting results. The chilled chromosome suspension prevents the flow karyotype from drifting during long sorting procedures. Also, keeping the sorted chromosomes at 4 °C prevents the chromosomal DNA from degradation.
10. On the flow karyotype, chromosomes 9, 10, 11, and 12 tend to cluster together and hard to distinguish. The same phenomena is also observed among chromosomes 14, 15, and 16. Therefore, you can only estimate their corresponding distribution on the flow karyotype. It might be necessary to collect more than six plates for these particular chromosomes.
11. During single chromosome amplification, it is very important to prevent DNA contamination by disinfecting the tissue culture hood, pipettors, PCR cooler, and Microarray Microplate Centrifuge with 70% ethanol. Always use filter tips during this procedure, and change pipet tips between handling each sample even when dispensing reagents.
12. Measure DNA concentration using Qubit dsDNA HS assay kit as follows: First make the Qubit working solution by diluting the Qubit dsDNA HS reagent 1:200 in Qubit dsDNA HS buffer. For samples, 198 μL of Qubit working solution is

mixed with 2 μL of DNA sample in a Qubit assay tube. For standards, 190 μL of Qubit working solution is mixed with 10 μL of Qubit dsDNA HS standard #1 or #2 in the Qubit assay tube. All tubes are mixed by vortexing 2–3 s and incubated at room temperature for 2 min. Finally, DNA concentration is measured by using Qubit 2.0 Fluorometer.

13. In order to determine the optimal length of time for DNA fragmentation, a time course study was conducted using the same amount of DNA (160 ng) under various lengths of time (e.g., 15, 20, 25, and 30 min). The size distribution of DNA fragments is then analyzed using Agilent 2100 Bioanalyzer.

14. DNA purification using AMPure XP magnetic beads: First, add the indicated amount of AMPure XP magnetic beads to the sample in a 1.5 mL DNA LoBind tube. Mix thoroughly on a vortex mixer or by pipetting up and down at least ten times. Incubate the tube at room temperature for 10 min to allow DNA to bind to the beads. Capture the beads by placing the tube on an appropriate magnetic stand at room temperature for 10 min or until the liquid is completely clear. Carefully remove and discard the supernatant without disturbing the beads. Wash the beads twice with freshly made 80% ethanol for at least 30 s to remove contaminants. Allow the beads to dry at room temperature for 10 min, and resuspend the beads thoroughly in the indicated amount of elution buffer. Incubate at room temperature for 2 min to release DNA from the beads. Capture the beads again by placing the tube on the magnetic stand at room temperature for 10 min or until the liquid is completely clear. Recover DNA in the supernatant and transfer to a new 1.5 mL DNA LoBind tube.

15. There are total 96 indexed Adaptor_truS_B primers. Each has a unique barcode [15]. A mixed library with different indexes can be pooled together for Illumina sequencing.

16. Sometimes, the collected fragment size might be a little bit off from the DNA ladder control. Therefore, it is helpful to collect several tubes near the preferred fragment size, which is determined by the length of paired-end sequencing reads you want to perform. Always remember that there are ~53–58 bps of adaptor sequence on both ends of the fragment.

17. PCR cycle number is determined by the amount of input DNA template. For example, five cycles of PCR using 50 ng DNA template will get ~500 ng PCR products, which is enough for Illumina HiSeq sequencing system. Try not to over-amplify DNA to reduce duplicate reads.

18. As PicoPLEX™ WGA introduces a linker sequence into the ends of the amplicons, a portion of the reads might have low diversity in the beginning bases, which affects base calling in the Illumina platform. This can be alleviated by the use of Phix spike-in at 5–10%. Alternatively, an addition of a random

1-mer, 2-mer, 3-mer, and 4-mer, (equally mixed), can be introduced to the beginning of all inserts during library construction (modified from the random nonamer approach used in Yang et al. [11]) to serve as a base-balancing spacer.

Acknowledgment

This work was supported by NIH grant R01HG007834.

References

1. Venter JC (2010) Multiple personal genomes await. Nature 464:676–677
2. Tewhey R, Bansal V, Torkamani A, Topol EJ, Schork NJ (2011) The importance of phase information for human genomics. Nat Rev Genet 12:215–223
3. Browning SR, Browning BL (2011) Haplotype phasing: existing methods and new developments. Nat Rev Genet 12:703–714
4. Glusman G, Cox HC, Roach JC (2014) Whole-genome haplotyping approaches and genomic medicine. Genome Med 6:73–88
5. Kong A, Masson G, Frigge ML, Gylfason A, Zusmanovich P, Thorleifsson G, Olason PI, Ingason A, Steinberg S, Rafnar T, Sulem P, Mouy M, Jonsson F, Thorsteinsdottir U, Gudbjartsson DF, Stefansson H, Stefansson K (2008) Detection of sharing by descent, long-range phasing and haplotype imputation. Nat Genet 40:1068–1075
6. Stephens M, Smith NJ, Donnelly P (2001) A new statistical method for haplotype reconstruction from population data. Am J Hum Genet 68:978–989
7. Roach JC, Glusman G, Hubley R, Montsaroff SZ, Holloway AK, Mauldin DE, Srivastava D, Garg V, Pollard KS, Galas DJ, Hood L, Smit AF (2011) Chromosomal haplotypes by genetic phasing of human families. Am J Hum Genet 89:382–397
8. Roach JC, Glusman G, Smit AF, Huff CD, Hubley R, Shannon PT, Rowen L, Pant KP, Goodman N, Bamshad M, Shendure J, Drmanac R, Jorde LB, Hood L, Galas DJ (2010) Analysis of genetic inheritance in a family quartet by whole-genome sequencing. Science 328:636–639
9. Ma L, Xiao Y, Huang H, Wang QW, Rao WN, Feng Y, Zhang K, Song Q (2010) Direct determination of molecular haplotypes by chromosome microdissection. Nat Methods 7:299–301
10. Fan HC, Wang JB, Potanina A, Quake SR (2010) Whole-genome molecular haplotyping of single cells. Nat Biotechnol 29: 51–57
11. Yang H, Chen X, Wong WH (2011) Completely phased genome sequencing through chromosome sorting. Proc Natl Acad Sci U S A 108:12–17
12. Kaper F, Swamy S, Klotzle B, Munchel S, Cottrell J, Bibikova M, Chuang HY, Kruglyak S, Ronaghi M, Eberle MA, Fan JB (2013) Whole-genome haplotyping by dilution, amplification, and sequencing. Proc Natl Acad Sci U S A 110:5552–5557
13. Kuleshov V, Xie D, Chen R, Pushkarev D, Ma ZH, Blauwkamp T, Kertesz M, Snyder M (2014) Whole-genome haplotyping using long reads and statistical methods. Nat Biotechnol 32:261–266
14. Zong C, Lu S, Chapman AR, Xie XS (2012) Genome-wide detection of single-nucleotide and copy-number variations of a single human cells. Science 338:1622–1626
15. Kozarewa I, Turner DJ (2011) 96-plex molecular barcoding for the Illumina genome analyzer. Methods Mol Biol 733:279–298

Part V

Genome Wide Haplotyping

Chapter 11

Long Fragment Read (LFR) Technology: Cost-Effective, High-Quality Genome-Wide Molecular Haplotyping

Mark A. McElwain, Rebecca Yu Zhang, Radoje Drmanac, and Brock A. Peters

Abstract

In this chapter, we describe Long Fragment Read (LFR) technology, a DNA preprocessing method for genome-wide haplotyping by whole genome sequencing (WGS). The addition of LFR prior to WGS on any high-throughput DNA sequencer (e.g., Complete Genomics Revolocity™, BGISEQ-500, Illumina HiSeq, etc.) enables the assignment of single-nucleotide polymorphisms (SNPs) and other genomic variants onto contigs representing contiguous DNA from a single parent (haplotypes) with N50 lengths of up to ~1 Mb. Importantly, this is achieved independent of any parental sequencing data or knowledge of parental haplotypes. Further, the nature of this method allows for the correction of most amplification, sequencing, and mapping errors, resulting in false-positive error rates as low as 10^{-9}. This method can be employed either manually using hand-held micropipettors or in the preferred, automated manner described below, utilizing liquid-handling robots capable of pipetting in the nanoliter range. Automating the method limits the amount of hands-on time and allows significant reduction in reaction volumes. Further, the cost of LFR, as described in this chapter, is moderate, while it adds invaluable whole genome haplotype data to almost any WGS process.

Key words Long fragment, Whole-genome amplification (WGA), Haplotype, Phasing, Next-generation sequencing (NGS), Molecular barcode

1 Introduction

Currently, almost all WGS technologies result in short reads (<500 bases) and typically cannot "phase" variants, that is, determine the order of variants that are inherited as contiguous blocks from each parent (haplotypes) [1]. In humans, accurate haplotype information is critical to the emerging field of clinical whole genome sequencing (WGS). This can be demonstrated by a hypothetical

---

Electronic supplementary material: The online version of this chapter (doi:10.1007/978-1-4939-6750-6_11) contains supplementary material, which is available to authorized users.

Irene Tiemann-Boege and Andrea Betancourt (eds.), *Haplotyping: Methods and Protocols*, Methods in Molecular Biology, vol. 1551, DOI 10.1007/978-1-4939-6750-6_11, © Springer Science+Business Media LLC 2017

example in which a patient's genome carries two deleterious mutations at different loci in a single gene. Two combinations of alleles are possible and are associated with very different clinical outcomes: in the first, the patient carries both mutations on one allele and thus also has a wild-type copy of the gene, which may be sufficient for healthy functioning. In the second scenario, both copies each contain one mutation resulting in a protein with reduced or complete loss of function, which is more likely to lead to a disease state. This important information is, to date, lacking from most WGS data, despite the rapid advances in technology development over the past 10 years.

Traditional methods for incorporating haplotype information into WGS data rely on additional information from parental sequencing and/or population-based inference, but de novo mutations and, in the case of population inference, rare inherited SNPs, cannot be phased with these techniques [2]. More recent approaches, which do not rely on external data, are either variants of our technique, or are based on metaphase chromosome separation, fosmid cloning, or proximity ligation. These methods have been described in detail [3–7] and have advantages and disadvantages, but a detailed discussion is beyond the scope of this chapter.

Our Long Fragment Read (LFR) technology [8, 9], as outlined below and illustrated in Fig. 1, is an efficient and inexpensive process for generating extremely high-quality long haplotypes. The method depends upon first diluting long-fragment genomic DNA to a low enough concentration such that it can be aliquoted across a 384-well plate at approximately 0.1 haploid genomes per well. At this dilution factor, there is a 10% probability of any particular locus being aliquoted into any particular well (resulting in ~38 wells per locus). The probability of two fragments of any locus being co-aliquoted into any well is $10\% \times 10\% = 1\%$ (resulting in ~4 wells carrying two fragments). Half of the these wells will remain informative, as the two fragments will be both maternally inherited or both paternally inherited, leaving only two "uninformative" wells carrying both maternally and paternally inherited fragments from any locus. Therefore, on average, ~36 wells (38 minus 2) carry haplotype information for each locus, with ~18 wells carrying only maternally inherited fragments and ~18 wells carrying only paternally inherited fragments. The diluted genomic DNA is then amplified, further fragmented, ligated to 384 barcoded adapters, and finally subjected to WGS to generate haplotyped sequence data.

We have previously demonstrated some of the advantages of our method compared to other WGS and haplotyping methods. First, the technique preserves very long physical DNA fragments during extraction, aliquoting, and low-bias whole genome amplification, resulting in contigs with N50 lengths of ~500–1000 kb. Second, aliquots are separately amplified, barcoded, and then further amplified in bulk, providing information that can be used to

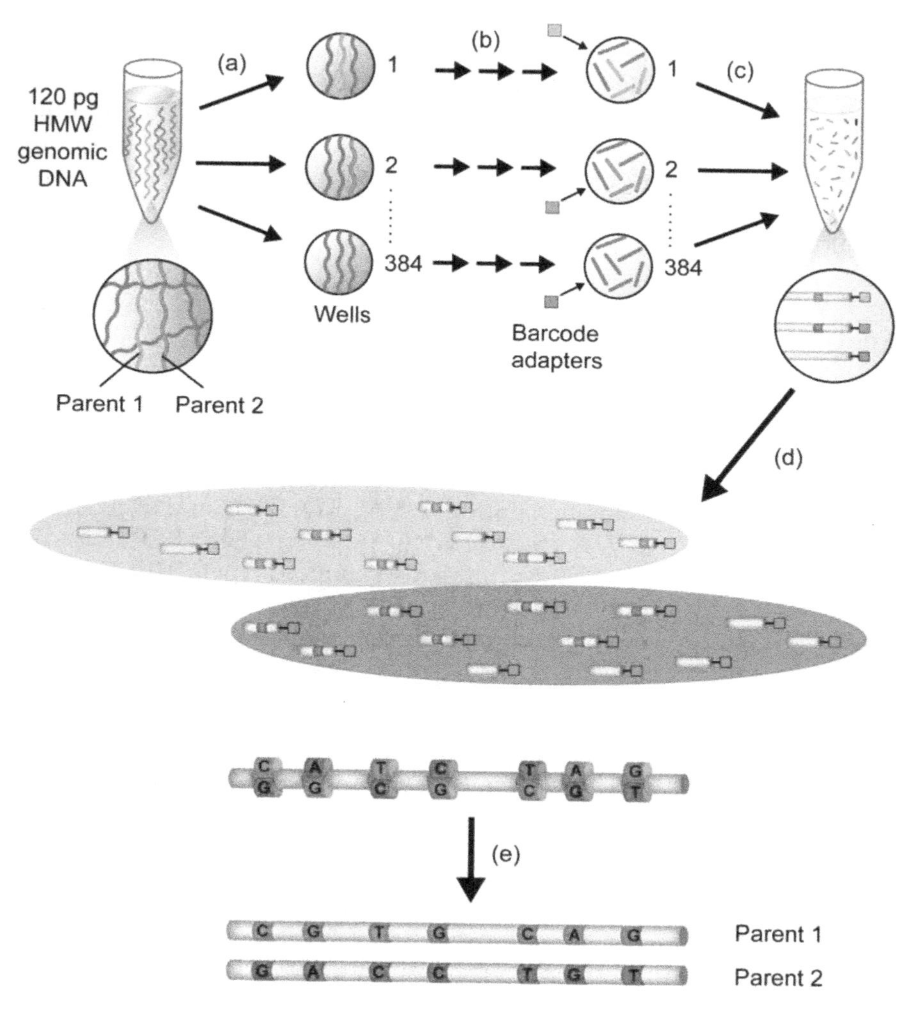

Fig. 1 The LFR technology. An overview of the LFR technology and controlled random enzymatic fragmenting is shown. (**a**) First, 100–130 pg of high molecular mass (HMM) DNA is physically separated into 384 distinct wells; (**b**) through several steps, all within the same well without intervening purifications, the genomic DNA is amplified, fragmented, and ligated to unique barcode adapters; (**c**) all 384 wells are combined, purified, and sequenced by any WGS platform; (**d**) mate-paired reads are mapped to the genome and barcode sequences are used to group tags into haplotype contigs; and (**e**) the final result is a diploid genome sequence. This figure is reproduced from Ref. [9]

computationally correct amplification errors [9, 10]. This is because an LFR library is generated from the amount of DNA contained in ~10 cells, so if a variant is associated with only one or two barcodes, it is likely an error and the reads associated with the remaining ~40 barcodes (resulting from denaturing and sequencing DNA from ~10 diploid cells) likely represents the true sequence. Third, our method can efficiently identify de novo mutations and assign them to parental haplotypes [10, 11]—impossible to do in most cases using approaches that process DNA in bulk and rely on parental sequence data for haplotyping. Fourth, our process is efficient and streamlined due to our ability to perform five consecutive enzymatic reactions in the same multi-well plate without intervening DNA purifications. Lastly, a unique advantage of our method is the ability to generate high-quality, phased WGS data from as few as five human cells, which is not possible with fosmid cloning, proximity ligation, or chromosome separation.

2 Materials

2.1 Reagents

1. Highly pure, high-molecular weight genomic DNA: Purified by Agilent RecoverEase DNA Isolation kit (or similar), 5 ng at 1 ng/μL.
2. Denature buffer: 1.6 M KOH, 10 mM EDTA. Add 50 μL EDTA (500 mM) to 224 mg pellet KOH. Add deionized H_2O to 2.5 mL total volume. Prepare 100 μL single-use aliquots and store at −20 °C (*see* **Note 1**). Expiration date is 1 month from manufacture.
3. Phosphorothioate-protected random 8mer. Oligonucleotide is resuspended at 1 mM in deionized H_2O (*see* **Note 2**).
4. 10× phi29 reaction buffer: 500 mM Tris–HCl pH 7.5, 100 mM $MgCl_2$, 100 mM $(NH_4)_2SO_4$, 40 mM DTT.
5. dNTP mix (25 mM total).
6. dUTP (1 mM).
7. phi29 polymerase (Enzymatics, 10 U/μL).
8. Uracil-DNA Glycosylase, Heat-Labile (UDG), (Enzymatics, 1 U/μL).
9. Human apurinic/apyrimidinic endonuclease 1 (APE1), (NEB, 10 U/μL).
10. DNA Polymerase I (*E. coli*) (NEB, 10 U/μL).
11. Recombinant Shrimp Alkaline Phosphatase (rSAP), (USB, 1 U/μL).
12. 384 uniquely barcoded sequencing adapters appropriate for your sequencing platform (stored at 5 μM in 10 mM Tris-HCl, 0.1 mM EDTA, pH 8.0—*see* **Note 3**).
13. T4 ligase (Enzymatics, 600 U/μL).
14. 10× ligase buffer: 300 mM Tris–HCl (pH 7.8), 60 mM $MgCl_2$, 25% PEG8000, 10 mM ATP.
15. AMPure XP beads (Agencourt).
16. 0.5% solution of Tween-20 (expires 1 week after dilution to 0.5%) made fresh from a 10% Tween-20 stock.

17. 70% ethanol.
18. TE: 10 mM Tris–HCl, 0.1 mM EDTA, pH 8.0.
19. Molecular biology-grade deionized water.

2.2 Plasticware and Instrumentation

1. 96 well full skirt PCR microplates (Axygen)—for automated liquid handling setup.
2. 384-well PCR plates (Armadillo; Thermo Scientific)—*see* **Note 4**.
3. Automated Microplate Heat Sealer (Thermo Scientific ALPS 3000 or similar)—*see* **Note 5**.
4. Liquid handling workstation, equipped with tip cutter (Hamilton Microlab Star).
5. Liquid handling system (TTP Mosquito HTS or similar)—*see* **Note 6**.
6. Liquid handling system (Formulatrix Tempest or similar)—*see* **Note 7**.
7. KingFisher Flex Magnetic Particle Processor (Thermo Scientific).
8. 96-well Assay Block, V-bottom, 2 mL (Costar)—for manual liquid handling.
9. 96-Well Polypropylene Microplates, round bottom, 355 μL (Greiner).
10. MPS 1000 Mini Plate Spinner (Labnet).
11. Reagent reservoirs.

3 Methods

Set up all reactions at room temperature. Use wide-bore pipet tips for all genomic DNA pipetting steps (*see* **Note 8**). The steps listed below are described as they are currently performed at Complete Genomics, Inc., using robotic liquid handling instruments. Alternatively, one could hand-pipet all reagents if desired (*see* **Note 9**). These steps describe batch processing of 24 DNA samples in parallel with automated pipetting, and eight samples in parallel with manual pipetting. The robotic liquid handling steps in Subheadings 3.1 and 3.3 (Reaction plate buffer preseed and multiple displacement amplification) are the rate limiting steps; therefore, they are performed in three partially overlapping "sub-batches" of eight samples. Each "sub-batch" can be offset by 3–5 min.

3.1 Reaction Plate Buffer Preseed (Automated Pipetting Method)

If performing the process manually, skip this step.

1. For each batch of 24 plates, mix 1 mL 10× phi29 reaction buffer with 9 mL molecular biology-grade deionized water for a final buffer concentration of 1×.

2. Use liquid handling system to dispense 0.6 μL 1× phi29 reaction buffer into each well of twenty-four 384-well PCR plates (*see* **Note 7**).
3. Centrifuge plates for 20 s in the Mini Plate Spinner, and then place upside-down in a plastic bag to prevent excessive evaporation.

3.2 Cell Lysis/ Denaturation

1. If starting with cells, rather than bulk isolated genomic DNA, set up 96-well full skirt PCR microplates (or 96-well Assay Block,V-bottom if performing steps manually) for cell lysis and DNA denaturation: add 10 cells in a 10 μL volume to each well of column 3. Add 30 μL (or 1 mL if performing steps manually) of molecular biology-grade water to each well of column 4.
2. Add 1 μL un-diluted denature buffer to the side of each well of column 3 and tap plate to add droplet to cell suspension. Incubate 2 min.
3. Add 20 μL (or 47.8 μL if performing steps manually) random 8mer (1 mM) to each denatured cell sample in column 2. Incubate 2 min.
4. Transfer 19 μL (or 736 μL if performing steps manually) molecular biology-grade water from column 4 to column 3.
5. If performing steps with robotic liquid handling, use multichannel pipettor to transfer denatured DNA/random 8mer mixtures from column 3 of the 96-well full skirt PCR microplate into odd-numbered columns of a 384-well PCR plate at 3 μL/well (Fig. 2). Centrifuge plates for 20 s in Mini Plate Spinner and proceed to **step 6** of Subheading 3.3.
6. If performing steps manually, proceed to **step 6** of Subheading 3.4.

3.3 Multiple Displacement Amplification (MDA) of Long-Fragment Genomic DNA (Automated Pipetting Method)

1. Set up the 96-well full skirt PCR microplate for DNA denaturation and primer addition: mix 100 μL denature buffer with 900 μL molecular biology-grade water and aliquot 110 μL per well in column 1. Add 5 μL each DNA sample (1 ng/μL) to each well in column 2 (i.e., 1 sample per row). Add 26 μL random 8mer (1 mM) to each well of column 3. Add 60 μL molecular biology-grade water to each well of column 4.
2. Use the liquid handling workstation to transfer 95 μL diluted denature buffer to each well containing DNA sample (*see* **Note 10**). Mix twice at a flow rate of 5 μL/s. Incubate 2 min.
3. Use the liquid handling workstation to transfer 7 μL denatured DNA from column 2 to column 3 (*see* **Note 10**). Do not mix. Incubate 2 min.
4. Use the liquid handling workstation to transfer 32 μL water from column 4 to column 3 (*see* **Note 10**). Mix three times at a flow rate of 5 μL/s.

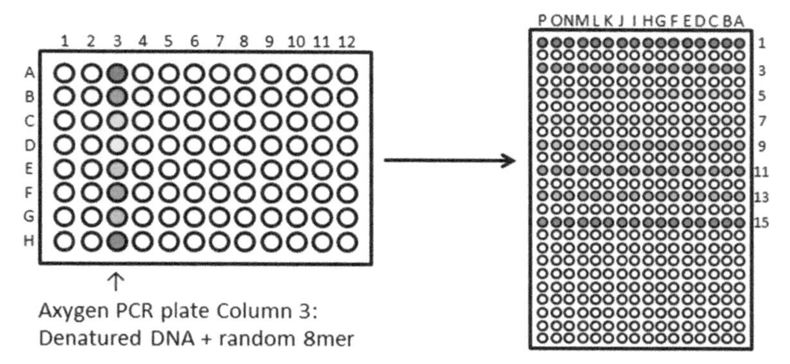

Fig. 2 Setup for aliquoting genomic DNA/8mer mixtures. After denaturing genomic DNA and mixing with 8mers, a liquid handling workstation is used to aliquot the DNA/8mer mixture from a 96-well Axygen PCR plate into odd-numbered columns of a 384-well Thermo Scientific Armadillo plate. Each well in column 3 of the Axygen plate represents a different DNA sample. Sample A03 is aliquoted into each well of column 1 of the Armadillo plate, sample B03 into column 3, etc. This plate will in turn be used as a source plate for the TTP Mosquito to aliquot 0.1 μL of DNA/8mer into each well of the reaction plate

5. Set up DNA aliquoting source plate: use the liquid handling workstation to transfer denatured DNA/random 8mer mixtures from column 3 of the 96-well full skirt PCR microplate into odd-numbered columns of a 384-well PCR plate at 3.5 μL/well. DNA sample 1 (well A03 of 96-well full skirt PCR microplate) will be transferred to column 1 of 384-well PCR plate, DNA sample 2 (well B03 of the 96-well full skirt PCR microplate) to column 3 of the 384-well PCR plate, etc. (*see* **Note 10**) (Fig. 2). Centrifuge the 384-well PCR plate 20 s in Mini Plate Spinner.
6. Aliquoting DNA into reaction plates: Use liquid handling system (*see* **Note 6**) to transfer 0.1 μL denatured DNA/8mer mix from source plate (384-well PCR plate generated in **step 5** of Subheading 3.3) into each well of the destination/reaction plates (384-well PCR plates generated in Subheading 3.1).
7. While the DNA aliquoting on the liquid handling system is ongoing, prepare MDA reagent: mix 2.5 mL molecular biology-grade deionized water, 493.1 μL 10× phi29 reaction buffer, 164.4 μL dNTP mix (25 mM), and 137.0 μL dUTP (1 mM). Just before proceeding with the next step, add 205.4 μL phi29 polymerase (10 U/μL).
8. Use liquid handling system to add 0.2 μL MDA reagent to each well of reaction plates generated in **step 6** of Subheading 3.2. Centrifuge reaction plates 20 s in Mini Plate Spinner.

9. Seal plates. When using an automated Microplate Heat Sealer use the following settings: 134 °C for 3.5 s (*see* **Note 11**).
10. Incubate reaction plates UPSIDE DOWN 45 min in a 37 °C incubator and 5 min in a 70 °C incubator (*see* **Note 12**).
11. Allow plates to cool to room temperature and centrifuge briefly (pulse to 160 × *g*).
12. If not proceeding immediately with subsequent processing, place plates UPSIDE DOWN at −20 °C.

3.4 Multiple Displacement Amplification (MDA) of Long-Fragment Genomic DNA (Hand Pipetting Method)

1. Set up 96-well Assay Block, V-bottom for denaturation and primer addition: mix 100 μL denature buffer with 900 μL molecular biology-grade water and aliquot 110 μL per well in column 1. Add 5 μL each DNA sample (1 ng/μL) to each well in column 2 (i.e., 1 sample per row). Add 51 μL random 8mer (1 mM) to each well of column 3. Add 1 mL molecular biology-grade water to each well of column 4.
2. Using wide-bore pipet tips, transfer 95 μL diluted denature buffer to each well containing DNA sample. Mix twice at a flow rate of 5 μL/s. Incubate 2 min.
3. Using wide-bore pipet tips, transfer 4.57 μL denatured DNA from column 2 to column 3. Do not mix. Incubate 2 min.
4. Using wide-bore pipet tips, transfer 794 μL water from column 4 to column 3. Mix three times at a flow rate of 5 μL/s.
5. Set up DNA aliquoting source plate: using wide-bore pipet tips, transfer 67 μL denatured DNA/random 8mer mixtures from column 3 of the 96-well Assay Block, V-bottom to each well of a row in a 96-well full skirt PCR microplate.
6. Aliquoting DNA into reaction plates: Use wide-bore pipet tips and a 12-channel pipettor to transfer 2 μL denatured DNA/8mer mix from source vessel (96-well full skirt PCR microplate generated in **step 5** of Subheading 3.4 or reagent reservoir filled with DNA solution generated in **step 6** of Subheading 3.2) into each well of the reaction plate (384-well PCR plate).
7. Prepare MDA reagent: mix 2.2 mL molecular biology-grade deionized water, 1106 μL 10× phi29 reaction buffer, 110.6 μL dNTP mix (25 mM), and 92.2 μL dUTP (1 mM). Just before proceeding with the next step, add 138.2 μL phi29 polymerase (10 U/μL).
8. Use 12-channel pipettor to add 1 μL MDA reagent to each well of reaction plates generated in **step 6** of Subheading 3.4. Centrifuge reaction plates 20 s in Mini Plate Spinner.
9. Seal plates using Microplate Heat Sealer (settings: 134 °C for 3.5 s) (*see* **Note 11**).
10. Incubate reaction plates UPSIDE DOWN 45 min in a 37 °C incubator and 5 min in a 70 °C incubator (*see* **Note 12**).

11. Allow plates to cool to room temperature and centrifuge briefly (pulse to 1000 rpm).
12. If not proceeding immediately with subsequent processing, place plates UPSIDE DOWN at −20 °C.

3.5 Fragmentation Step 1: Uracil Excision

1. If needed, thaw reaction plates and centrifuge briefly (pulse to 1000 rpm).
2. Prepare UDG/APE1 reagent according to Table 1: Uracil excision recipes.
3. Peel sealing film off reaction plates.
4. Use the liquid handling system to add 0.2 μL (or a multi-channel pipettor to add 1.0 μL) of UDG/APE1 reagent per well to each well of 24 reaction plates (or eight plates, if hand-pipetting). Centrifuge plates 20 s in Mini Plate Spinner.
5. Seal plates using Microplate Heat Sealer (settings: 134 °C for 3.5 s) (*see* **Note 11**).
6. Incubate reaction plates UPSIDE DOWN 2 h in a 37 °C incubator and 15 min in a 70 °C incubator (*see* **Note 12**).
7. Allow plates to cool to room temperature and centrifuge briefly (pulse to 1000 rpm).
8. If not proceeding immediately with subsequent processing, place plates UPSIDE DOWN at −20 °C.

3.6 Fragmentation Step 2: Nick Translation

1. If needed, thaw reaction plates and centrifuge briefly (pulse to 1000 rpm).
2. Prepare nick translation reagent according to Table 2: Nick translation recipes.
3. Peel sealing film off reaction plates.
4. Use liquid handling system to add 0.2 μL (or multi-channel pipettor to add 1.0 μL) of nick translation reagent per well to each well of 24 reaction plates (or 8 plates, if hand-pipetting). Centrifuge plates 20 s in Mini Plate Spinner.

Table 1
Uracil excision recipes

Robotic		Manual
2.4 mL	dH_2O	3.1 mL
291.8 μL	10× phi29 reaction buffer	368.6 μL
227.6 μL	UDG (1 U/μL)	157.1 μL
45.6 μL	APE1 (10 U/μL)	41.9 μL
0.2	Vol. added per well (μL)	1.0

Table 2
Nick translation recipes

Robotic		Manual
2.2 mL	dH_2O	3.2 mL
259.6 μL	10× phi29 reaction buffer	368.6 μL
103.8 μL	DNA Pol1 (10 U/μL)	114.5 μL
0.2	Vol. added per well (μL)	1.0

Table 3
Dephosphorylation recipes

Robotic		Manual
2.3 mL	dH_2O	2.9 mL
299.5 μL	10× phi29 reaction buffer	368.6 μL
374.4 μL	rSAP (1 U/μL)	368.6 μL
0.2	Vol. added per well (μL)	1.0

5. Seal plates using Microplate Heat Sealer (settings: 134 °C for 3.5 s) (*see* **Note 11**).
6. Incubate reaction plates UPSIDE DOWN 1 h at room temperature and 15 min in a 70 °C incubator (*see* **Note 12**).
7. Allow plates to cool to room temperature and centrifuge briefly (pulse to 1000 rpm).
8. If not proceeding immediately with subsequent processing, place plates UPSIDE DOWN at −20 °C.

3.7 Dephosphorylation

1. If needed, thaw reaction plates and centrifuge briefly (pulse to 1000 rpm).
2. Prepare dephosphorylation reagent according to Table 3: Dephosphorylation recipes.
3. Peel sealing film off reaction plates.
4. Use the liquid handling system to add 0.2 μL (or multi-channel pipettor to add 1.0 μL) of dephosphorylation reagent per well to each well of 24 reaction plates (or 8 plates, if hand-pipetting). Centrifuge plates 20 s in Mini Plate Spinner.
5. Seal plates using the Microplate Heat Sealer (settings: 134 °C for 3.5 s) (*see* **Note 11**).
6. Incubate reaction plates UPSIDE DOWN 1 h in a 37 °C incubator and 15 min in a 70 °C incubator (*see* **Note 12**).

7. Allow plates to cool to room temperature and centrifuge briefly (pulse to 1000 rpm).
8. If not proceeding immediately with subsequent processing, place plates UPSIDE DOWN at −20 °C.

3.8 Barcoded Adapter Ligation and Pooling

1. Dilute 5 μM sequencing adapters in molecular biology-grade water to 1 μM final concentration.
2. Peel sealing film off reaction plates.
3. Use the liquid handling workstation to add 2.5 μL sequencing adapters to each well of reaction plate (one unique barcode per well).
4. Prepare ligation reagent according to Table 4: Ligation recipes.
5. Use Formulatrix Tempest to add 1 μL (or multi-channel pipettor to add 1.5 μL) of ligation reagent to each well of 24 reaction plates (or 8 plates, if hand-pipetting). Centrifuge plates 20 s in Labnet MPS 1000 Mini Plate Spinner.
6. Seal plates using Microplate Heat Sealer (settings: 134 °C for 3.5 s)
7. Incubate reaction plates right side up for 2 h at room temperature.
8. Centrifuge plates in a benchtop centrifuge at 650 × g for 1 min.
9. Use the liquid handling workstation to pool reaction products from each of the 384 wells of a reaction plate into one well of a 96-well deep-well plate (i.e., contents of 24 reaction plates [1 batch] are transferred to 24 wells of a deep-well plate). Alternatively, if hand-pipetting, use a multi-channel pipettor to pool into a reagent reservoir; or invert over a collection vessel and centrifuge.
10. Use 1× ligase buffer to normalize volumes of pooled samples to 1400 μL per well.

3.9 Sample Purification

1. Prepare beads: mix 35.8 mL AMPure XP beads and 25.2 μL 10% Tween-20.
2. Prepare wash buffer (70% ethanol) and wash plates: mix 140 mL 200 proof ethanol and 60 mL deionized water. Aliquot 950 μL per well in two 96-well Assay Blocks, V-bottom.

Table 4
Ligation recipes

Robotic		Manual
4.9 mL	dH_2O	1.25 mL
4.7 mL	10× Ligase buffer	3.7 mL
833.2 μL	T4 Ligase (600 U/μL)	593.3 μL
1.0	Vol. added per well (μL)	1.5

3. Prepare elution buffer and elution plates: mix 5 mL TE and 10 μL 0.5% Tween-20. Aliquot 45 μL elution buffer per well in a 96-Well Polypropylene Microplate, round bottom.
4. Split 1400 μL of pooled ligation products into four 96-well Assay Blocks, V-bottom (350 μL per well per plate).
5. Add 350 μL AMPure bead/Tween-20 mixture, prepared in **step 1** of Subheading 3.9, to each well.
6. Perform automated AMPure bead purification, using the Magnetic Particle Processor. General parameters are:
   - Use comb to mix beads: 1 min fast mixing each plate, followed by 3 min medium-speed mixing each plate, and finally one more round of 3 min of medium-speed mixing of each plate.
   - Collect DNA-bound beads from each plate.
   - Wash twice in 70% ethanol plates: 1 min of slow mixing in the first plate and 1 min of medium-speed mixing in the second plate.
   - Air-dry beads for 3 min.
   - Release beads into elution plate; mix slowly for 10 min and then collect beads.

3.10 PCR and Further Preparation of Sequencing Libraries

The remaining steps for library preparation and sequencing depend greatly on the sequencing platform. On Complete Genomics' platform, these library prep steps for processing an LFR library are identical to a standard, non-haplotyped workflow.

4 Notes

1. There is a concern that dissolution of atmospheric CO_2 in the denature buffer could decrease the pH and render the alkaline denature step incomplete. Thus, we consider 100 μL aliquots to be single-use and try to avoid uncapping the tubes more than once.
2. Two phosphorothioate bonds are incorporated at the 3′ end to protect the oligo from the strong 3′ exonuclease activity of Phi29 polymerase. For example, when ordering from Integrated DNA Technologies, the sequence should be entered as NNNNNN*N*N.
3. The oligonucleotide sequences for the adapters we use on Complete Genomics' platform are as follows (all sequences are 5′ – 3′):

   Ad1 Top Strand (X represents specific base in barcode):

   /5Phos/ACTGCTGACGTACTGXXXXXXXXXXAGCACG AGACGTTCTCGACA

Full list of 384 barcoded adapters available at (Supplementary Material).

Ad1 Bottom Strand:

TACGTCAGCAG/3dT-Q/

Ad2 Top Strand:

/5phos/TCTGCTGAGTCGAGAACGT/3ddC/

Ad2 Bottom Strand:

CGACTCAGCAG/3dA-Q/

4. Other plates may be acceptable alternatives, for example Bio-Rad Hard-Shell 384-well plates. The most important criterion is that the plate must not warp during multiple cycles from room temperature to 37–65 °C; the plate needs to remain flat for accurate pipetting with multi-channel liquid handling robots. However, we have further observed differences between well plate manufacturers with respect to amplification yield and genome bias. Our experiences suggest the Armadillo plates best suit our needs.
5. Whatever plate sealing system is used, it is important to use sealing conditions that allow a plate to tolerate five cycles of heat sealing and peeling.
6. We strongly recommend the TTP Mosquito HTS for this step, as it shears the DNA less than other systems we have investigated. We use this instrument only to aliquot DNA. The most important criteria for choosing an instrument for this task are that it does not mechanically fragment the long DNA fragments, and has a capability for 100 nL-range pipetting.
7. This instrument is used for dispensing most of the reaction buffers. It must be able to reproducibly dispense volumes in the 200–1000 nL range, and handle solutions of varying viscosities. Take note that, in this chapter, we have reported reagent volumes that vary with reagent types due to idiosyncrasies in our laboratory setup. Therefore, an operator wishing to translate this process into their own lab may need to reoptimize dead volumes and overages. We dedicate specific channels on the Formulatrix Tempest to each pipetting step. Lastly, we engineer wash steps into the end of each pipetting program, in which the channels are washed with 10 mL of molecular biology-grade deionized water.
8. Long genomic DNA fragments can shear if aspirated and dispensed through typical narrow-bore pipet tips. It is not necessary to purchase wide bore tips; rather, we have experienced minimal DNA shearing when using a razor blade to manually cut off tips until the bore diameter is roughly 1 mm or larger. However, it is critical to aspirate and dispense slowly to reduce shearing even with wide-bore tips.

9. Pipetting all reagents by hand is acceptable, provided all volumes are scaled up appropriately to reduce pipetting error. We use reagent reservoirs and multi-channel pipettors to add reagents to the reaction plates for slightly higher throughput. Performing the MDA reaction in a larger volume may require optimization of incubation time and other parameters to prevent over-amplification. In our experience, genome bias increases rapidly with higher amplification levels. This process is further adaptable to a hand-pipetting in a 96-well plate format rather than 384-well plates, provided no more than 20 haploid genome equivalents are used.
10. Any steps in which the liquid handling workstation pipets genomic DNA should use pipet tips cut to approximately a 1 mm bore size to reduce DNA shearing. Hamilton sells an automated tip cutter that we incorporate into our workflow, cutting the tips 5 mm from the end.
11. Do not stack plates until sealing film has cooled to room temperature. Heat from the film can easily inactivate the heat labile enzymes used in this process; phi29 is especially sensitive to heat inactivation.
12. Make sure NOT to stack the plates in the incubator, to ensure heat transfer and air circulation are optimal. The operator may find it useful to pause after adding reagents to 8 or 12 plates to place completed plates in the incubator, and process the remaining one or two "sub-batches" at 3-5 min intervals.

Acknowledgments

We would like to acknowledge the ongoing contributions and support of all Complete Genomics employees, in particular the many highly skilled individuals who work in the libraries, reagents, and sequencing groups that make it possible to generate high-quality whole genome data. Specifically, we would like to thank Robert Chin, Ramya Srinivasan, Daniel Hayden, and Joseph Peterson for help in improving the conditions and automation processes of this protocol.

References

1. McKernan KJ, Peckham HE, Costa GL et al (2009) Sequence and structural variation in a human genome uncovered by short-read, massively parallel ligation sequencing using two-base encoding. Genome Res 19(9): 1527–1541
2. Abecasis GR, Auton A, Genomes Project C et al (2012) An integrated map of genetic variation from 1,092 human genomes. Nature 491(7422):56–65
3. Kitzman JO, Mackenzie AP, Adey A et al (2011) Haplotype-resolved genome sequencing of a Gujarati Indian individual. Nat Biotechnol 29(1):59–63
4. Suk EK, McEwen GK, Duitama J et al (2011) A comprehensively molecular haplotype-resolved

genome of a European individual. Genome Res 21(10):1672–1685

5. Selvaraj S, R Dixon J, Bansal V et al (2013) Whole-genome haplotype reconstruction using proximity-ligation and shotgun sequencing. Nat Biotechnol 31(12):1111–1118
6. Kuleshov V, Xie D, Chen R et al (2014) Whole-genome haplotyping using long reads and statistical methods. Nat Biotechnol 32(3):261–266
7. Amini S, Pushkarev D, Christiansen L et al (2014) Haplotype-resolved whole-genome sequencing by contiguity-preserving transposition and combinatorial indexing. Nat Genet 46(12):1343–1349
8. Drmanac R (2006) Nucleic acid analysis by random mixtures of non-overlapping fragments. United States Patent WO 2006/138284 A2, 2006
9. Peters BA, Kermani BG, Sparks AB et al (2012) Accurate whole-genome sequencing and haplotyping from 10 to 20 human cells. Nature 487(7406):190–195
10. Peters BA, Kermani BG, Alferov O et al (2015) Detection and phasing of single base de novo mutations in biopsies from human in vitro fertilized embryos by advanced whole-genome sequencing. Genome Res 25(3): 426–434
11. Schaaf CP, Gonzalez-Garay ML, Xia F et al (2013) Truncating mutations of MAGEL2 cause Prader-Willi phenotypes and autism. Nat Genet 45(11):1405–1408

Chapter 12

Contiguity-Preserving Transposition Sequencing (CPT-Seq) for Genome-Wide Haplotyping, Assembly, and Single-Cell ATAC-Seq

Lena Christiansen, Sasan Amini, Fan Zhang, Mostafa Ronaghi, Kevin L. Gunderson, and Frank J. Steemers

Abstract

Most genomes to date have been sequenced without taking into account the diploid nature of the genome. However, the distribution of variants on each individual chromosome can (1) significantly impact gene regulation and protein function, (2) have important implications for analyses of population history and medical genetics, and (3) be of great value for accurate interpretation of medically relevant genetic variation. Here, we describe a comprehensive and detailed protocol for an ultra fast (<3 h library preparation), cost-effective, and scalable haplotyping method, named Contiguity Preserving Transposition sequencing or CPT-seq (Amini et al., Nat Genet 46(12):1343–1349, 2014). CPT-seq accurately phases >95 % of the whole human genome in Mb-scale phasing blocks. Additionally, the same workflow can be used to aid de novo assembly (Adey et al., Genome Res 24(12):2041–2049, 2014), detect structural variants, and perform single cell ATAC-seq analysis (Cusanovich et al., Science 348(6237):910–914, 2015).

Key words Haplotyping, Phasing, Human genome, Contiguity preserving transposition, Combinatorial indexing, CPT-seq, Assembly, Single cell ATAC-seq

1 Introduction

The rapid evolution of next-generation sequencing technologies [1] in both throughput and accuracy has allowed an expansion of many important applications ranging from population sequencing, fetal and cancer genome analysis, and metagenomics. Standard next-generation library preparation methods typically fragment the genome into several hundred base long library elements, effectively removing long-range genomic and haplotyping information [2]. As such, most genomes sequenced to date use methods that do not account for the diploid nature of the genome. However, the distribution of variations on each individual chromosome can have a significant impact on gene regulation, protein function, and clinical manifestations of disease. Haplotype-resolved genome sequencing,

Irene Tiemann-Boege and Andrea Betancourt (eds.), *Haplotyping: Methods and Protocols*, Methods in Molecular Biology, vol. 1551, DOI 10.1007/978-1-4939-6750-6_12, © Springer Science+Business Media LLC 2017

either experimentally or computationally, or a combination of the two, could greatly improve the accuracy of interpretation of medically relevant genetic variants [2].

Here, we describe an experimental haplotyping approach using a novel method of library creation. This method retains long-range information during the fragmentation process by the use of Contiguity Preserving Transposition sequencing (CPT-seq). The fundamental approach to deriving haplotype information from a genomic sample is by diluting intact genomic DNA (gDNA) to sub-haploid representation across a large number of compartments, and subsequently creating indexed libraries from the individual compartments [3]. The CPT-seq method accomplishes this compartmentalization and indexing using two key innovations: contiguity preserving transposition and combinatorial indexing. The first innovation relies on transposition to tagment gDNA, allowing short-insert sequencing libraries to be created while at the same time tracking library elements that originate from the same contiguous piece of DNA. The second key innovation is the introduction of a combinatorial indexing scheme wherein thousands of virtual compartments can be created from a limited number of physical compartments. This combinatorial indexing occurs at two different steps, with the first occurring during the initial tagmentation event, and the second during PCR amplification of individual physical compartments.

The assay starts with unsheared DNA purified from the sample, and then continues with two tiers of indexing carried out in 96-well plates. The first round comprises a 96-plex indexed transposition, in which indices from transposomes—barcoded DNA molecules recognized and inserted by the transposase—introduced into gDNA. The tagmented gDNA, still physically associated with the transposase, is then pooled, diluted, and randomly split into a second 96-well plate such that <1 copy of each genomic region is contained in each of the 96 wells. Each contiguous piece of DNA from the first tegmentation step will then be moved intact into the second 96-well plate, but usually to a different well than other molecules containing the homologous region. As a result, each of the 96 physical compartments now contains fractions of the genome in the form of long transposase-indexed DNA molecules. The transposon-tagged DNA is then fragmented by the protein denaturant SDS (Sodium Dodecyl Sulfate) through the release of the transposase, resulting in each well enriched for neighboring pieces of DNA. Subsequently, the fragments are further indexed by 96-plex PCR leading to a total index space of $96 \times 96 = 9216$ combinatoric indices (Fig. 1), and sequencing libraries are prepared from this material. Sequencing reads with the same index share genomic neighborhood information and can be assembled into the long continuous blocks to infer haplotype information or facilitate genome assembly [3, 4]. We have demonstrated that with this

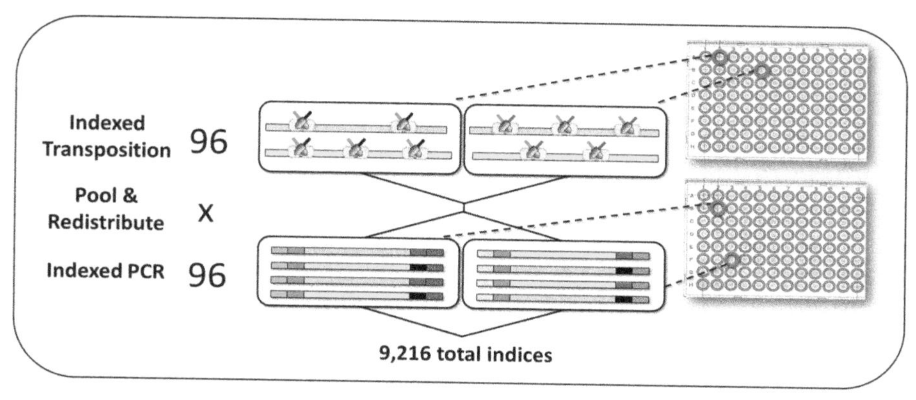

Fig. 1 CPT-seq utilizes two tiers of indexing carried out in 96-well plates. The first round comprises a 96-plex indexed transposition, in which the indices from the transposomes are introduced into the DNA molecule. After pooling, dilution, and randomly splitting into the second 96-well plate, the transposed DNA are fragmented by SDS through the release of the transposase, resulting in neighboring-enriched libraries per PCR compartment. The fragments are further indexed by 96-plex PCR leading to a total index space of 96 × 96 = 9216 combinatoric indices

rapid and scalable workflow, the human genome can be assembled into long, accurate Mb-scale haplotype blocks with >95 % heterozygous variants covered [3]. Additionally, the current protocol supports de novo assembly [4] and single-cell ATAC-seq applications [5]. Here, we describe a comprehensive and detailed protocol for CPT-seq.

2 Materials

Prepare all solutions using Super-Q purified water and store at room temperature (unless indicated otherwise).

2.1 Genomic DNA Quality Assessment

1. 1× TE buffer: 10 mM Tris–HCl, 1 mM EDTA, pH 8.0.
2. 0.5× pulsed field gel electrophoresis Tris-Borate EDTA (TBE) buffer: 5 mM Tris-Borate, 0.5 mM EDTA, pH 8.0.
3. Certified Megabase Agarose.
4. CHEF-DR II Chiller System for pulsed field gel electrophoresis. Instruction manuel catalog number M1703729, http://www.bio-rad.com/webroot/web/pdf/lsr/literature/M1703729B.pdf.
5. GeneRuler High Range DNA Ladder (10,171 bp–48,502 bp).
6. 6× DNA loading dye: 10 mM Tris–HCl; pH 7.6, 0.03 % bromophenol blue, 0.03% xylene cyanol FF, 60 % glycerol, and 60 mM EDTA.
7. Gentra Puregene Cell Kit (catalog number 158745) (QIAGEN).

8. Qubit dsDNA HS Assay Kit (catalog number Q32854) (Life Technologies).
9. SYBR Gold Nucleic Acid Gel Stain (catalog number S-11494) (Life Technologies).

2.2 Transposome Formation Components

1. Twelve (12) i7 METS transposon oligonucleotides (IDT, standard desalted and scaled to 100 nmol) resuspended in resuspension buffer (RSB): 10 mM Tris, pH 8.5 (Illumina) (*see* **Note 1**).
2. Eight (8) i5 METS transposon oligonucleotides (IDT, standard desalted and scaled to 100 nmol) resuspended in resuspension buffer (RSB): 10 mM Tris, pH 8.5 (Illumina) (*see* **Note 2**).
3. 5′ phosphorylated pMENTs oligonucleotide (IDT, standard desalted and scaled to 25 nmol) (*see* **Note 3**).
4. Annealing buffer: 10 mM Tris–HCl, 1 mM EDTA, 25 mM NaCl, pH 8.0.
5. Standard storage buffer: 50 mM Tris–HCl, 0.1 mM EDTA, 1 mM DTT, 100 mM NaCl, 0.1 % Triton X-100, 50% glycerol.
6. EZ-Tn5 transposase (Epicentre).
7. Novex 8% TBE gel (catalog number EC6215BOX) (ThermoFisher Scientific).
8. 100 bp DNA ladder.
9. Thin-walled 0.2 mL polypropylene 8-well PCR strip-tubes.
10. Resuspension buffer (RSB): 10 mM Tris, pH 8.5 (Illumina).

2.3 Transposition Reagents

1. Qubit dsDNA HS Assay Kit.
2. 2× Nextera TD buffer (part of Nextera DNA Library Preparation kit catalog number FC-121-1031) (Illumina).
3. Wide-orifice pipet tips.
4. Plastic solution basin.
5. Hard-shell 96-well PCR plate.
6. Microseal Adhesive Seals.
7. EDTA.
8. BSA.
9. SDS.
10. Rocking Platform Shaker.

2.4 PCR Indexing Reagents

1. NPM (part of Nextera DNA Library Preparation kit catalog number FC-121-1031) (Illumina).
2. Nextera Index Kit (catalog number FC-121-1012) (Illumina).
3. Truseq Index Plate Fixture Kit (catalog number FC-130-1005) (Illumina).

4. Zymo DNA Clean & Concentrator–500 kit (catalog number D0043) (Zymo Research).
5. 2% agarose E-gel (catalog number G5018-02) (ThermoFisher Scientific).
6. Agilent High Sensitivity DNA Kit (5067-4626) (Agilent).

2.5 Clustering Reagents

1. Truseq PE Cluster Kit v3 (catalog number PE-401-3001) (Illumina).
2. Custom Read 1 primer "C15ME" (IDT, standard desalted and scaled to 25 nmol) (*see* **Note 4**).
3. cBot (catalog number SY-301-2002) (Illumina).

2.6 Sequencing Reagents

1. Truseq SBS Kit v3 (catalog number FC-401-3001) (Illumina).
2. Custom Index 1 primer "D15'ME'" (IDT, standard desalted and scaled to 25 nmol) (*see* **Note 5**).
3. Custom Read 2 primer "D15ME" (IDT, standard desalted and scaled to 25 nmol) (*see* **Note 6**).
4. 3 mL HT2 Wash Buffer (part of Truseq PE Cluster Kit v3 catalog number PE-401-3001) (Illumina).
5. 8 mL 0.1 N NaOH.
6. Hiseq 2000 (catalog number SY-401-1001) (Illumina).

3 Methods

Carry out all procedures at room temperature.

3.1 Input DNA Quality Assessment

1. Prepare genomic DNA using Qiagen's Gentra™ Puregene Cell Kit on cultured cells following supplier's preparation protocol. Quantify genomic DNA using high sensitivity dsDNA Qubit reagent and store genomic DNA in 1× TE buffer at 4 °C (no freeze thaw cycles) until ready for quality assessment (https://tools.thermofisher.com/content/sfs/manuals/Qubit_dsDNA_HS_Assay_UG.pdf).
2. In a 500 mL glass flask, prepare 1% agarose gel solution in 0.5× TBE buffer by mixing 2 g of pulsed-field Certified Megabase agarose with 200 mL 0.5× TBE buffer. Warm agarose and buffer solution in microwave for 2 min or until agarose has dissolved.
3. Pour agarose solution into molding tray for a Bio-Rad pulsed field gel electrophoresis system. Allow agarose solution to solidify at room temperature.
4. Prepare 0.5× TBE buffer by mixing 1 L Super-Q and 1 L 1× TBE Buffer. Turn on Bio-Rad pulsed field gel electrophoresis system and fill with 0.5× TBE buffer. Allow system to sit at room temperature for 30 min.

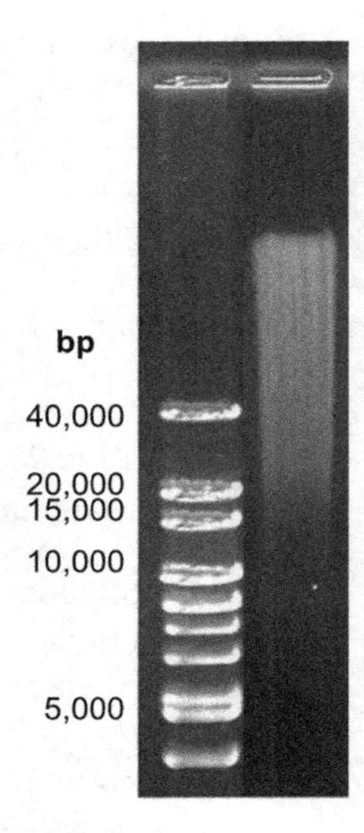

Fig. 2 PFGE image for gDNA sample NA12878 prepared using the Gentra Puregene Cell Kit. Sample was loaded on a 1 % agarose gel and run on a Bio-Rad Pulsed-Field Gel Electrophoresis System for 16 h at 14 °C at 170 V with a switch time starting at 1 s and progressing to 6 s

5. Slow down flow of system to load genomic DNA sample. Gently mix 6 μL 100 ng genomic DNA sample with 4 μL 6× loading dye and load into a well of the agarose gel. Load GeneRuler High Range DNA Ladder in an addition well. Allow samples to sit in the well for 10 min before turning up flow rate to 170 V.
6. Run at 170 V for 16 h at 14 °C with a switch time ranging from 1 to 6 s.
7. Stain gel for 30 min with 1 mg/mL SYBR Gold and image on Gel Doc to confirm high molecular weight (HMW) genomic DNA (Fig. 2) (*see* **Note** 7).

3.2 Assembly of 96 Indexed Transposome Complexes

1. In strip tubes, anneal 20 transposon oligos (METS), each containing the Tn5 Mosaic End (ME) sequence at their 3′-end, to its universal 5′-phosphorylated, 19 bp ME complimentary oligo (pMENTs). Mix oligos in a 1:1 M ratio in annealing buffer at a final stock concentration of 100 μM and anneal using the following thermocycling condition: 95 °C for 5 min followed by a slow ramp to 25 °C at 0.1 °C/s (*see* **Note 8**).
2. Mix 12.5 μL of the 20 unique annealed transposons at 100 μM with 15.62 μL 80 μM EZ-Tn5 transposase and 71.8 μL standard storage buffer to make a final stock concentration of 12.5 μM. Incubate mixtures at 37 °C for 1 h (*see* **Note 9**).

3. In order to make 96 unique indexed transposome complexes, the 8 i5 and 12 i7 transposome complexes were aliquoted, in a 1:1 ratio, into columns 1–12 and rows A-H, respectively, of a 96-well PCR plate.
4. Prepare a working stock of 96 transposome complexes diluted to 2.5 μM in standard storage buffer and store at −20 °C. Store original 12.5 μM stock plate at −20 °C.

3.3 Transposition of HMW Genomic DNA

1. Thaw 2× Nextera TD buffer at room temperature and store on ice until ready to use.
2. Quantify genomic DNA with High Sensitivity dsDNA Qubit reagent following Life technologies's supplier protocol (https://tools.lifetechnologies.com/content/sfs/manuals/Qubit_dsDNA_HS_Assay_UG.pdf).
3. Prepare 120 μL of genomic DNA at 1 ng/μL in 1× TE buffer in a 1.7 mL microcentrifuge tube. Thoroughly mix genomic DNA and buffer by gently swirling with a wide-orifice pipette tip.
4. Make the following master mix for 110 transposition reactions in a plastic solution basin. Add reagents in the following order and mix thoroughly by gently rocking solution basin back and forth: 1100 μL 2× Nextera tagmentation buffer, 880 μL Super-Q water and 110 μL 1 ng/μL genomic DNA.
5. Aliquot 19 μL master mix into 96-well PCR plate with wide-orifice pipet tips.
6. Place 96-well plate on ice. Add 1 μL of each indexed transposome complex (2.5 pmol) to each well with standard 10 μL pipet tips, and mix by gently swirling pipet tip (*see* **Note 10**).
7. Cover PCR plate with Microseal "B" adhesive seal and spin at 500 × *g* for 10 s.
8. Incubate samples in thermocycler at 55 °C for 10 min (*see* **Note 11**).
9. Remove plate from thermocycler and with wide-orifice pipet tips add 20 μL 40 mM EDTA to tagmentation reactions to stop the reaction.
10. Cover plate with sealing film and spin at 500 × *g* for 10 s.
11. Incubate EDTA-treated samples in thermocycler at 37 °C for 15 min. During incubation, thaw BSA, NPM, and Nextera index primers at room temperature and keep on ice until ready to use.
12. After incubation, remove sealing film and pool 20 μL from all 96 wells into plastic basin with wide-orifice pipet tips.
13. Cover plate with sealing film, and store remaining tagmentation products at 4 °C.

14. Cover solution basin with sealing film and gently rock for 5 min at room temperature at 2 rpm (VWR Rocking Platform Shaker). A low rocking speed is used to prevent damage to genomic DNA and to prevent spilling of sample from solution basin.
15. Dilute 25 pg/μL sample to 1 pg/μL with 1× TE buffer in new plastic solution basin: 1410 μL TE buffer, 30 μl 100 ng/μL BSA, and 60 μl 25 pg/μL pooled samples (use wide-orifice pipet tip).
16. Cover solution basin with sealing film and gently rock for an additional 5 min at room temperature at 2 rpm.
17. While sample is rocking, add 10 μL of Nextera i7 index primer (Nextera Index Kit) to a new 96-well low DNA-binding PCR plate according to Nextera Index Default Layout (*see* **Note 12**).
18. Using standard 10 μL pipet tips, transfer 10 μL of 1 pg/μL dilution to 96-well plate (10 pg/well) containing the i7 primer. Mix thoroughly by swirling pipet tip.
19. Incubate samples at room temperature for 5 min. Cover solution basin containing leftover pool, and place aside on lab bench.
20. Add 2 μL of 1% SDS to each well and mix by swirling pipet tip: 10 μL 1 pg/μL diluted sample with BSA, 10 μL Nextera i7 index primer, and 2 μL 1% SDS.
21. Cover plate with sealing film and spin at 500 × *g* for 10 s.
22. Incubate the 96-well plate at 55 °C for 15 min in a thermocycler then keep plate on ice until ready to continue.

3.4 PCR Indexing of Transposed Genomic DNA

1. After incubation, add 10 μL Nextera i5 indexed primers from Nextera Index Kit according to Nextera Index Default Layout (*see* **Notes 13** and **14**).
2. Make a PCR master mix for 105 reactions. Mix 3150 μL of NPM and 3990 μL of Super-Q water and add 68 μL to all 96 wells of the PCR plate on ice for a final volume of 100 μL.
3. Mix thoroughly by pipetting sample up and down.
4. Cover plate with sealing film, and spin at 500 × *g* for 10 s.
5. Use the following PCR parameters: initial extension at 72 °C for 3 min followed by an initial denaturation at 98 °C for 30 s, 15 cycles of denaturation at 98 °C for 10 s, annealing at 63 °C for 30 s, and extension at 72 °C for 3 min.
6. Post PCR, pool 50 μL from all 96 reactions into a plastic solution basin and combine with 9.6 mL Zymo binding buffer, and add to a 500 μg Zymo-Spin column. Wash column two times with 10 mL Zymo wash buffer and elute in 2 mL Zymo elution buffer.
7. QC samples on a High Sensitivity chip on BioAnalyzer according to Agilent user manual (http://www.chem.agilent.

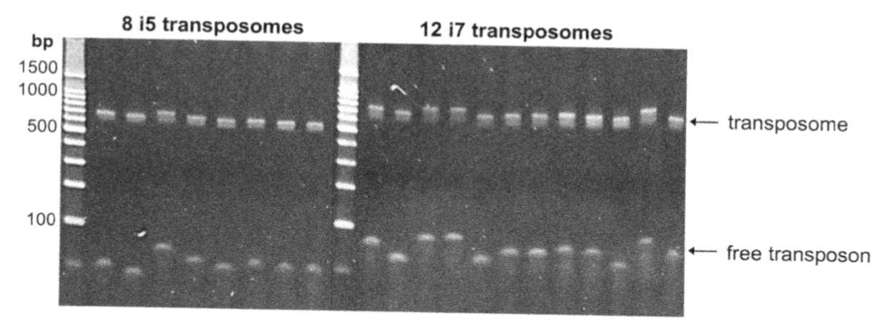

Fig. 3 (**a**) Electropherogram trace for CPT-Seq sequencing library using Agilent's High Sensitivity DNA Kit. (**b**) CPT-seq sequencing library run on 2% agarose E-gel alongside 100 bp ladder

com/library/usermanuals/Public/G2938-90321_SensitivityDNA_KG_EN.pdf) (Fig. 3). In replace of a BioAnalyzer, samples can also be run on a 2% agarose E-gel (Invitrogen) to confirm accurate library size (250–1000 bp) and quantified using HS dsDNA Qubit reagent.

3.5 Clustering CPT-Seq Library

1. Thaw Truseq v3 PE Cluster Kit in cold water bath for 1 h.
2. Dilute sequencing library to 2 nM in RSB (Truseq v3 PE Cluster Kit).
3. Mix the following and let sit at room temperature for 5 min: 10 μL 2 nM library and 10 μL 0.1 N NaOH.
4. After 5 min, add 980 μL of chilled HT1 (samples are at 20 pM) and place denatured library on ice.
5. Dilute denatured library to 8 pM in HT1 and aliquot 150 μL into each lane of an eight-well strip-tube for clustering.
6. Dilute Read 1 sequencing primer in HT1 to 0.5 μM final concentration and aliquot 150 μL into an eight-well strip tube.
7. Prepare cBot for clustering according to Illumina's standard cBot protocol.
8. Proceed with standard Illumina clustering protocol with custom Read 1 sequencing primer "C15ME" and cBot recipe "PE_Amp_Lin_Block_TubeStripHyb_v8.0" (*see* **Note 15**).
9. After clustering, wash cBot and store flowcell at 4 °C until ready to use.
10. Set up clustered flowcell on an Illumina HiSeq 2000 using a custom sequencing recipe and the custom sequencing primers for Index 1 and Read 2 (*see* **Note 16**).

3.6 Sequencing CPT-Seq Library on Hiseq2000

1. Perform maintenance wash on Hiseq before setting up sequencing run. Wash will take a couple of hours.
2. Prepare reagents for 200 cycle PE Hiseq run. Thaw SRE, ICB, and CMR (Illumina Truseq SBS Kit v3, catalog number

FC-401-3001) in cold water bath. When reagents are thawn, keep on ice.

3. When ICB is thawn, add four tubes of LFN and one tube of EDP to solution. Mix thoroughly by inverting conical and keep on ice until use.
4. Thaw PE regents in cold water bath. When reagents are thawn, keep on ice. Thaw an extra tube of wash buffer (HT2).
5. In a 15 mL conical tube, prepare 4 mL of 0.5 μM Read 2 sequencing primer D15ME in HT1 and label "position 16." This tube will replace the standard HP11 PE reagent. Keep primer on ice until ready to use.
6. In a 15 mL conical tube, prepare 4 mL of 0.5 μM i7 Index 1 sequencing primer in HT1 and label "position 17." This tube will replace the standard HP12 PE reagent. Keep primer on ice until ready to use.
7. Prepare 8 mL of 0.1 N NaOH in 15 mL conical tube. This tube will be added to position 18 on the PE reagent rack.
8. After thoroughly mixing reagents, add SBS and PE reagents to Hiseq racks according to Hiseq User Manuel. Replace standard PE reagents in positions 16 and 17 with custom Read 2 and Index 1 reagents, respectively.
9. Top off position 19 with 3 more mL of HT2.
10. Follow instructions on Hiseq monitor to set up sequencing run.
11. Proceed with standard Illumina sequencing protocol with custom Read 2 sequencing primer "D15ME" and i7 Index 1 sequencing primer "D15'ME'" and custom sequencing recipe "NCP_2x51."

3.7 Data Analysis

Detailed protocols for data analysis can be found elsewhere [3]. The general approach we have taken for phasing is described below.

1. Prepare hg19 human genome as the reference for sequence alignment, and a VCF file containing highly reliable heterozygous SNPs for the sample DNA from the standard variant calling workflow.
2. Fastq file extraction for genomic sequence and index sequence was carried out by bcl2fastq2 software (Illumina). Every genomic fragment is demultiplexed according to its index and distributed into individual fastq files for each index.
3. For each index, the fastq files are aligned against hg19 human genome to generate bam file by BWA or BOWTIE. After removing the duplicate reads, the unique reads are sorted using samtools according to their genomic positions and grouped into the island structure when the distance between

them is shorter than 15 kb threshold, or can be numerically estimated from the histogram of the distances between the neighboring reads (*see* ref. 3, Fig. 3).

4. Heterozygous SNPs from the reference VCF file are assigned to the islands obtained from **step 3**. The islands containing less than 2 SNPs are not informative for haplotyping and therefore filtered out. SNPs that are linked by only one data point or showing conflicting calls by multiple islands are removed as well. The remaining islands are used as input to RefHap for phasing [6, 7].
5. 1000 Genomics Project or other SNP panels can be used to impute additional phased SNPs [8].

4 Notes

1. 12 i7 indexed METS oligonucleotide sequences:

   P7-PCR_i7-1-B15Ext-Clpx_i7_D15_ME: GTCTCGTGG GCTCGGCTGTCCCTGTCCCGAGTAATCACCGTCT CCGCCTCAGATGTGTATAAGAGACAG

   P7-PCR_i7-2-B15Ext-Clpx_i7_D15_ME: GTCTCGTGG GCTCGGCTGTCCCTGTCCTCTCCGGACACCGTCT CCGCCTCAGATGTGTATAAGAGACAG

   P7-PCR_i7-3-B15Ext-Clpx_i7_D15_ME: GTCTCGTGG GCTCGGCTGTCCCTGTCCAATGAGCGCACCGTCT CCGCCTCAGATGTGTATAAGAGACAG

   P7-PCR_i7-4-B15Ext-Clpx_i7_D15_ME: GTCTCGTGG GCTCGGCTGTCCCTGTCCGGAATCTCCACCGTCT CCGCCTCAGATGTGTATAAGAGACAG

   P7-PCR_i7-5-B15Ext-Clpx_i7_D15_ME: GTCTCGTGG GCTCGGCTGTCCCTGTCCTTCTGAATCACCGTCT CCGCCTCAGATGTGTATAAGAGACAG

   P7-PCR_i7-6-B15Ext-Clpx_i7_D15_ME: GTCTCGTGG GCTCGGCTGTCCCTGTCCACGAATTCCACCGTCT CCGCCTCAGATGTGTATAAGAGACAG

   P7-PCR_i7-7-B15Ext-Clpx_i7_D15_ME: GTCTCGTGG GCTCGGCTGTCCCTGTCCAGCTTCAGCACCGTCT CCGCCTCAGATGTGTATAAGAGACAG

   P7-PCR_i7-8-B15Ext-Clpx_i7_D15_ME: GTCTCGTGG GCTCGGCTGTCCCTGTCCGCGCATTACACCGTCT CCGCCTCAGATGTGTATAAGAGACAG

   P7-PCR_i7-9-B15Ext-Clpx_i7_D15_ME: GTCTCGTGG GCTCGGCTGTCCCTGTCCCATAGCCGCACCGTCT CCGCCTCAGATGTGTATAAGAGACAG

   P7-PCR_i7-10-B15Ext-Clpx_i7_D15_ME: TCTCGTGG GCTCGGCTGTCCCTGTCCTTCGCGGACACCGTCT CCGCCTCAGATGTGTATAAGAGACAG

P7-PCR_i7-11-B15Ext-Clpx_i7_D15_ME: GTCTCGTGGGCTCGGCTGTCCCTGTCCGCGCGAGACACCGTCTCCGCCTCAGATGTGTATAAGAGACAG

P7-PCR_i7-12-B15Ext-Clpx_i7_D15_ME: GTCTCGTGGGCTCGGCTGTCCCTGTCCCTATCGCTCACCGTCTCCGCCTCAGATGTGTATAAGAGACAG

2. 8 i5 indexed METS oligonucleotide sequences:

   P5-PCR_i5-1-A14Ext-Clpx_i5_C15_ME: TCGTCGGCAGCGTCTCCACGCTATAGCCTGCGATCGAGGACGGCAGATGTGTATAAGAGACAG

   P5-PCR_i5-2-A14Ext-Clpx_i5_C15_ME: TCGTCGGCAGCGTCTCCACGCATAGAGGCGCGATCGAGGACGGCAGATGTGTATAAGAGACAG

   P5-PCR_i5-3-A14Ext-Clpx_i5_C15_ME: TCGTCGGCAGCGTCTCCACGCCCTATCCTGCGATCGAGGACGGCAGATGTGTATAAGAGACAG

   P5-PCR_i5-4-A14Ext-Clpx_i5_C15_ME: TCGTCGGCAGCGTCTCCACGCGGCTCTGAGCGATCGAGGACGGCAGATGTGTATAAGAGACAG

   P5-PCR_i5-5-A14Ext-Clpx_i5_C15_ME: TCGTCGGCAGCGTCTCCACGCAGGCGAAGGCGATCGAGGACGGCAGATGTGTATAAGAGACAG

   P5-PCR_i5-6-A14Ext-Clpx_i5_C15_ME: TCGTCGGCAGCGTCTCCACGCTAATCTTAGCGATCGAGGACGGCAGATGTGTATAAGAGACAG

   P5-PCR_i5-7-A14Ext-Clpx_i5_C15_ME: TCGTCGGCAGCGTCTCCACGCCAGGACGTGCGATCGAGGACGGCAGATGTGTATAAGAGACAG

   P5-PCR_i5-8-A14Ext-Clpx_i5_C15_ME: TCGTCGGCAGCGTCTCCACGCGTACTGACGCGATCGAGGACGGCAGATGTGTATAAGAGACAG

3. 5′ phosphorylated pMENTS oligonucleotide sequence:
   /5Phos/CTGTCTCTTATACACATCT

4. Custom Read 1 sequencing primer "C15ME" sequence:
   GCGATCGAGGACGGCAGATGTGTATAAGAGACAG

5. Custom Index 1 sequencing primer "D15'ME'" sequence:
   CTGTCTCTTATACACATCTGAGGCGGAGACGGTG

6. Custom Read 2 sequencing primer "D15ME" sequence:
   CACCGTCTCCGCCTCAGATGTGTATAAGAGACAG

7. Average expected size of high molecular weight genomic DNA is 100–200 kb.

8. Oligos were purchased from IDT (standard desalting). Eight of these oligonucleotides (i5 oligonucleotides) have adapter sequences to make them compatible with the P5 Illumina sequencing end, and the other 12 have the adapter for the P7 side (i7 oligonucleotides). Transposon indexing (which intro-

	1	2	3	4	5	6	7	8	9	10	11	12
A	N501 N701	N501 N702	N501 N703	N501 N704	N501 N705	N501 N706	N501 N707	N501 N708	N501 N709	N501 N710	N501 N711	N501 N712
B	N502 N701	N502 N702	N502 N703	N502 N704	N502 N705	N502 N706	N502 N707	N502 N708	N502 N709	N502 N710	N502 N711	N502 N712
C	N503 N701	N503 N702	N503 N703	N503 N704	N503 N705	N503 N706	N503 N707	N503 N708	N503 N709	N503 N710	N503 N711	N503 N712
D	N504 N701	N504 N702	N504 N703	N504 N704	N504 N705	N504 N706	N504 N707	N504 N708	N504 N709	N504 N710	N504 N711	N504 N712
E	N505 N701	N505 N702	N505 N703	N505 N704	N505 N705	N505 N706	N505 N707	N505 N708	N505 N709	N505 N710	N505 N711	N505 N712
F	N506 N701	N506 N702	N506 N703	N506 N704	N506 N705	N506 N706	N506 N707	N506 N708	N506 N709	N506 N710	N506 N711	N506 N712
G	N507 N701	N507 N702	N507 N703	N507 N704	N507 N705	N507 N706	N507 N707	N507 N708	N507 N709	N507 N710	N507 N711	N507 N712
H	N508 N701	N508 N702	N508 N703	N508 N704	N508 N705	N508 N706	N508 N707	N508 N708	N508 N709	N508 N710	N508 N711	N508 N712

Fig. 4 TBE gel image of 8 i5 and 12 i7 individual transposomes alongside 100 bp ladder

duces a pair of indices on either side of the genomic insert) is achieved by using only 20, i.e., 8 i5 (i.e., P5-side index) plus 12 i7 (i.e., P7-side index), index-containing oligonucleotides, creating 8 × 12 = 96 different index combinations.

9. To confirm formation of transposomes, run 1 μL of a 1:60 dilution of 12.5 μM transposome stock with 4 μL 6× loading dye and 7 μL water on a Novex 8 % TBE gel for 1 h at 100 V (*see* Fig. 4).
10. Do not pipet sample up and down in order to minimize mechanical shearing and damage to HMW genomic DNA. Manually check that all pipet tips are on equally firmly to ensure even reagent delivery.
11. Preheat block to 55 °C before transferring to thermocycler.
12. The i7 primer is used as a nonspecific surface blocker, preventing any potential DNA loss post-SDS treatment.
13. It is recommended to use the Truseq Index Plate Fixture Kit (Illumina catalog number FC-130-1005) to assist in correctly arranging index primers for appropriate delivery to wells.
14. Similar to transposon-level indexing, PCR-level indexing is also generated with 8 i5 and 12 i7 PCR primers, yielding 96 unique PCR index combinations (*see* Fig. 5).
15. Before loading the cluster plate onto the cBot, tap reagent plate on hard surface to ensure there are no air bubbles at the bottom of the plate.

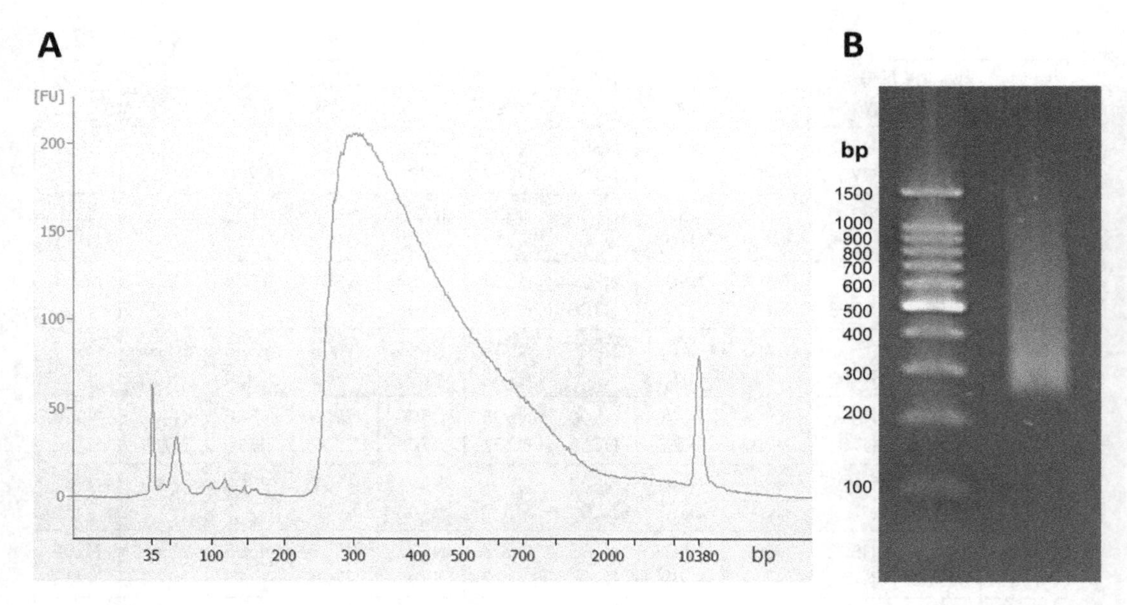

Fig. 5 Nextera Index default layout for PCR-level indexing using Illumina's Nextera Index Kit. 8 i5 and 12 i7 PCR primers yield 96 unique PCR index combinations

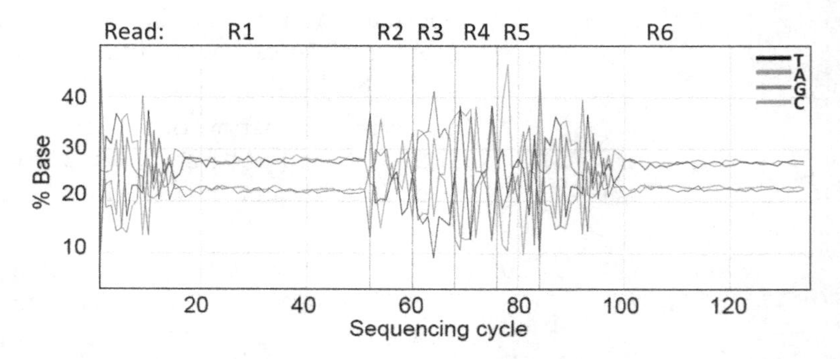

Fig. 6 A typical % base versus sequence cycle for a custom 2x51 CPT-seq v3 Hiseq 2000 run. The order of sequencing reads is as follows: gDNA read 1 (cycles 1–51), index 1 (transposon i7, cycles 52–59, and PCR i7, cycles 60–67), index 2 (PCR i5, cycles 68–75, and transposon i5, cycles 76–83), and gDNA read 2 (cycles 84–134)

16. For the custom sequencing recipe, the order of sequencing reads is as follows: genomic DNA read 1 (cycles 1–51), index 1 (transposon i7, cycles 52–59, followed by 27 dark cycles, and PCR i7, cycles 60–67), index 2 (PCR i5, cycles 68–75, followed by 21 dark cycles, and transposon i5, cycles 76–83), and genomic DNA read 2 (cycles 84–134) (*see* Fig. 6).

Acknowledgments

We would like to thank the research, development, software, and engineering departments at Illumina for the sequencing technology development.

References

1. Bentley DR et al (2008) Accurate whole human genome sequencing using reversible terminator chemistry. Nature 456(7218):53–59
2. Snyder MW, Adey A, Kitzman JO, Shendure J (2015) Haplotype-resolved genome sequencing: experimental methods and applications. Nat Rev Genet 16:344–358 and references cited therein
3. Amini S, Pushkarev D, Christiansen L, Kostem E, Royce T, Turk C, Pignatelli N, Adey A, Kitzman JO, Vijayan K, Ronaghi M, Shendure J, Gunderson KL, Steemers FJ (2014) Haplotype-resolved whole-genome sequencing by contiguity-preserving transposition and combinatorial indexing. Nat Genet 46(12): 1343–1349
4. Adey A, Kitzman JO, Burton JN, Daza R, Kumar A, Christiansen L, Ronaghi M, Amini S, Gunderson KL, Steemers FJ, Shendure J (2014) In vitro, long-range sequence information for de novo genome assembly via transposase contiguity. Genome Res 24(12):2041–2049
5. Cusanovich DA, Daza R, Adey A, Pliner HA, Christiansen L, Gunderson KL, Steemers FJ, Trapnell C, Shendure J (2015) Multiplex single-cell profiling of chromatin accessibility by combinatorial cellular indexing. Science 348(6237):910–914
6. Duitama J, Huebsch T, McEwen G, Suk EK, Hoehe M (2010) ReFHap: a reliable and fast algorithm for single individual haplotyping. In: Proceedings of the first ACM international conference on bioinformatics and computational biology (BCB '10). AMC, New York, NY, pp 160–169
7. Xie M, Wang J, Jiang T (2012) A fast and accurate algorithm for single individual haplotyping. BMC Syst Biol 6(Suppl 2):S8
8. The 1000 Genomes Project Consortium (2012) An integrated map of genetic variation from 1,092 human genomes. Nature 491:56–65

Chapter 13

A Fosmid Pool-Based Next Generation Sequencing Approach to Haplotype-Resolve Whole Genomes

Eun-Kyung Suk, Sabrina Schulz, Birgit Mentrup, Thomas Huebsch, Jorge Duitama, and Margret R. Hoehe

Abstract

Haplotype resolution of human genomes is essential to describe and interpret genetic variation and its impact on biology and disease. Our approach to haplotyping relies on converting genomic DNA into a fosmid library, which represents the entire diploid genome as a collection of haploid DNA clones of ~40 kb in size. These can be partitioned into pools such that the probability that the same pool contains both parental haplotypes is reduced to ~1 %. This is the key principle of this method, allowing entire pools of fosmids to be massively parallel sequenced, yielding haploid sequence output. Here, we present a detailed protocol for fosmid pool-based next generation sequencing to haplotype-resolve whole genomes including the following steps: (1) generation of high molecular weight DNA fragments of ~40 kb in size from genomic DNA; (2) fosmid cloning and partitioning into 96-well plates; (3) barcoded sequencing library preparation from fosmid pools for next generation sequencing; and (4) computational analysis of fosmid sequences and assembly into contiguous haploid sequences.

This method can be used in combination with, but also without, whole genome shotgun sequencing to extensively resolve heterozygous SNPs and structural variants within genomic regions, resulting in haploid contigs of several hundred kb up to several Mb. This method has a broad range of applications including population and ancestry genetics, the clinical interpretation of mutations in personal genomes, the analysis of cancer genomes and highly complex disease gene regions such as MHC. Moreover, haplotype-resolved genome sequencing allows description and interpretation of the diploid nature of genome biology, for example through the analysis of haploid gene forms and allele-specific phenomena. Application of this method has enabled the production of most of the molecular haplotype-resolved genomes reported to date.

Key words Haplotype-resolving genomes, Molecular haplotypes, Phasing, Clone-based haplotyping, Fosmid library, Fosmid pools, Fosmids, Next generation sequencing, Haplotype assembly, Phasing algorithm

1 Introduction

Human genomes are diploid by nature. Thus, to fully understand human biology and link genetic variation to gene function and phenotype, it is essential to determine both parental sequences of an individual genome independently [1, 2]. Present technologies,

Irene Tiemann-Boege and Andrea Betancourt (eds.), *Haplotyping: Methods and Protocols*, Methods in Molecular Biology, vol. 1551, DOI 10.1007/978-1-4939-6750-6_13, © Springer Science+Business Media LLC 2017

however, routinely read out "mixed diploid" sequences. Therefore, they cannot distinguish between the unique combinations of variants on each of the two chromosomal homologues, the haplotypes. Over the past few years, a number of experimental methods to haplotype-resolve genomes have been developed. Among those, clone-based haplotyping, in particular fosmid pool-based next generation sequencing (NGS), has enabled phasing by far the largest number of genomes to date, over 30 [3–9]. Here, we present the principle and concrete steps of this method.

The key principle is to convert human genomic DNA into a library of fosmids, haploid DNA fragments ~40 kilobases (kb) in size, and partition this library into pools of fosmids such that the probability that both parental haplotypes co-occur is reduced to ~1 % [10]. Thus, multiple pools can be massively parallel sequenced to generate redundant coverage of both haploid genomes of an individual. In our original report introducing this principle [10], we have established "haploid clone pools" of ~5000 fosmids, random mixtures of DNA fragments representing ~5 % of a haploid genome. These result from partitioning a library of ~1.44×10^6 fosmid clones, ensuring 7× coverage of each haploid genome, into 3 × 96-well plates. In order to increase the throughput, we have chosen to combine these plates into one 96-well plate, each well containing a super-pool of ~15,000 fosmids. These super-pools are barcoded and subjected to NGS. Analysis of the NGS data showed that only 1.31 % of SNP calls, on average, were heterozygous per super-pool, confirming that the method works as expected [3]. To estimate the number of super-pools necessary to be sequenced in order to reach a sufficient coverage level, simulation studies were performed. Accordingly, 40 pools were estimated to result in an average haploid read coverage of ~12× (diploid read coverage of ~24×) and 85 % of heterozygous SNPs phased; 48 pools were required to achieve a read coverage of 14.5× and ~29×, respectively, and 92 % of SNPs phased.

In the molecular genetics part of our protocol, we describe the following steps (overview in Fig. 1): (1) Extraction of high molecular weight (HMW) genomic DNA (gDNA), mechanical shearing, and gel-based selection of DNA fragments of ~40 kb in size; (2) ligation of size-selected fragments into pEpiFos vector; (3) phage packaging and mass transfection of *E. Coli* to obtain a total of 1.44×10^6 fosmid clones; (4) partitioning of these fosmid clones into 3 × 96 deep well plates to generate pools of ~5000 fosmids and amplifying those in liquid culture; (5) combining the 3 × 96 well plates to generate super-pools of 15,000 fosmid clones; (6) amplification of fosmid clones per well on agar plates; (7) isolation of fosmid DNA from amplified clones; (8) preparation of barcoded sequencing libraries per super-pool, and (9) processing them for NGS. We also indicate where the protocol can be adapted to newer sequencing platforms, so that it is clear which steps to modify. This

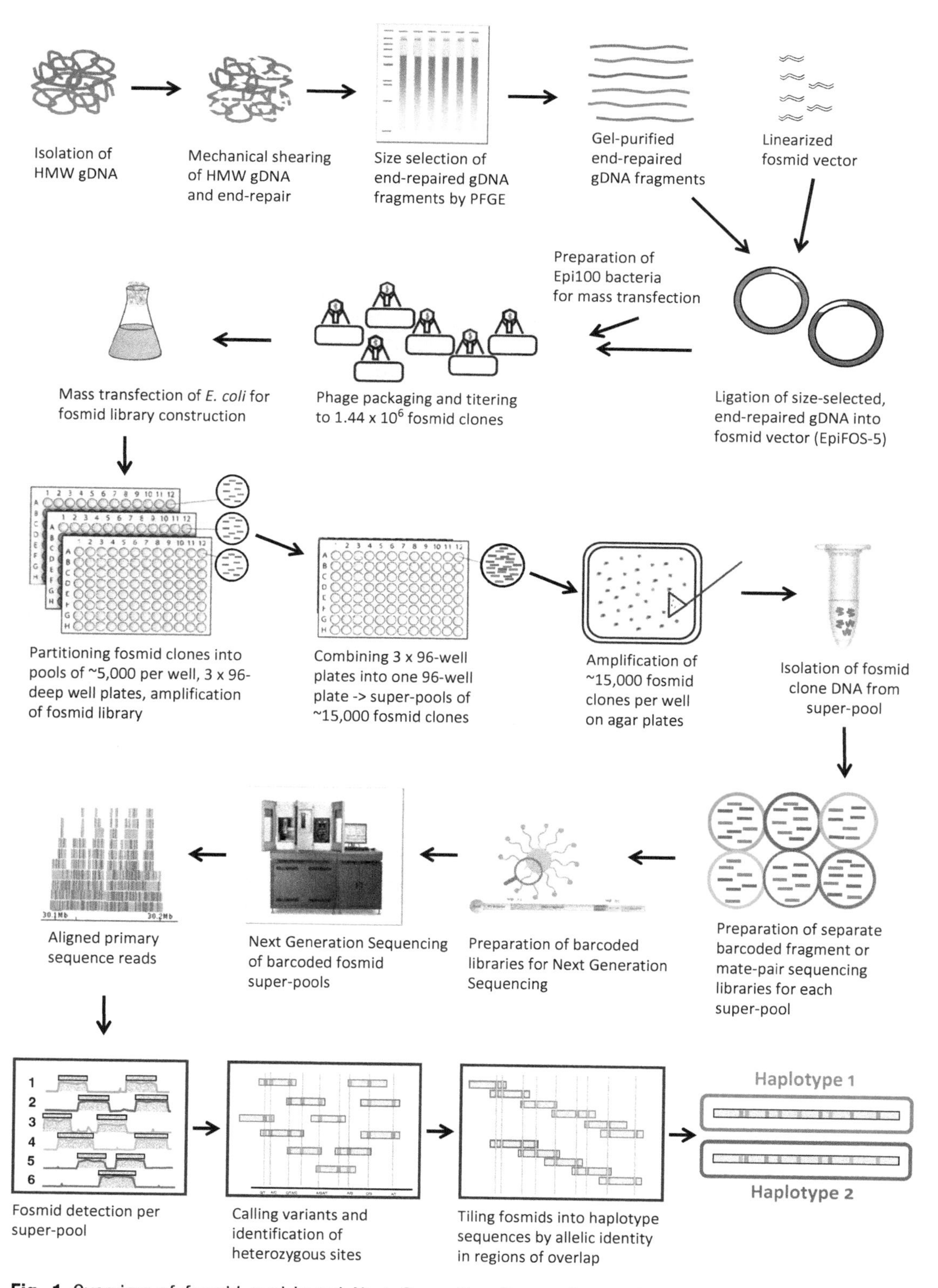

Fig. 1 Overview of fosmid pool-based Next Generation Sequencing method to haplotype-resolve whole genomes. The principle and molecular genetics and computational key steps of this method, as applied to haplotype-resolving one individual human genome, are shown

protocol can be complemented by any routine protocol for whole genome shotgun sequencing (WGS), as we have done to generate our first haplotype-resolved genome, "Max Planck One" (MP1) [3]. This will add to the quality of heterozygote detection and help to resolve larger structural variants. Successive production of a set of 12 genomes has shown that assembly of haplotype contigs solely from fosmid pool-based NGS is feasible [9].

In the computational part of our protocol, we describe the steps involved in assembling a haplotype-resolved genome from the sequence read output in a user-oriented mode, providing downloadable algorithms and scripts. In particular, we make our specifically designed heuristic algorithm "ReFHap" [11] available, which has proven to be particularly efficient in terms of computing time and represents an optimal compromise between accuracy, completeness, and computational resources [5]. Our computational tools should be usable by any scientist with experience in NGS data analysis. Haplotype assembly includes the following steps (*see* Fig. 6 in the bioinformatics part): (1) alignment of fosmid pool sequence reads and sorting these alignments per pool; (2) merging sorted alignments to detect phase-informative, heterozygous variants; (3) fosmid detection and allele calling per pool; and (4) phasing by the use of ReFHap to generate haplotype contigs of phased variants per chromosome.

Our method was empirically corroborated by application to haplotyping HapMap trio child NA12878 [5]. For this sample, whole genome sequencing of the family trio had resulted in the resolution of ~80% of all heterozygous positions [12]. Where comparable, the trio-based and molecular phasing data were entirely identical, showing that fosmid pool-based haplotyping can produce highly accurate results, even at low coverage. Our method, however, allowed resolution of a much higher number of heterozygous SNPs, ~98% in total, highlighting the power of our approach. Moreover, our method enabled generation of the most comprehensively haplotype-resolved human genome to date, MP1, with over 99% of all heterozygous SNPs and virtually all individual and rare SNPs phased into long haplotype blocks with an N50 of ~1 Megabase (Mb), i.e., 50% of the haplotype-resolved sequence were within blocks of at least ~1 Mb. The maximum block length achieved was ~6.3 Mb [3]. Finally, we have applied our protocol to phase an entire set of 16 human genomes [9], equivalent to about half of the published production of clone pool-based haplotype-resolved genomes.

In sum, haplotype-resolved genome sequencing is applicable to a broad range of scenarios, including population and anthropological genetics and the analysis of human diseases, for example the determination of the molecular haplotypes underlying highly variable and complex disease gene regions such as MHC, or GWAS regions. Moreover, knowledge of phase is critical for clinical interpretation of mutations such as compound heterozygosity or pharmacogenetically relevant variants. The same is true for the accurate and actionable description and interpretation of personal genomes as a whole. Finally, haplotype-resolved genome sequencing provides key information on genome function, for example through resolving haploid gene forms and their regulatory environments, and allele-specific phenomena in general. These include gene expression, the regulation of transcription, methylation, and other epigenetic mechanisms.

2 Materials

2.1 Fosmid Library Construction

2.1.1 High Molecular Weight (HMW) Genomic DNA (gDNA) Isolation

1. DNA Extraction Kit for HMW gDNA (Stratagene, Cat. Nr. 200600).
2. Centrifuge (12,500 ×*g*) for 15 ml/50 ml tubes.
3. Centrifuge (refrigerated, 14,000 ×*g*) for 1.5/2.0 ml tubes and 96-well plates.
4. Shaking Water bath.
5. Nanodrop ND100.
6. Rotating tube roller.
7. 50 ml reaction tubes.
8. 1.5/2 ml reaction tubes.
9. Pipettors and large-bore pipet tips.
10. Sterile glass rod.
11. Absolute EtOH p.A.
12. 70% EtOH.
13. TE buffer: 10 mM Tris–HCl (pH 7.5), 1 mM EDTA.
14. Pulsed Field Gel Electrophoresis System (CHEF DR II).
15. Ultrapure Low Melting Point Agarose (Thermo Fisher).
16. 10× TBE: [890 mM Tris, 890 mM Boric acid, 20 mM EDTA (pH 8.0)]; 0.5× TBE.
17. Lambda DNA-Mono Cut Mix DNA Ladder (NEB).
18. Loading dye.
19. Ethidium bromide.
20. UV-illuminator.

2.1.2 Mechanical Shearing of HMW gDNA and End-Repair

1. HydroShear shearing device (Digilab Genomic Solutions).
2. Sterile water.
3. Sterile disposable 1 ml syringes, 23G × 1¼ injection needle.
4. Agarose gel electrophoresis system.
5. 100% Isopropanol p.A.

6. TE buffer: 10 mM Tris–HCl (pH 7.5), 1 mM EDTA.
7. EpiFOS Fosmid Library Production Kit (Epicentre, Cat. No. FOS0901).
8. 1.5 ml reaction tubes.
9. Thermomixer.

2.1.3 Size Selection of End-Repaired gDNA Fragments by PFGE

1. Pulsed Field Gel Electrophoresis System (CHEF DR II).
2. 1 % Low melting agarose.
3. 0.5× TBE.
4. 6× Loading Dye.
5. Lambda DNA-Mono Cut Mix DNA Ladder (NEB).
6. Sterile water.
7. Gel chamber (14 cm wide, 13 cm long) with preparative comb (4 wells; 1.5 mm thick, 27 mm width, 14 mm length, 200 μl volume/well).
8. SYBR® Gold Nucleic Acid Gel Stain 10,000× (Molecular Probes).
9. Dark Reader Transilluminator (e.g., Clare Chemical Lab) or Safe Imager Blue Light Transilluminator (Molecular Probes).
10. Razor blade (sterile).
11. Gelase enzyme (1 U/μl) (Epicentre).
12. Thermomixer.
13. Ice bath.
14. 1.5 ml reaction tubes.
15. Centrifuge (refrigerated, 14,000 × *g*) for 1.5/2.0 ml tubes.
16. 3 M Sodium acetate (pH 7).
17. 100 % Isopropanol p.A.
18. 70 % EtOH.
19. TE buffer: 10 mM Tris–HCl (pH 7.5), 1 mM EDTA.
20. 0.8 % agarose gel.
21. 0.5× TBE.
22. Fosmid Control (FC) DNA (included in EpiFOS Fosmid Library Production Kit).
23. Ethidium bromide.
24. UV-illuminator.
25. Aluminum foil.

2.1.4 Preparation of Epi100 Bacteria for Mass Transfection

1. EPI100 bacteria (included in EpiFOS Fosmid Library Production Kit).
2. 10 cm petri dishes.
3. LB broth: 10 g/l Bacto-Tryptone, 10 g/l NaCl, 5 g/l Yeast Extract.

4. LB agar: LB-Broth, 15 g/l Bacto-Agar.
5. 10 mM $MgSO_4$.
6. Autoclave.
7. Incubator (37 °C).
8. 50 ml Erlenmeyer flask.
9. UV-Spectrophotometer (Eppendorf).

2.1.5 Ligation of Size-Selected gDNA into pEpiFOS-Vector, Phage Packaging, and Testing Titer

1. Fast-Link Ligase and Buffer (included in EpiFOS Fosmid Library Production Kit).
2. 200 μl Microtubes.
3. 1.5 ml reaction tubes.
4. Thermomixer.
5. Sterile water.
6. Ice bath.
7. MaxPlax-Lambda Packaging Extract (included in EpiFOS Fosmid Library Production Kit).
8. Chloroform.
9. Epi100 bacteria (included in EpiFOS Fosmid Library Production Kit).
10. LB-Agar plates with 12.5 μg/ml chloramphenicol (10 cm petri dishes).
11. Incubator (37 °C).

2.1.6 Mass Transfection of E. Coli

1. Laminar flow hood.
2. LB broth with 10 mM $MgSO_4$ and 12.5 μg/ml Cloramphenicol.
3. 96-deep well plates.
4. Erlenmeyer flask.
5. EPI100 bacteria (included in EpiFOS Fosmid Library Production Kit).
6. Shaking incubator (37 °C).

2.1.7 Partitioning Fosmid Clones into Pools and Amplification of Fosmid Library

1. 96-deep well plates.
2. LB broth.
3. Multipette (Eppendorf).
4. Breathable seal.
5. Shaking incubator (37 °C).
6. LB Agar plates with 12.5 μg/ml Chloramphenicol.
7. Sterile glycerol (90%).
8. Eight-channel pipette.
9. Aluminum foil.
10. −80 °C freezer.

2.2 Isolation of Fosmid DNA from Fosmid Clone Pools

2.2.1 Plating and Scraping Fosmid Super-Pools

1. Large LB agar plates (22 cm × 22 cm) with 12.5 μg/ml Chloramphenicol.
2. LB broth, and LB broth with 12.5 μg/ml Cloramphenicol.
3. Inoculating loop.
4. Drigalski spatula.
5. 50 ml reaction tubes (Falcon).

2.2.2 Extraction of Fosmid Super-Pool DNA

1. Ice bath.
2. Centrifuge (12,500 × g) for 15 ml/50 ml Falcon tubes.
3. Centrifuge (refrigerated, 12,500 × g) for 1.5/2.0 ml tubes and 96-well plates.
4. QIAGEN Large-Construct Kit.
5. Fluted filters.
6. QIAGEN-tip 500 (to be purchased in addition, 2 per extraction required).
7. ATP disodium salt (AppliChem).
8. Water bath.
9. 100% Isopropanol p.A.
10. 70% EtOH.
11. 1.5 ml reaction tubes.
12. TE buffer: 10 mM Tris–HCl (pH 7.5), 1 mM EDTA.
13. Agarose gel electrophoresis system.

2.3 Fosmid Pool-Based Next Generation Sequencing Library Preparation

2.3.1 Barcoded Preparation of Fragment Libraries

1. Covaris S2 Sonicator (Covaris).
2. Covaris microTUBEs (Covaris).
3. Low TE buffer (Applied Biosystems).
4. Covaris G7 adaptor (Covaris).
5. Lonza Flash Gel System (Lonza).
6. 2.2% Lonza FlashGel Cassette (Lonza).
7. GeneRuler Low Range DNA Ladder, ready-to-use with 6× Orange DNA Loading Dye (Fermentas).
8. MinElute Gel Extraction Kit (QIAGEN).
9. 1.5 ml DNA LoBind tubes (Eppendorf).
10. End-It DNA End-Repair Kit (Epicentre).
11. Thermoshaker.
12. Qubit Fluorometric Quantitation (Thermo Scientific).
13. Qubit dsDNA HS Assay Kit (Thermo Scientific).
14. 4% Reliant NuSieve 3:1 agarose gel (Lonza).
15. UV Multibright Transilluminator (Intas).

16. SOLiD Fragment Library Oligo Kit (Applied Biosystems).
17. Quick Ligation Kit (NEB).
18. MinElute Reaction Cleanup Kit (QIAGEN).
19. Gel electrophoresis system (Bio-Rad).
20. Ethidium bromide (AppliChem).
21. DNA Polymerase I (*E. coli*) (10 U/μl) (NEB).
22. 100 mM dNTP-Mix (GeneAmp).
23. Microcentrifuge 5417R (Eppendorf).
24. Thermomixer.

2.3.2 Barcoded Preparation of Mate-Paired Libraries

1. SOLiD Mate-Paired Library Oligo Kit (Applied Biosystems).
2. HydroShear (Genomic Solutions, Inc.).
3. HydroShear Standard Shearing Assembly 1–5 kb (Genomic Solutions, Inc.).
4. QIAquick Gel Extraction Kit (QIAGEN).
5. End-It DNA End-Repair Kit (Epicentre).
6. 500 mM EDTA.
7. Quick Ligation Kit (NEB).
8. Microcentrifuge 5417R (Eppendorf).
9. 10× TAE (Applied Biosystems).
10. Agarose-LE (Applied Biosystems).
11. Gel electrophoresis system (any supplier).
12. Ethidium bromide (AppliChem).
13. 1 Kb DNA Ladder (Invitrogen).
14. Gel imaging system (any supplier).
15. Plasmid-Safe ATP-Dependent DNase (Epicentre).
16. DNA Polymerase I (*E. coli*) (10 U/μl) (NEB).
17. 100 mM dNTP Mix (GeneAmp).
18. T7 Exonuclease (10 U/μl) (NEB).
19. S1 Nuclease (400–1500 U/μl) (Invitrogen).
20. 3 M Sodium chloride, 5 M Sodium chloride.
21. Tris–HCl (500 mM, pH 7.5).
22. 1 M Magnesium chloride (Ambion).
23. Streptavidin Dynabeads, Dynal MyOne C1 (Thermo Fisher Scientific).
24. SOLiD Buffer Kit (including 1× Bead Wash Buffer, 1× Bind & Wash Buffer, 1× Low Salt Binding Buffer, Low TE Buffer, 1× TEX Buffer, 2-Butanol) (Applied Biosystems).
25. 100× BSA (NEB).

26. Six Tube Magnetic Stand (Applied Biosystems).
27. Vortexer (any supplier).
28. Rotator for 1.5–2.0 ml tubes (any supplier).

2.3.3 Large-Scale PCR of Fragment and Mate-Paired Libraries

1. PCR SuperMix (Invitrogen).
2. Cloned Pfu polymerase (2.5 U/μL) (Stratagene).
3. 4% Reliant NuSieve 3:1 agarose gel (Lonza).
4. GeneRuler Low Range DNA Ladder, ready-to-use with 6× Orange DNA Loading Dye (Fermentas).
5. MinElute Reaction Cleanup Kit (QIAGEN).
6. QIAquick Gel Extraction Kit (QIAGEN).
7. Gel electrophoresis system (any supplier).
8. Gel imaging system (any supplier).
9. SOLiD Mate-Paired Library Oligo Kit (Applied Biosystems).
10. SOLiD Fragment Library Oligo Kit (Applied Biosystems).
11. 96-Well GeneAmp PCR System 9700 (Applied Biosystems).
12. Lonza Flash Gel System (Lonza).
13. 2.2% Lonza FlashGel Cassette (Lonza).
14. Ethidium bromide (AppliChem).
15. 15 ml conical tubes (Falcon).
16. Microcentrifuge (12,500 × *g*).
17. Qubit Fluorometric Quantitation (Thermo Scientific).
18. Qubit dsDNA HS Assay Kit (Thermo Scientific).
19. 1.5 ml DNA LoBind tubes (Eppendorf).
20. Six Tube Magnetic Stand (Applied Biosystems).

2.3.4 Preparation of Fragment and Mate-Paired Sequencing Libraries for Emulsion PCR

1. Qubit Fluorometric Quantitation (Thermo Scientific).
2. Qubit dsDNA HS Assay Kit (Thermo Scientific).
3. SOLiD Buffer Kit (including 1x Bead Wash Buffer, 1× Bind & Wash Buffer, 1× Low Salt Binding Buffer, Low TE Buffer, 1× TEX Buffer, 2-Butanol) (Applied Biosystems).

2.4 Processing Next Generation Sequencing Libraries for Instrument Run

2.4.1 Emulsion PCR

1. 96-Well GeneAmp PCR System 9700 (Applied Biosystems).
2. SOLiD Buffer Kit (Applied Biosystems).
3. SOLiD ePCR Kit (Applied Biosystems).
4. 1 ml glass pipet (any supplier).
5. 5 ml glass pipet (any supplier).
6. ULTRA-TURRAX Tube Drive (IKA).
7. SOLiD ePCR Tubes and Caps (IKA).

8. Covaris S2 Sonicator (Covaris).
9. Covaris-2 Series Machine Holder for 1.5-ml microcentrifuge tube (Covaris).
10. Covaris-2 Series Machine Holder for 0.65-ml microcentrifuge tube (Covaris).
11. 15 ml tube (Falcon).
12. 50 ml tube (Falcon).
13. Vortexer (any supplier).
14. Semi-automated Xstream pipettor (Eppendorf).
15. Repeater plus pipette (Eppendorf).
16. MicroAmp Optical 96-Well Reaction Plates (Applied Biosystems).
17. MicroAmp Optical Adhesive Film (Applied Biosystems).
18. Six Tube Magnetic Stand (Applied Biosystems).
19. NanoDrop ND1000 Spectrophotometer (Thermo Scientific).
20. Nuclease-free water.

2.4.2 Breaking the Emulsion PCR

1. SOLiD Buffer Kit (Applied Biosystems).
2. Repeater plus pipette (Eppendorf).
3. 50 ml tube (Falcon).
4. SOLiD Emulsion Collection Tray Kit (Applied Biosystems).
5. Fume hood (any supplier).
6. NanoDrop ND1000 Spectrophotometer (Thermo Scientific).
7. Six Tube Magnetic Stand (Applied Biosystems).
8. 1.5 ml LoBind Tubes (Eppendorf).

2.4.3 Enrichment of Templated Beads

1. SOLiD Bead Enrichment Kit (Applied Biosystems).
2. SOLiD Buffer Kit (Applied Biosystems).
3. Microcentrifuge (12,500 × *g*).
4. 15 ml tubes (Falcon).
5. Six Tube Magnetic Stand (Applied Biosystems).
6. Covaris S2 Sonicator (Covaris).
7. Covaris-2 Series Machine Holder for 1.5-ml microcentrifuge tube (Covaris).
8. Covaris-2 Series Machine Holder for 0.65-ml microcentrifuge tube (Covaris).
9. 0.5 ml LoBind Tubes (Eppendorf).
10. 1.5 ml LoBind Tubes (Eppendorf).
11. 2.0 ml LoBind Tubes (Eppendorf).

2.4.4 3′-End Modification of Enriched Templated Beads

1. Six Tube Magnetic Stand (Applied Biosystems).
2. SOLiD Buffer Kit (Applied Biosystems).
3. Covaris S2 Sonicator (Covaris).
4. Covaris-2 Series Machine Holder for 1.5-ml microcentrifuge tube (Covaris).
5. SOLiD Bead Deposition Kit (Applied Biosystems).
6. 1.5 ml LoBind Tubes (Eppendorf).
7. NanoDrop ND1000 Spectrophotometer (Thermo Scientific).

2.4.5 Bead Deposition on SOLiD Sequencing Slide and Instrument Run

1. SOLiD Bead Deposition Kit (Applied Biosystems).
2. Covaris S2 Sonicator (Covaris).
3. 1.5 ml LoBind Tubes (Eppendorf).
4. Six Tube Magnetic Stand (Applied Biosystems).
5. Covaris-2 Series Machine Holder for 1.5-ml microcentrifuge tube (Covaris).
6. SOLiD Slide Kit (Applied Biosystems).
7. SOLiD Bead Deposition Kit (Applied Biosystems).
8. SOLiD Fragment Library Sequencing Kit (Applied Biosystems).
9. SOLiD Mate-Paired Library Sequencing Kit (Applied Biosystems).
10. SOLiD Instrument Buffer Kit (Applied Biosystems).
11. SOLiD Deposition Chamber 1, 4, 8 well (Applied Biosystems).

2.5 Computational Analysis of Fosmid Sequences and Haplotype Assembly

2.5.1 Hardware Requirements

The computational requirements for primary and secondary sequence analysis include high-end computing servers with > 1 Terabyte (TB) disk space (RAID system) and > 32 Gigabyte (GB) Memory (RAM), and > 4 CPU cores per server or cluster node.

2.5.2 Software Requirements

The following list of software tools is used at different (generally successive) stages of this process:

For SOLiD NGS Analysis: BioScope™ Software
For other NGS systems such as Illumina:
Bwa [13]: http://bio-bwa.sourceforge.net/
Bowtie2 [14]: http://bowtie-bio.sourceforge.net/bowtie2/manual.shtml
Picard: http://broadinstitute.github.io/picard/
NGSEP [15]: http://sourceforge.net/projects/ngsep/
GATK [16]: https://www.broadinstitute.org/gatk/
Samtools [17]: http://samtools.sourceforge.net/
Fosmid detector [3] and ReFHap [5, 11]: http://www.molgen.mpg.de/~genetic-variation/SIH/Data/algorithms. This address includes a README.txt file with detailed instructions to run both the fosmid detector and ReFHap.

3 Methods

The following procedures apply to haplotype-resolving one individual genome.

3.1 Fosmid Library Construction

To establish a high quality individual fosmid library, two kits proved to work well and efficiently in conjunction, the DNA Extraction Kit from Stratagene and the EpiFOS Fosmid Library Production Kit from Epicentre. The first one, specifically, provides three solutions almost ready to use, and a protease mixture to digest cellular proteins and RNase to eliminate RNA. Its protocol is applicable to isolate HMW gDNA from blood, whole tissue, and cultured cells. The second kit can be applied to establish a fosmid library from sheared gDNA, and provides all components and reagents required to end-repair and ligate the sheared gDNA into a single copy fosmid vector, phage-package the fosmid clone DNA, and transfect *E. coli* cells (*see* **Note 1**).

3.1.1 High Molecular Weight Genomic DNA Isolation

The preparation of fosmid clones requires very high quality HMW gDNA.

1. Start with 8 ml EDTA blood in a 50 ml reaction tube for the extraction of a total of 40–50 μg HMW gDNA per individual sample. Add 42 ml 1× Solution 1 according to the manufacturer's protocol, and incubate the sample on ice for 2 min, followed by spinning it at 350 × *g* and 4 °C for 15 min. Discard the supernatant carefully, because the pellet is very instable. Resuspend the pellet in 11 ml of Solution 2, add 5 μl pronase (225 mg/ml) to a final concentration of 100 μg/ml and incubate the sample in a shaking water bath for 1 h at 60 °C. Transfer it on ice for 10 min, add 4 ml of Solution 3, and invert the tube several times (*see* **Note 2**). Incubate the sample for 5 more min on ice to precipitate the cellular proteins. Spin the tube at 2000 × *g* and 4 °C for 15 min and transfer the supernatant to a sterile 50 ml tube by using a large-bore pipet tip (*see* **Note 3**).
2. For RNA digestion, add RNase (10 mg/ml) to a final concentration of 20 μg/ml and incubate the sample in a water bath for 15 min at 37 °C.
3. The HMW gDNA can be precipitated after adding two volumes of absolute EtOH and gently inverting the tube. Use a sterile glass rod to spool the gDNA and rinse it with 70% EtOH. Dry the spooled DNA, transfer it into a sterile 50 ml tube, and carefully resuspend it in 500 μl pre-warmed TE buffer (*see* **Note 4**). Avoid any vortexing or pipetting, which might degrade the HMW gDNA.

4. Dissolve the pellet on a slowly rotating tube roller at 4 °C (*see* **Note 5**). Calculate the yield and concentration of your sample by measuring the OD_{260}. Store the DNA at 4 °C.
5. Control the quality of the extracted HMW gDNA: run a Pulsed Field Gel Electrophoresis (PFGE) with a 1% agarose gel in 0.5× TBE, load 3 μl of the HMW gDNA, use λ DNA-Mono Cut Mix DNA Ladder in the outer lanes as reference, run PFGE at 6 V/cm, switch time 0.2–2 s, 13–16 h (*see* **Note 6**).

3.1.2 Mechanical Shearing of HMW gDNA and End-Repair

To provide the basis for the fosmid library construction, the extracted HMW gDNA needs to be fragmented into ~40 kb segments and end-repaired to be cloned into a fosmid vector. DNA fragments larger than ~60 kb or smaller than ~20 kb can prevent phage-packaging at a later stage. Moreover, the use of fragments <20 kb might result in the formation of chimeric clones.

1. Dilute 8 μg HMW gDNA to a final concentration of 20 ng/μl.
2. Shear the DNA using a HydroShear shearing device; use the 4–40 kb (LARGE) shearing assembly. Since every shearing assembly has slightly different shearing properties, test different speed codes at the outset: prepare three aliquots of 8 μg DNA (400 μl) and test them with speed codes "16," "17," or "18," retraction speed "20," and 25 shearing cycles. Check and compare the results on an agarose gel.
3. Alternatively, if a HydroShear device is not available, the DNA can be sheared manually using a sterile 1 ml disposable syringe with a 23G 1¼ needle, aspirating 400 μl of the diluted HMW gDNA and pulling the syringe up and down for 50 s (12 times) (*see* **Note 7**).
4. After controlling the shearing results on by PFGE (*see* Subheading 3.1.1, **step 5**), precipitate the sheared DNA with isopropanol (100%) (*see* Subheading 3.1.3, **step 5**) and resuspend the sheared gDNA in 26 μl TE.
5. To produce blunt-end gDNA fragments, mix 4 μl End-Repair 10× Buffer, 4 μl of 2.5 mM dNTPs and 4 μl of 10 mM ATP into a 1.5 ml tube, add 26 μl of sheared gDNA and 2 μl End-Repair Enzyme (final volume 40 μl), incubate the mix for 60 min at room temperature, and transfer the tube into a preheated thermomixer (70 °C) for 10 min to inactivate the enzymes (*see* **Note 8**).

3.1.3 Size Selection of End-Repaired gDNA Fragments by PFGE

Size-selection by PFGE is performed to guarantee suitable DNA fragments for library production.

1. Prepare a 1% low melting point (LMP) agarose gel (with 200 ml 0.5× TBE), using a preparative comb with four slots (*see* **Note 9**). Mix 40 μl end-repaired gDNA with 7 μl 6× loading dye, and slowly pipet gDNA sample into one preparative

slot. Keeping one well empty between gDNA sample and size ladder, load 1 μl λ DNA-Mono Cut Mix DNA ladder (mixed with 1 μl 6× loading dye and 4 μl dH_2O) on both sides of the gDNA sample. Run a PFGE at 6 V/cm; initial sweep time 0.5; final sweep time 2.0 s; for 20–22 h.

2. Use Sybr Gold Dye to stain the agarose gel to avoid the need for UV exposure. Prepare a staining solution by pipetting the 10 μl Sybr Gold Stock Solution into 100 ml of 0.5× TBE (in a 100 ml glass flask wrapped in aluminum foil), pour the staining solution into a plastic tray, lift the gel into the staining bath, and leave for at least 30 min (*see* **Note 10**).
3. Visualize stained gDNA fragments on a Dark Reader Transilluminator. Ensure that the bulk of sheared gDNA has migrated within the correct size range (20–50 kb). Make a hole with a pipet tip at the 30 kb and 48 kb band of the size ladder, turn off illumination, use a sterile razor blade to select sheared gDNA by making a cut parallel to the 30 and 48 kb size ladder band (using the holes as marks) and excising the gel pieces with size-selected DNA fragments by vertical cuts (*see* Fig. 2 and **Note 11**).
4. To recover the size-selected gDNA, warm Gelase 50× Reaction Buffer in a thermomixer at 45 °C, set a second thermomixer (or a water bath) to 70 °C, push excised gel slice into a 2 ml tared tube, and weigh the gel piece. Translate solid gel weight into volume of molten agarose (1 mg of solid gel will result in 1 μl molten agarose), calculate volume of 50× Gelase Buffer required to yield a 1× Buffer, and calculate Gelase enzyme units (1 U/μl; 1 U Gelase enzyme per 100 μl molten agarose) needed after the subsequent step. Heat tube with weighed gel

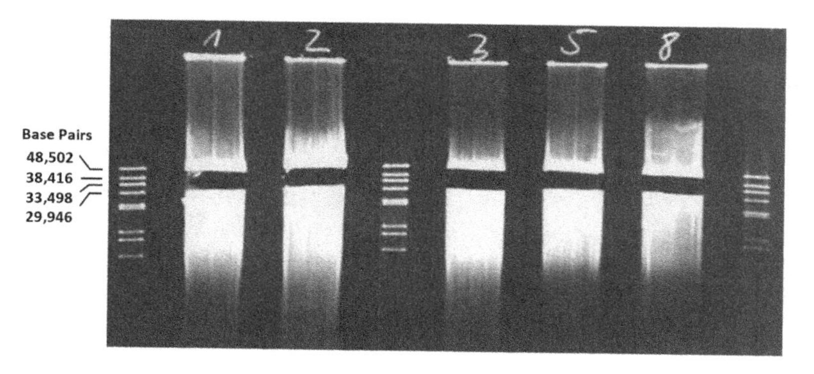

Fig. 2 Size-selection of end-repaired gDNA fragments after PFGE. Sheared gDNA from five different individuals resolved by PFGE on a 1 % low melting point (LMP) agarose gel (200 ml 0.5× TBE) together with a λ DNA-Mono Cut Mix DNA Ladder; gel stained with Sybr Gold Dye. Gel pieces with fragments between ~30 and 48 kb in size were cut out

slices in a 70 °C thermomixer (or water bath), keep for 10–15 min until the agarose is molten, quickly transfer tubes with molten agarose to 45 °C thermomixer, add appropriate volume of pre-warmed 50× Gelase Buffer and appropriate units of Gelase enzyme, gently mix the solution, and incubate for at least 60 min at 45 °C. Heat inactivate Gelase enzyme at 70 °C, then chill the mix for 15 min in an ice bath. Centrifuge the reaction mix for 20 min at maximum speed (11,000 × *g*), and transfer most of supernatant to a new 1.5 ml tube, making sure to not aspirate the gelatinous pellet.

5. Purify size-selected gDNA after gelase digest, add 1/10 volume 3 M sodium acetate (pH 7), mix gently, add 1 volume isopropanol, gently invert the tube, leave for 10 min at room temperature, centrifuge for 20–30 min at full speed, remove supernatant and wash pellet with 1 ml 70% ethanol twice, air-dry pellet for 10 min, and resuspend in 11 μl TE.
6. To control for quality and quantity of size-selected gDNA, run an 0.8% TBE agarose gel (minigel of 7–10 cm length), load 1 μl of size-selected gDNA, use 1 μl Fosmid Control (FC) DNA from the EpiFOS Fosmid Library Production Kit (Epicentre) (size 40 kb, 100 ng/μl) on both sides of the sample, additionally load 1 μl each of a 1:2 and 1:4 dilution of the FC DNA, run the gel at 6 V/cm for 40–60 min, stain with ethidium bromide, and visualize on an UV-illuminator and check the quantity and size of the size-selected gDNA by comparing the sample to the FC DNA dilutions.

3.1.4 Preparation of Epi100 Bacteria for Mass Transfection

The EpiFOS Fosmid Library Production Kit (Epicentre) provides a glycerol stock of EPI100 bacteria suitable for mass transfection.

1. Use multiple 10 cm petri dishes, prepare LB agar [without chloramphenicol (CA)], autoclave and pour warm into dishes, and use solidified LB agar plates to streak Epi100 plating strain (*see* **Note 12**). Incubate at 37 °C overnight, seal and store the plate at 4 °C. The day before the phage-packaging procedure, pick a single colony from the plated strain and inoculate a 50 ml Erlenmeyer flask with fresh LB broth and 10 mM $MgSO_4$. Shake the flask overnight at 37 °C, and use 2 ml of the overnight culture for the next step (Subheading 3.1.4, **step 2**). The overnight culture can be stored at 4 °C (*see* **Note 13**).
2. On the day of phage packaging, prepare an Erlenmeyer flask with fresh 50 ml LB broth and 10 mM $MgSO_4$, then inoculate with 2 ml of the Epi100 overnight culture from **step 1**; shake at 37 °C until OD_{600} reaches 0.85, measuring every 20 min.
3. Prepare multiple LB plates with CA in 10 cm petri dishes; these will be used to determine the titer of packaged fosmid clones (phage particles) in Subheading 3.1.5. Seal and store LB plates at 4 °C (*see* **Note 14**).

3.1.5 Ligation of Size-Selected gDNA into EpiFOS-Vector, Phage Packaging, and Testing Titer

In order to establish an individual fosmid library that ensures a ~7× coverage of the haploid and ~15× coverage of the diploid genome, a total of 1.44×10^6 fosmid clones need to be generated.

1. For the ligation reaction, combine 1 μl Fast-Link Ligation 10× Buffer, 1 μl 10 mM ATP, 1 μl pEpiFOS-Vector (500 ng), and 250 ng of the size-selected gDNA (corresponding to a 10:1 M ratio of vector:insert) in a new 200 μl microtube, add 1 μl Fast-Link Ligase (2 U/μl) and dH_2O to a final reaction volume of 10 μl, seal the tube and incubate the reaction for 2 h at room temperature, inactivate by placing micro tube at 70 °C for 10 min. In order to control the fosmid library production procedure, prepare a control reaction, use 2.5 μl of 40 kb FC insert DNA provided in the kit instead of size-selected gDNA. The control ligation reaction used to determine the packaging efficiency (*see* Subheading 3.1.5, **step 3**) should yield more than 1×10^7 colony forming units (cfu)/ml of control insert DNA.
2. For phage packaging, thaw one tube of MaxPlax-Lambda Packaging Extract on ice, immediately transfer half of the extract (25 μl) into a new 1.5 ml tube, and store the remaining half of packaging extract at −80 °C freezer. Add 10 μl of ligated DNA and mix by pipetting the reaction repeatedly (do not introduce air bubbles), spin down the tube, and incubate the reaction mixture for 90 min in a thermomixer at 30 °C. Five min before the end of the incubation period, take the remaining half of the packaging extract out of the freezer and put it on ice to thaw it. Then add it to the reaction mixture and incubate the reaction again for 90 min at 30 °C. Finally, add 940 μl Phage Dilution Buffer (PDB) to a total volume of 1 ml, mix gently, then add 25 μl of chloroform to the reaction, and store at 4 °C.
3. Phage titer testing: To determine the number of phage particles that contain fosmid clones, use 10 μl of the 1 ml phage packaging reaction (*see* **step 2**), add 990 μl PDB to generate a $1{:}10^2$ phage dilution, then use 100 μl of the $1{:}10^2$ dilution and 900 μl of PDB to prepare a $1{:}10^3$ dilution. Processing two dilutions in parallel is useful to be able to determine the efficiency of the phage packaging reaction as precisely as possible. Then mix 10 μl of each phage dilution with 100 μl of prepared Epi100 bacteria (*see* Subheading 3.1.4) into a 1.5 ml tube each, incubate both mixtures for 20 min at 37 °C by the use of a thermomixer, then plate the total volumes of 110 μl of transfected Epi100 cells each on an LB plate with CA (petri dish), and incubate at 37 °C overnight. To make sure that the packaging efficiency is sufficient to establish a complex fosmid library, count the numbers of colonies grown from each of the phage dilutions to calculate the total expected number of circular fosmid clones packaged by phages (*see* **step 2**) as follows: (Number of colonies x dilution factor × 1000 μl) divided by

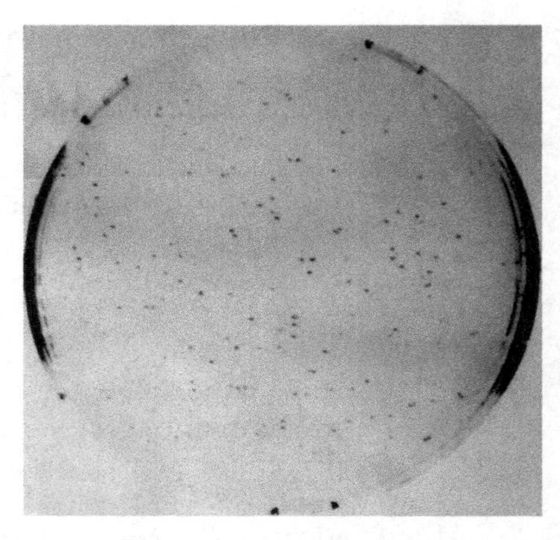

Fig. 3 Titer testing to determine the total number of fosmid clones expected from mass transfection. A petri dish (agar plate with chloramphenicol) with multiple bacterial colonies is shown; these were grown overnight from Epi100 bacteria transfected with a 1:10^2 phage dilution. 166 colonies are counted, allowing calculation of the total number of expected fosmid clones to nearly 1.7×10^6, sufficient to generate an individual fosmid library from a single cloning procedure

(volume of plated phage particles in µl) (*see* **Note 15** and Fig. 3). If the expected total number of fosmid clones is below 1.44×10^6—that is, if less than 150 fosmid clones are counted from plating the 1:10^2 phage dilution—start a second fosmid cloning round beginning with the first step, shearing of 8 µg HMW gDNA. Store the packaged phages in PDB at 4 °C until the required number of fosmid clones has been obtained.

3.1.6 Mass Transfection of E. Coli

Make sure that all the following steps are performed in one workflow under sterile conditions. Specifically, work under a sterile laminar flow hood and use sterile media only.

1. To begin with, prepare 440 ml LB broth with 10 mM $MgSO_4$ and 12.5 µg/ml CA, aliquot the prepared LB into 3×96-deep well plates, 1.5 ml per well.
2. For mass transfection, prepare all phage packaging reactions required to generate 1.5×10^6 fosmid clones and determine their total volume (*see* Subheading 3.1.5, **step 3**). Fill an Erlenmeyer flask with the prepared Epi100 bacteria (*see* Subheading 3.1.4) to bring the final transfection volume up to 30 ml. For example, if one phage packaging reaction of 1 ml has proven sufficient to achieve a total of 1.5×10^6 phages, start with 29 ml of Epi100 cells.
3. Add the phages to the Epi100 cells and let the phage particles adsorb for 20 min at 37 °C. Preserve an aliquot of 20 µl from the mass transfection mix, expected to contain ~1000 fosmid clones, in order to control the clone pool complexity; store on ice until use in Subheading 3.1.7, **step 2**.

3.1.7 Partitioning Fosmid Clones into Pools and Amplification of Fosmid Library

Partitioning the fosmid clones into 3×96-deep well plates, i.e., 288 wells, allows generation of "haploid clone pools" that contain 5000 fosmid clones per well. This library format minimizes the probability that complementary parental haplotypes co-occur in one well. Thus, multiple fosmid pools, each representing a random mixture of ~40 kb haploid DNA segments covering 5 % of the haploid genome, can be sequenced to saturate both haploid genomes. The partitioning and amplification of an entire fosmid library with agar plates is very labor-, and cost-intensive. To simplify the procedure we have chosen to use liquid cultures at this stage as described below (*see* **Note 16**).

1. Dispense 100 μl of the mass transfection into each well of the 3×96-deep well plates, expected to correspond to 5000 fosmid clones per well, and cover with breathable seal. Each well contains prepared LB broth (*see* Subheading 3.1.6, **step 1**).
2. After partitioning fosmid clones into deep wells, incubate for 20 h at 37 °C at 200 rpm in a shaking incubator to amplify the clones.
3. Take the 20 μl sample preserved in Subheading 3.1.6, **step 3**, and plate 5 μl (expected to correspond to ~250 colonies) and 2.5 μl (expected to correspond to ~125 colonies) each on a 10 cm LB plate with CA and incubate at 37 °C overnight.
4. The next day, count the number of grown clones. Determine the obtained clone pool complexity as follows: (Number of cfu/Volume of plated fosmid clones) = (Number of cfu per well/Dispensed mass transfection volume). For example, if 260 cfu are counted on an LB plate, the number of cfu, or fosmid clones that have been obtained per well in the library is calculated as (260 cfu/5 μl) = (Number of cfu per well/100 μl), and is 5200 in this case
5. For long-term storage of fosmid libraries, aliquot 250 μl of glycerol (90 %) into each of the wells of the 3×96-well plates to a final volume of 1.85 ml, mix and store at −80 °C..
6. In order to prepare working plates, thaw glycerol stock deep well plates and pipet 150 μl per well from the deep well stock plate into a 96-well working plate; make sure to pipet deep well content up and down at least once before removing 150 μl to capture all individual clones. Use a plate with wells of at least 250 μl of capacity.
7. To prepare super-pool plates with 15,000 fosmid clones per well, combine the three 96-well working plates into a single 96-well plate. Transfer 50 μl from each 96-well working plate into a single, new 96-well plate; use an 8-channel pipette to always combine the wells with identical positions on the working plates. Cover the super-pool plate with breathable seal, pack into aluminum foil and store in −80 °C freezer.

3.2 Isolation of Fosmid DNA from Fosmid Clone Pools

Due to the usage of the EpiFos single-copy vector, each transfected Epi100 bacterium will only contain one fosmid clone. To isolate sufficient amounts of fosmid DNA from single fosmid super-pools for sequencing library preparation, while preserving library complexity, the use of LB agar plates for this amplification step (*see* **Note 16**) has proven to be indispensable. As a consequence of such large-scale cultivation of Epi 100 cells, large amounts of Epi100 genomic DNA and proteins must be removed, requiring additional purification steps. The fosmid clones scraped from the agar plate are divided into two portions for more efficient exonuclease digest of bacterial genomic DNA, and the proteins are filtrated before extraction of fosmid insert DNA (*see* **Note 17**).

3.2.1 Plating and Scraping Fosmid Super-Pools

As outlined in the Introduction, routinely, 40–48 super-pools per fosmid library are sequenced to be able to phase approximately 85–92% of the heterozygous SNPs. Thus, this number of super-pools was amplified for subsequent DNA isolation, preparing two plates per super-pool.

1. Prepare two large (22×22 cm) LB agar plates per super-pool with CA. Thaw 96-well super-pool plates with 15,000 fosmid clones per well on ice. Scratch thawed fosmid-glycerol mixture with the inoculating loop to yield approximately 3 μl fosmid-glycerol mixture per well, and mix with 1 ml LB medium and CA in a 1.5 ml tube. Pipet 500 μl of the LB-clone mix onto one LB plate and the other half of the LB-clone mix onto the second LB plate, spread with Drigalski spatula, and incubate at 37 °C overnight.
2. To scrape the clones, rinse 3×5 ml LB medium over the first incubated LB plate, carefully scrape colonies from agar plate with a Drigalski spatula, and pipet clones from LB plate into a single 50 ml Falcon tube. Repeat for the second LB plate containing the second half of the super-pool, and pipet collected clones into a new 50 ml Falcon tube.

3.2.2 Extraction of Fosmid Super-Pool DNA

1. Chill centrifuge to 4 °C, and centrifuge the two 50 ml Falcon tubes each containing half of the clones from a single super-pool at 6000 ×*g* for 15 min. For extraction, the QIAGEN Large-Construct Kit is used, which contains most of components, reagents, and buffers. Resuspend pellet in 10 ml P1 Buffer. Add 10 ml P2 Buffer and mix by inverting rigorously (5×) to lyse cells (but do not vortex), leave reaction for 2 min at room temperature, stop cell lysis reaction by adding 10 ml prechilled (4 °C) P3 buffer, and cool reaction 10 min on ice. Centrifuge the Falcon tubes for 30 min at 4 °C at a minimum of 12,500×*g* (*see* **Note 18**).
2. Prepare two fluted filters, wet the filter paper with dH_2O and put each filter onto a new 50 ml Falcon tube, pipet the supernatant (containing fosmid DNA) from the first centrifuged

Falcon tube onto one pre-wetted fluted filter, and the supernatant from the second Falcon tube onto a second pre-wetted fluted filter, wait until supernatant has been filtrated.

3. Equilibrate two QIAGEN-tips 500 (from the QIAGEN Large-Construct Kit) with 10 ml QBT Buffer, transfer the clear filtrate from each Falcon tube to one QIAGEN-tip, wash the tip twice with 30 ml QC Buffer, elute fosmid super-pool DNA with 15 ml preheated (65 °C) QF Buffer. Pool the eluates from both Falcon tubes, which are from a single fosmid super-pool.
4. Precipitate fosmid super-pool DNA by adding 0.7 volumes (21 ml) isopropanol, mix and centrifuge for 40 min at 12,500 × *g*. Wash pellet with 5 ml ethanol (70 %), centrifuge for 30 min at 12,500 × *g*, air-dry pellet at room temperature, and redissolve DNA in 9.5 ml EX Buffer.
5. To remove contaminating genomic DNA from the EPI100 bacteria and nicked or damaged large-construct DNA, an exonuclease digestion is strongly recommended. Therefore, prepare first 100 mM ATP solution: Dissolve 2.75 g ATP (adenosine 5′triphosphate disodium salt, anhydrous, MW 551.14) in 40 ml distilled water, adjust the pH to 7.5 with 10 M NaOH, bring the volume up to 50 ml, distribute into 300 μl aliquots, and store at −20 °C. Secondly, prepare ATP-dependent Exonuclease by resuspending 80 μg ATP-Dependent Exonuclease in 225 μl Exonuclease Solvent. Then add 200 μl ATP-dependent Exonuclease and 300 μl of the 100 mM ATP solution to dissolved DNA to remove noncircular DNA from the bacterial genome, mix thoroughly, and incubate reaction mix at 37 °C overnight.
6. Equilibrate one QIAGEN-tip 500 with 10 ml QBT buffer, mix the exonuclease-digested fosmid super-pool DNA with 10 ml QS buffer, and transfer the whole DNA mix to QIAGEN-tip 500. After the DNA has passed the tip, wash two times with 30 ml QC Buffer, elute and precipitate fosmid super-pool DNA as described in **steps 3** and **4**, resuspend air-dried DNA pellet in 300 μl TE, and dissolve at room temperature overnight. Transfer fosmid DNA into a new 1.5 ml Eppendorf tube and rinse Falcon tube with additional 100 μl TE to collect all fosmid super-pool DNA. Check the quality and quantity of extracted fosmid super-pool DNA on a 1.5 % agarose gel, run at 120 V for 20 min (*see* Fig. 4).

3.3 Fosmid Pool-Based Next Generation Sequencing Library Preparation

At the time of developing and applying our method to production, we used a SOLiD System from ABI/Life Technologies. The extracted super-pool fosmid DNA can also be analyzed in conjunction with any other NGS technology following the manufacturer's protocols for NGS sequencing library preparation. All reagents have been provided with the SOLiD system sequencing kits, if not

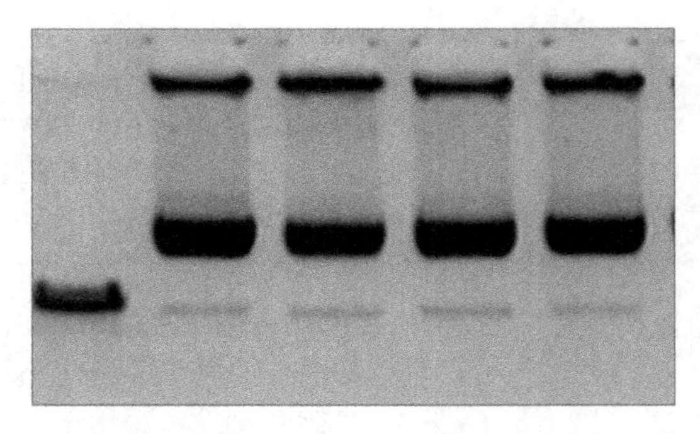

Fig. 4 Fosmid super-pool DNA controlled after extraction. The DNA isolated from four super-pools is shown resolved on a 1.5 % agarose gel stained with ethidium bromide; on the *left*, (linearized) fosmid control (FC) DNA, the faint DNA bands of the same size as FC DNA seemingly represent linear fosmid DNA, which has escaped purification. The larger, thick bands represent the circular fosmid DNA

specified otherwise. The fosmid DNA isolated from a single fosmid super-pool is used as input to prepare a sequencing library; thus, up to 96 sequencing libraries could be generated if one would want to utilize all super-pools from an individual fosmid pool library. To allow parallel processing of multiple NGS sequencing libraries, we created barcode tags by modification of the P1-adaptor sequence. With the barcoded sequences, up to 16 super-pools could be multiplexed later in the clonal amplification step and the sequencing run.

3.3.1 Barcoded Preparation of Fragment Libraries

In this case, the isolated fosmid super-pool DNA is used as input to create a library of short DNA sequencing template molecules 100–150 bp in length.

1. At least 30 min ahead of time, prepare Covaris S2 Sonicator for DNA shearing. Fill with deionized water to mark level 12 (water should cover glass microtube), degas for 30 min while continuously keeping the pump on, then cool down to 4 °C. Make sure that no bubbles appear at the bottom of the tube after degasing has finished (*see* **Note 19**).
2. Use 3 μg of extracted fosmid pool DNA in 100 μl Low TE buffer, place capped Covaris microtube into Covaris G7 adaptor. Transfer 100 μl DNA into microtube using extra thin pipet tip to avoid damaging of microtube presplit septa (cap of Covaris microtube) and make sure to not introduce an air bubble at the bottom of the microtube; shear the input DNA for 360 s ("DC" 20 %, intensity 5, cycles burst 200, time 60 s, six cycles, frequency sweeping mode) to generate fragments between 100 and 150 bp in size.
3. Control the shearing result by running a Lonza FlashGel System with a 2.2 % Lonza FlashGel Cassette: mix 3 μl of sonicated DNA with 3 μl 1× Loading Dye (diluted from 6× Orange DNA Loading Dye), pipet 2.5 μl Low Range DNA Ladder in

the first and last lane, run gel at 275 V for 6 min, visualize DNA while migrating. If the bulk of sheared DNA is above a size range of 100–150 bp, repeat Covaris shearing as described above for 180 s, and control fragment sizes again on a Lonza gel. Concentrate sheared DNA with QIAGEN MinElute columns (from MinElute Gel Extraction Kit, QIAGEN) as per manufacturer's protocol, elute twice in 21 μl elution buffer, take 1.5 μl of eluted sample to measure final DNA concentration, and transfer sheared DNA to a 1.5 ml LoBind tube.

4. End-repair of fragments for blunt-end ligation: mix 40 μl of sheared fosmid pool DNA, 6 μl End-Repair Buffer, 6 μl 10× dNTPs, 6 μl ATP, 2 μl End-Repair enzyme from the End-It DNA End-Repair Kit (Epicentre), incubate at 21 °C in a thermoshaker for 30 min, purify with MinElute Columns as per protocol, elute twice in 20 μl elution buffer, quantify with Qubit per manufacturer's protocol.
5. Size selection of end-repaired fragments: run a 4% Reliant NuSieve 3:1 agarose gel with ethidium bromide (*see* **Note 20**), mix total volume of end-repaired DNA with 7 μl Loading Dye (6×), load three slots each with 15 μl DNA, keep one well on both sides of the samples empty, use 5 μl Low Range DNA ladder at both sides of the gel, run for 40 min at 110 V (210 mA). Visualize sample on an UV Multibright Transilluminator shortly, use UV protection shield to excise the DNA between bands 100 and 150 bp (*see* Fig. 5a), transfer gel piece into one 2 ml tube and cut into smaller pieces, weigh

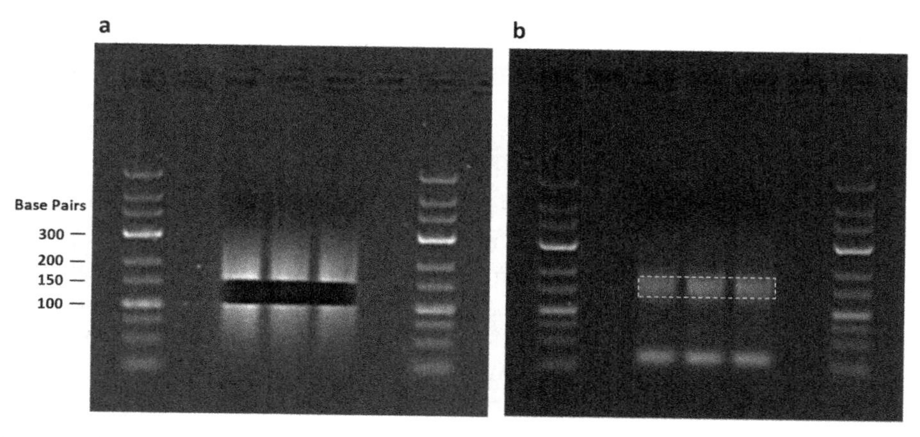

Fig. 5 Size-selection of sheared fosmid super-pool DNA for preparation of fragment sequencing libraries. (**a**) Three aliquots of sheared, end-repaired fosmid clone DNA fragments before ligation of adaptor sequences are shown resolved on a 4% Reliant NuSieve 3:1 agarose gel (Lonza) with ethidium bromide together with a Low Range DNA Ladder, and a gel piece containing DNA fragments of 100–150 bp in size has been cut out. (**b**) The excised DNA fragments are shown resolved after ligation of adaptor sequences, and are now slightly larger, in the range of 150–200 bp in size. These ligation products are again excised, as indicated by the *white dashed lines*, for subsequent purification and nick translation. The bands at the bottom represent excess adaptor sequences that were not ligated

tared tube and digest gel as described by the manufacturer's protocol, *see* also Subheading 3.1.3, **step 4**; then purify with QIAGEN MinElute columns, elute twice in 25 μl elution buffer, and use 1 μl to determine DNA concentration with Qubit (*see* **Note 21**).

6. The barcoding of a fosmid super-pool DNA sample is achieved by ligating a modified P1 adaptor sequence containing a unique barcode sequence tag of 6 bp (*see* **Note 22** including a list of these unique barcode sequences) to the DNA fragments instead of the universal adaptors provided by the SOLiD Fragment Library Oligo Kit. First, the necessary picomoles (pmol) of adaptor required for ligation are determined according to the following formula:

   1 μg DNA × 10^6 pg/μg × 1 pmol/660 pg × 1/(Average insert fragment size) = 17 pmol

   (# μg DNA) × (17 pmol DNA) = (# pmol DNA for adaptor ligation)

   (# pmol DNA for adaptor ligation) × (30) = (# pmol adaptors needed)

   (# pmol adaptors needed)/(# pmol/μl stock) = (# μl adaptor needed)

   With an average size of 125 bp for sheared DNA fragments, the volume of universal adaptors can be calculated as follows: (Total amount of purified DNA in pg/660)/125 × (30/50).

7. Ligation reaction: pipet 47–49 μl of purified fosmid pool DNA into a 1.5 ml LoBind tube, add 100 μl 2× Quick Ligation Buffer (from the Quick Ligation Kit, NEB), P1 and P2 adaptors (50 pmol/μl) as calculated in **step 6**, and 5 μl Quick Ligase, fill with dH_2O to a total reaction volume of 200 μl, incubate exactly 10 min in a thermomixer preheated to 21 °C, and stop the reaction instantly by adding 600 μl ERC buffer. Purify the reaction with MinElute columns, using the QIAGEN MinElute Reaction Cleanup Kit according to the manufacturer's protocol, and elute twice in 20 μl elution buffer to a final sample volume of 40 μl.

8. Size-selection of ligation products: mix the total sample volume with 7 μl 6× Loading Dye, load sample in three aliquots of 16 μl into three wells of a Reliant 4% NuSieve 3:1 agarose gel (with ethidium bromide), keep one well on each side empty, use Low Range DNA Ladder on both sides of gel, run for 40 min at 110 V, visualize on UV Transilluminator, size-select as described earlier (Subheading 3.3.1, **step 5**) the band between 150 bp and 200 bp (size corresponds to fosmid pool DNA plus ligated adaptor sequences; *see* Fig. 5b), and weigh gel slices in tared 2.0 ml tube. If gel slice weighs more than 400 mg, divide gel pieces into two tubes (*see* **Note 23**).

9. Purification of size-selected ligation products with MinElute Gel Extraction Kit (QIAGEN) according to the manufacturer's protocol: add 6 volumes QG buffer to 1 volume (equivalent to 100 mg or 100 μl) of gel, dissolve the gel by vortexing the tube at room temperature, vortex tube every 2–3 min for minimally 15 min until gel has completely dissolved. Add 1 gel volume of isopropanol, invert the sample, apply to MinElute column and centrifuge for 1 min, discard flow-through, add 500 μl QG buffer and centrifuge 1 min, discard flow-through, wash with 750 μl PE buffer, centrifuge 1 min, discard flow-through and centrifuge again 1 min, place the MinElute column in a new 1.5 ml LoBind tube, elute DNA with 18 μl of elution buffer twice, and use 1 μl for DNA concentration measurement with Qubit (*see* **Note 24**).
10. Nick translation: transfer 34 μl of adaptor-ligated DNA into a new 1.5 ml tube, add 4 μl 10× NEBuffer 2 (NEB), 0.8 μl dNTP-Mix (100 mM) and 1 μl DNA Polymerase I (10 U/μl) (NEB), fill with water to 40 μl, mix and incubate reaction for 30 min at 16 °C in a thermomixer; if necessary, inactivate the enzyme at 65 °C for 10 min and store reaction at 4 °C until large-scale PCR setup (*see* Subheading 3.3.3).

3.3.2 Barcoded Preparation of Mate-Paired Libraries

Mate pairs are defined as a pair of DNA fragments that originate from the two ends of the same genomic DNA fragment, the distance between the two mates depends upon the size of the original genomic DNA fragment (insert size), and can range from ~100 bp to several kb. To create mate-paired tags, a linear DNA fragment needs to be circularized by the use of LMP CAP adaptors connected to an internal biotinylated adaptor. These adaptors are provided by the SOLiD Mate-Paired Library Oligo Kit from Applied Biosystems (*see* also Subheading 2.3.2). The resulting DNA circle has one nick in each strand due to the LMP CAP adaptor. During nick translation, the length of the fragments that will be released after T7 exonuclease and S1 nuclease digest depends on the time and temperature of the DNA polymerase reaction. The largest part of the inserted, circularized DNA is cut out enzymatically, which leaves the two ends of the inserted DNA as mate-paired tags. After ligation of universal adaptors and subsequent PCR amplification, the mated tags (2 × 50 bp) can be sequenced together, allowing computational detection of larger structural genome variants. Each mate-paired library can also be indexed by using modified P2-adaptors containing a unique barcode sequence.

1. Shear 5 μg of extracted fosmid pool DNA in 125 μl nuclease-free water with the HydroShear (standard assembly) to generate gDNA fragment sizes of 1–2 kb (speed code 5, 20 cycles), 3–4 kb (speed code 13, 20 cycles), and 5–10 kb (speed code 15, 5 cycles). Control the shearing results on a 0.8 % agarose gel (*see* **Note 25**).

2. Purify sheared fosmid pool DNA with QIAquick Gel Extraction Kit: add 3 volumes QG buffer and 1 volume isopropanol to sheared DNA, transfer 750 μl of DNA-QG buffer mix to QIAquick column (maximum DNA binding capacity 10 μg per column), wait for 2 min at room temperature, centrifuge 1 min at minimally 13,000 × *g* and discard flow-through, repeat the steps until the entire sample has been loaded, wash column with 750 μl PE buffer, centrifuge for 2 min at minimally 10,000 × *g*, repeat centrifugation, air-dry column for 2 min, transfer column to a new 1.5 ml LoBind tube, elute by adding 30 μl EB buffer to column, let stand for 2 min, centrifuge column at minimally 10,000 × *g* for 1 min, repeat elution and centrifugation step.
3. End-repair with End-It DNA End-Repair Kit (Epicentre) as per manufacturer's protocol; the total reaction volume is 10-fold the amount of input DNA (in μg). Purify with QIAquick columns (*see* **step 2**).
4. Ligation of end-repaired sheared fosmid pool DNA to LMP Cap Adaptors resulting in a nick on each DNA strand during circularization: first calculate the molarity of each DNA fragment size (insert) based on the formula $((10^6/660) \times (1/\text{average insert size})) = X$ pmol/μg DNA. Use the molarity to calculate the amount of adaptor needed for the ligation reaction in μl ((total amount of input DNA × pmol of DNA fragment × 100)/(50 pmol) = μl adaptor). Combine 150 μl 2× Quick Ligase Reaction Buffer (NEB), the calculated volume of LMP cap adaptor, 7.5 μl Quick Ligase enzyme (NEB), the total amount of input DNA and add dH_2O to have a 300 μl reaction mix, then incubate the mix at room temperature for 10 min. Purify with QIAquick columns (*see* **step 2**).
5. Size selection: To remove unbound CAP adaptors, prepare a 0.8 % TAE agarose gel (Applied Biosystems) with ethidium bromide, mix ligated DNA with 10x loading dye, load 11 μl of sample per well, keep one well empty on each side of the DNA sample, load 1 μl of the 1 kb size DNA Ladder at both sides of the gel, run the gel at 120 V/cm, visualize the gel on an UV illuminator, excise a gel slice corresponding narrowly to the insert size and extract DNA with QIAquick Gel Extraction Kit as per protocol.
6. Circularize LMP CAP adaptor-ligated DNA with a biotinylated internal adaptor: for circularization, the components are calculated per μg of size-selected insert DNA: for a 1–2 kb insert size use 182.5 μl of 2× Quick Ligase Buffer, 1.5 μl Internal Adaptor (2 μM), 9 μl Quick Ligase, and add dH_2O to a volume of 365 μl. Accordingly, circularization of 3–4 kb insert sizes (or 5–10 kb insert sizes, respectively) require 280 μl (360 μl) of 2× Quick Ligase Buffer, 0.65 μl (0.4 μl)

Internal Adaptor, 14 μl (18 μl) Quick Ligase, then add dH_2O to a total volume of 560 μl (720 μl). Incubate the reaction for 10 min at room temperature, purify with QIAquick Gel Extraction Kit (*see* Subheading 3.3.2, **step 2**). The internal adaptor is biotin-labeled, enabling the specific binding of circularized DNA to streptavidin beads later in the protocol (*see* **step 11**) (*see* **Note 26**).

7. Isolate circularized DNA by digesting un-circularized DNA with Plasmid Safe ATP DNAse: combine 5 μl 25 mM ATP, 10 μl 10× Plasmid-Safe Buffer, the total volume of circularized DNA from **step 6**, add 0.33 μl Plasmid-Safe DNAse (10 U/μl) per μg of circularized DNA, then add dH_2O to a total reaction volume of 100 μl and incubate in a thermomixer at 37 °C for 40 min. The DNAse will digest only noncircularized DNA fragments. Purify with QIAquick Gel Extraction Kit Protocol (*see* Subheading 3.3.2, **step 2**) and quantify circularized DNA with Qubit; a minimum of 200 ng circularized DNA should be recovered to proceed with the next steps. For more complex genomes, 600 ng–1 μg circularized DNA is recommended for a high-complexity library.
8. The size of the mate-paired tags to be produced in this step critically depends on the reaction temperature and time of the nick-translation of circularized DNA: work on ice, per 1000 ng of circularized DNA add 5 μl of 100 mM dNTP mix, 50 μl 10× NEBuffer 2, circularized DNA, and bring up to a total volume of 490 μl with dH_2O; chill the reaction for at least 5 min by putting the tube into an ice-water bath, then quickly add 10 μl DNA polymerase I (10 U/μl); mix, incubate the reaction at 0 °C for 12–14 min, immediately stop the reaction by adding 3 volumes of Buffer QG and 1 volume of isopropanol and purify with QIAquick Gel Extraction protocol (*see* Subheading 3.3.2, **step 2**). The size of mate-paired tags created by DNA polymerase I can be controlled by reaction time and temperature (*see* **Note 27**).
9. T7 exonuclease digest to release the mated tags: per 1000 ng of circularized DNA combine 50 μl 10× NE Buffer 4 and circularized DNA, bring up to a total volume of 480 μl with dH_2O, add 20 μl T7 exonuclease (10 U/μl) and mix and incubate the reaction at 37 °C for 30 min. Purify with QIAquick Gel Extraction Kit according to the manufacturer's protocol (*see* Subheading 3.3.2, **step 2**). T7 exonuclease recognizes the nicks within the circularized DNA created in **step 8**, and digests the un-ligated DNA strand away from the tags creating a gap in the sequence. This allows cleavage of the mate-paired tags from the circularized template by S1 Nuclease (*see* the next step).
10. S1 Nuclease digest: Use S1 dilution buffer to prepare S1 Nuclease with an activity of 1 U/μl (Invitrogen); per 1000 ng

circularized DNA mix 50 μl 10× S1 Nuclease buffer, 25 μl 3 M sodium chloride, 50 μl 100 mM magnesium chloride, 20 μl of diluted S1 Nuclease, add dH_2O to a total volume of 500 μl, incubate the reaction at 37 °C for 30 min, immediately stop the reaction by adding 3 volumes of QG Buffer and 1 volume of isopropanol, and proceed with the QIAquick Gel Extraction Kit Protocol (*see* Subheading 3.3.2, **step 2**). S1 Nuclease removes the non-ligated part of the inserted DNA, leaving a linearized molecule with the ends (tags) of the inserted DNA attached to both sides of the molecule (mate-paired tags).

11. Binding of biotinylated library molecules to Streptavidin beads for purification: preparing Streptavidin Dynabeads for use after the end-repair step. Prepare the Streptavidin binding buffer by mixing 10 μl Tris–HCl (500 mM, ph 7.5), 200 μl 5 M sodium chloride, 1 μl 500 mM EDTA and add 289 μl dH_2O to a total volume of 500 μl. Prepare 1× Bead Wash Buffer (from the Solid Buffer Kit) by mixing 5 μl 100× BSA (NEB) and 495 μl dH_2O. Pre-wash Streptavidin Dynabeads: vortex bottle of Streptavidin Dynabeads, transfer 90 μl of beads into a 1.5 ml LoBind tube, add 500 μl 1× Bead Wash Buffer. Vortex and spin down, place tube into magnetic stand (Applied Biosystems) and wait until the solution clears. Aspirate the supernatant and discard it, add 500 μl 1× BSA, and vortex 15 s. Spin down, place tube into magnetic stand until solution clears, aspirate and discard supernatant. Add 500 μl of 1× Bind & Wash Buffer, vortex 15 s, spin down, place tube into magnetic stand to clear solution, aspirate and discard supernatant. Use pre-washed beads in **step 13**.
12. Combine 10 μl End-Repair buffer (10×), 10 μl ATP (10 mM), 10 μl dNTPs (2.5 mM each), 2 μl End-Repair Enzyme Mix, a total volume of S1-digested DNA from step 10, then add dH_2O to a total volume of 100 μl, incubate the reaction mix for 30 min at room temperature, stop the reaction by adding 5 μl EDTA (500 mM), 200 μl Streptavidin binding buffer (prepared in **step 11**) and 95 μl dH_2O.
13. Binding the library DNA molecules to Streptavidin beads: add the entire 400 μl reaction from **step 12** to the pre-washed Streptavidin beads (prepared in **step 11**), vortex, place the tube into a rotator for 30 min at room temperature, and spin down the tube.
14. Wash the DNA-Strepatavidin bead complex: place tube with DNA bound to beads into magnetic stand until solution clears. Remove and discard supernatant with pipet, remove tube from magnetic rack, and add 500 μl 1× Bead Wash Buffer (prepared in **step 11**). Transfer the suspension to a new 1.5 ml LoBind tube, vortex for 15 s, spin down, and resuspend in 500 μl 1× Bind & Wash Buffer. Vortex and spin down the solution and

clear on magnetic rack. Discard the supernatant again, and repeat washing step with another 500 μl 1× Bind & Wash Buffer. Resuspend the beads in 500 μl 1× Quick Ligase Buffer, vortex for 15 s, spin down and clear suspension on magnetic rack. Discard supernatant a last time, and resuspend beads in 97.5 μl 1× Quick Ligase Buffer.

15. Ligate P1 and P2 adaptors to end-repaired mate-paired molecules: calculate the necessary volume of P1 and P2 adaptors by using the formula in Subheading 3.3.1, **step 6**, with the size of circularized DNA corresponding to the average insert size. Combine 97.5 μl of the DNA-bead complex, the calculated volume of P1 and P2 Adaptor (50 pmol/μl each) and 2.5 μl Quick Ligase, and incubate the reaction for 15 min at room temperature (*see* **Note 28**).
16. Wash the DNA-bead complex as described in **step 14** but modified in the last step: add 500 μl 1× NEBuffer 2 (instead of Quick Ligase Buffer) and resuspend beads in 96 μl 1× NEBuffer 2.
17. Nick translation of library: Add 2 μl 100 mM dNTP mix and 2 μl DNA Polymerase I (10 U/μl) to the 96 μl DNA-bead complex in NEBuffer, incubate the reaction for 30 min at 16 °C, place tube into magnetic stand until solution clears, remove and discard supernatant, resuspend beads in 500 μl Buffer EB, place tube into magnetic stand until cleared, remove supernatant, and resuspend beads in 30 μl Buffer EB.

3.3.3 Large-Scale PCR of Fragment and Mate-Paired Libraries

The library is amplified with the Invitrogen PCR SuperMix. The number of PCR cycles should be as small as possible to avoid PCR-related biases due to differential amplification of library molecules.

Fragment Libraries:

1. To prepare a PCR master mix, pipet 400 μl PCR SuperMix into a 1.5 ml tube, add 10 μl of each primer, 40 μl dH_2O and 2.5 μl Pfu Polymerase (2.5 U/μl). Before adding the sample, transfer 50 μl of master mix without DNA sample into a new 200 μl tube (negative PCR control). Add 40 μl of nick-translated DNA (*see* Subheading 3.3.1, **step 10**) to the remaining master mix, vortex, and transfer four aliquots into new 200 μl PCR strips (*see* **Note 29**). Run large-scale PCR program (5 min at 95 °C, hold; few cycles (*see* **Note 30**) 15 s at 95 °C, 15 s at 62 °C, 60 s at 70 °C; 5 min at 70 °C, hold; 4 °C, forever, hold); the number of PCR cycles critically depends on the amount of input DNA as measured after purification of nick-translated DNA.
2. Purify PCR reaction with MinElute columns, elute in 60 μl, add 10 μl loading dye (6×), load the sample on a 4% Reliant NuSieve 3:1 agarose gel, run for 40 min at 110 V, size-select

PCR product between 150 and 200 bp and control for over-amplification; do not proceed, if PCR generates an over-amplification band. Gelase digest and extract DNA sample as described (*see* Subheading 3.3.3, **step 4**).

Mate-paired libraries:

1. Prepare a PCR master mix for four PCR reactions: mix 200 μl PCR SuperMix (Invitrogen), 8 μl of Library PCR Primer 1 (50 μM), 8 μl Library PCR Primer 2 (50 μM), 1 μl cloned Pfu DNA polymerase (2.5 U/μl), and 143 μl dH_2O. Aliquot 90 μl of PCR master mix to a 200 μl PCR tube, add 10 μl dH_2O as negative control. To the remaining 270 μl PCR master mix, add 27 μl DNA-bead complex, vortex, pipet 90 μl aliquots into three PCR tubes, run PCR program with initial denaturing step (10 min at 95 °C) followed by 5 cycles (15 s at 95 °C, 15 s at 62 °C, 60 s at 70 °C); 4 min at 60 °C, hold; 4 °C, forever, hold (*see* **Note 30**).
2. Control PCR reaction: Use a 4 μl aliquot of PCR reaction and mix with 1 μl 6× Orange DNA Loading Dye. Load sample on a 2.2% FlashGel System, load Low Range DNA Ladder in adjacent well, and run the gel for 6 min at 275 V. If fairly robust PCR amplification products are visible on the gel, pool all three PCR reactions into a new 1.5 ml LoBind tube. Otherwise, run two additional cycles on thermal cycler, control an aliquot using a Lonza FlashGel System, repeat thermal cycling steps until amplification is observed.
3. Place 1.5 ml tube on magnetic rack, carefully aspirate supernatant and transfer into a new 2.0 ml LoBind tube. Purify with QIAquick Gel Extraction protocol (*see* Subheading 3.3.2, **step 2**).
4. Gel purification step: use a Reliant 4% NuSieve 3:1 agarose gel with ethidium bromide, mix 6× Orange DNA Loading Dye to mate-paired library to a final 1× concentration. Load the entire dye-mixed sample DNA in aliquots of 11 μl per gel slot. Keeping one well on each side of sample DNA empty, load 2 μl Low Range DNA Ladder. Run the gel at 120 V, visualize and excise the sample DNA between bands 275 and 300 bp. Weigh gel slices in tared 15 ml conical tubes, add 6 volumes QG buffer per 1 volume gel, and dissolve the gel slices by vortexing at room temperature (do not heat). Add 1 gel volume isopropanol, and mix and invert. Apply 750 μl to a QIAquick column, let stand for 2 min at room temperature, centrifuge for 1 min at minimally 10,000 ×*g*. Repeat until the entire sample has been loaded on column. Wash the column by adding 500 μl QG buffer, centrifuge at 10,000 ×*g*, wash with 750 μl PE buffer, centrifuge, and air-dry the column for 2 min. Place column on a new 1.5 ml LoBind tube, add 30 μl EB buffer, wait 2 min, and centrifuge for 1 min. Quantify the eluted library with the Qubit system as per manufacturer's protocol.

3.3.4 Preparation of Fragment and Mate-Paired Sequencing Libraries for Emulsion PCR

After large-scale PCR purification, library quantification by the use of the Qubit system is sufficiently accurate to calculate the dilution required to obtain picogram concentrations of the library. Alternatively, TaqMan or SybrGreen Real Time PCR can be applied for quantification.

To prepare 500 pM fosmid DNA input samples, measure amplification products by real-time PCR or Qubit fluorometry as per manufacturer's protocol. Prepare several aliquots of 500 pM dilutions of amplified DNA as follows: dilute necessary volume of sample DNA in a volume of 30 μl low TE buffer to yield a final concentration of 5 ng/μl. Pipet 8 μl of the prepared dilution into a new 100 μl tube, and add 32 μl low TE buffer to prepare a 1 ng/μl dilution. Pipet 4.8 μl of 1 ng/μl dilution into a new tube, and add 75.2 μl low TE buffer to a final concentration of 60 pg/μl (500 pM for fragment libraries) (*see* **Note 31**).

3.4 Processing Next Generation Sequencing Libraries for Instrument Run

The SOLiD technology relies on sequencing library amplification on a solid support. To this end, an aqueous and an oil phase are mixed to form an emulsion. During emulsion PCR (ePCR), ideally, multiple copies of one single DNA template molecule are generated on one magnetic bead contained in one emulsion droplet. To avoid multi-clonal amplification, precise DNA quantification and input amounts are essential requirements. Un-templated beads are separated and discarded in a subsequent enrichment step. To ensure that the beads will deposit on the sequencing slide, a 3′modification of the enriched templated beads is performed. For control, a workflow analysis (WFA) run is carried out with a small portion of the templated beads to analyze the templated bead quality (e.g., multi-clonal beads result in poor sequencing outcome), and the bead concentration.

3.4.1 Emulsion PCR

1. Prepare oil phase (all components provided by the SOLiD ePCR Kit): pipet 35 ml of oil into a 50 ml Falcon tube, add 1.8 ml emulsion stabilizer 1 by using a 2.5 ml glass pipet, and add 400 μl of stabilizer 2 with a 1 ml glass pipet. Avoid air bubbles when aspirating stabilizer; in case bubbles have been drawn into pipet, dispense and re-aspirate. Fill Falcon tube with oil to 40 ml, close tube and vortex vigorously to emulsify oil components; open cap and let the Falcon tube degas for at least 30 min. Pipet 9 ml of oil phase with a 10 ml syringe into a SOLiD ePCR tube and cap, place tube on IKA Turrax.
2. During degasing of the oil phase, prepare aqueous phase containing the fosmid pool DNA sample. Mix reaction components depending on sequencing library molarity with a total reaction volume of 2720 μl. For a 1 pM sequencing library combine 280 μl 10× PCR Buffer, 392 μl 100 mM dNTP mix (25 mM per dNTP), 70 μl 1 M MgCl, 16.8 μl ePCR Primer 2 (500 μM), 11.2 μl ePCR Primer 1 dilution (10 μM), 5.6 μl

sequencing library template molecules (500 pM), 1,644.4 μl nuclease-free water and 300 μl AmpliTaq Gold (5 U/l) in a 15 ml Falcon tube, mix by gently inverting the closed tube.

3. Prepare P1 beads: vortex one vial of SOLiD P1 beads, spin down tube, place tube in magnetic rack, wait 1 min until the solution has cleared, discard supernatant, resuspend the beads in 200 μl 1× Bead Block Solution. Vortex, spin down, and place tube in Covaris S2 Sonicator. Run the Bead Block Declump program, place tube in magnetic rack, and wait until solution clears. Discard supernatant, resuspend in 200 μl 1× TEX buffer, vortex, and spin down.
4. Sonicate the prepared P1 beads with the Bead Declump program, immediately add 80 μl of sonicated P1 beads to aqueous phase (*see* **step 2**). Mix gently by swirling the reaction, and use an Eppendorf semiautomated Xstream pipettor with a 10 ml tip to aspirate 2.8 ml of aqueous phase with bead-DNA complex and P1 beads. Start the IKA Turrax to swirl the oil phase, make sure the program runs at least 5 min. When the IKA Turrax reaches full spinning speed, place Xstream pipettor tip in the middle of ePCR tube and dispense as programmed. Let the oil-aqueous phase mix swirl until IKA Turrax stops (after 5 min).
5. Use a 5 ml tip on Eppendorf Repeater Plus Pipette, dispense 100 μl emulsion PCR mix per well into a 96-well PCR reaction plate, check bottom of the 96-well plates for air bubbles, spin down if necessary, and run emulsion PCR in a thermal cycler with gold/silver block (program: 5 min at 95 °C, hold; 40 cycles 15 s at 93 °C, 30 s at 62 °C, 75 s at 72 °C; 7 min at 72 °C, hold; 4 °C, forever, hold).
6. After ePCR finishes, check wells for broken emulsion reactions. Instead of a homogeneous milky suspension, a broken emulsion is indicated by three different layers in a well: a milky suspension, followed by a clear liquid layer (an aqueous phase) and dark freckles from the beads at the bottom of the well. Remove the content of "broken wells" before the next step; do not proceed to the next step if more than four wells show broken emulsion reactions.

3.4.2 Breaking the Emulsion PCR

1. Under a fume hood, pipet 100 μl 2-butanol into each well of the 96-well ePCR plate. Pipet up and down to mix, then pipet the broken emulsion PCR suspensions into a 50 ml tray. Transfer into a new 50 ml Falcon tube, and rinse tray with additional 2-butanol to collect remaining beads from tray. Fill the Falcon tube to 30 ml with 2-butanol, cap and vortex, centrifuge for 5 min at 2000 × *g*, and carefully decant the supernatant (oil). Turn Falcon tube and place for 5 min onto a paper towel to drain remaining oil.

Alternatively use Emulsion Collection Tray: place blue metal adaptor on the ePCR 96-well plate, then place the Emulsion Collection Tray like a cap (bottom up) on the top of the ePCR 96-well plate with metal adaptor. Use parafilm around metal adaptor to seal the connected plates, and flip the plates so that the ePCR plate is upside-down over the Emulsion Collection Tray. Centrifuge the plate construction for 2 min at 550×*g* (centrifuge with 96-well plate adaptor), place the assembly under a fume hood, remove the ePCR plate, and add 10 ml of 2-butanol to the Collection Tray containing centrifuged ePCR emulsion. Pipette the mixture up and down until it appears homogeneous, transfer to a 50 ml Falcon tube, and rinse reservoir with 2-butanol to collect remaining beads. Cap and vortex Falcon tube, centrifuge at 2000×*g* for 5 min, decant the oil phase and place inverted tube on paper towel, and then wait 5–10 min.

2. Wash templated beads: Resuspend the beads in 600 μl 1× Bead Wash Buffer, carefully pipet up and down and transfer beads to a fresh 1.5 ml LoBind tube, rinse the remainder at the bottom of the 50 ml Falcon tube with additional 600 μl 1× Bead Wash Buffer and transfer to 1.5 ml LoBind tube, vortex, centrifuge at 21,000×*g* for 1 min. Remove the oil phase with pipet, change pipet tip, discarding supernatant. Repeat bead washing as described but use 150 μl to resuspend and transfer the beads, and 150 μl to transfer remaining beads, add 1 ml 1× Bead Wash Buffer, centrifuge at 21,000×*g* for 1 min, discard supernatant with pipet, resuspend in 200 μl 1× TEX buffer, put tube on magnetic rack until it clears, discard supernatant, add 200 μl 1× TEX buffer and sonicate the beads with program Declump 1. Determine the bead concentration using a 1:10 dilution on an UV-spectrophotometer (NanoDrop), and compare the color of the bead dilution to photographed colors for different bead concentrations. Optimally, the concentration reflects closely the input amount of beads, about 1.6 billion, indicated by a "medium" color. Adjust the bead concentration by adding 1× TEX in case the color is too dark, or by placing tube in magnetic rack and removing small volume of supernatant in case the color is too light, to match the bead suspension to required volume. Quantify with NanoDrop.

3.4.3 Enrichment of Templated Beads

1. Prepare buffers and enrichment beads (from the SOLiD Bead Enrichment Kit) for use in subsequent steps: for a single ePCR reaction, transfer 1.8 ml denaturing buffer into a 15 ml Falcon tube, add 200 μl denaturant, cap and vortex denaturing reagent. Prepare a fresh 60% glycerol solution, transfer 4 ml dH_2O into a 15 ml Falcon tube and add 3 ml glycerol twice, cap and vortex. For preparation of the enrichment beads, vortex the beads and immediately transfer 300 μl of the beads to

a new, 1.5 ml LoBind tube, centrifuge for 5 min at 21,000 ×*g*, discard the supernatant, resuspend in 900 μl 1× Bind and Wash Buffer, centrifuge for 5 min at 21,000 ×*g*, repeat Bind and Wash step, resuspend in 150 μl 1× Bind and Wash Buffer, add 1.5 μl 1 mM Enrichment Oligo, vortex and place on a rotator at room temperature for 30 min. Centrifuge beads for 5 min at 21,000 ×*g*, remove supernatant, resuspend in 900 μl 1× TEX buffer, repeat centrifugation and TEX wash step, and resuspend beads in 75 μl 1× Low Salt Binding Buffer.

2. Prepare templated beads for enrichment step: place tube with templated beads in magnetic rack, remove supernatant, resuspend in 300 μl freshly prepared denaturing buffer, wait 1 min, place tube in magnetic rack, remove supernatant, repeat resuspension and magnetic rack step, resuspend templated beads in 300 μl 1× TEX buffer, place in magnetic rack and remove supernatant, repeat resuspension in TEX and magnetic rack step. Resuspend the templated beads in 150 μl TEX and transfer to a new 0.5 ml LoBind tube. Declump beads with Program 1 on Covaris.

3. Pipet enrichment beads (from **step 1**) into 0.5 ml LoBind tube containing templated beads, vortex, spin down and sonicate the mixture using the Covalent Declump 3 program, pulse-spin and incubate at 61 °C, vortex and spin down the mixture every 5 min over a time period of 15 min. Cool beads on ice for 2 min. Use a new 1.5 ml LoBind tube and add 400 μl of the 60 % glycerol solution; do not vortex. Mix cooled beads by pipetting up and down, carefully load the total volume on the top of the glycerol-filled tube. Do not vortex at this point, but centrifuge for 3 min at 21,000 ×*g*. Prepare a 2.0 ml LoBind tube with 1 ml 1× TEX buffer, and transfer top layer of beads (swimming on glycerol) to the bottom of tube prepared with TEX (be cautious to not transfer the un-templated beads sitting at the bottom of the glycerol tube). Fill up to 2.0 ml with 1× TEX, vortex and centrifuge transferred templated beads at 21,000 ×*g* for 1 min. If the beads are not sufficiently pelleted (in case of a carry-over of too much glycerol), divide them into two halves and fill each one into a tube, add 500 μl 1× TEX buffer to each tube, vortex and centrifuge at 21,000 ×*g* for 1 min. Discard supernatant, and add 200 μl 1× TEX buffer to each tube. Resuspend the beads and pool the two halves again into one single tube. Otherwise, if the beads are pelleted, directly remove supernatant and add 400 μl 1× TEX buffer. Proceed with protocol as follows: vortex and centrifuge tube at 21,000 ×*g* for 1 min, remove supernatant, resuspend with 400 μl Denaturing Buffer (which is prepared fresh by mixing 1.8 ml Denaturation Buffer with 200 μl Denaturant and vortexing), let stand for 1 min, place

in magnetic rack, remove supernatant, repeat denaturing and magnetic clarification until all white enrichment beads are removed. Resuspend in 400 μl 1× TEX, place in magnetic rack, discard supernatant, repeat resuspension in TEX and magnetic rack step, resuspend templated beads in 200 μl 1× TEX, vortex, spin down and transfer into new 1.5 ml LoBind tube. Declump beads using program Declump 1, place tube in magnetic rack, remove supernatant and resuspend in 400 μl 1× TEX, place in magnetic rack, remove supernatant. Repeat the last two steps until supernatant is clear, and finally resuspend beads in 400 μl 1× TEX (*see* **Note 32**).

3.4.4 3′-End Modification of Enriched Templated Beads

The templates on the selected beads are subjected to a 3′-end modification to allow covalent bonding to the slide.

Sonicate enriched templated beads (*see* Subheading 3.4.3) with Declump 3, pulse-spin and prepare 500 μl 1× Terminal Transferase Reaction (TTR) mix (from the SOLiD Bead Deposition Kit) per ePCR reaction. Mix 55 μl 10× TTR, 55 μl 10× cobalt chloride, and 390 μl dH_2O to a total volume of 500 μl. Prepare 1 mM Bead Linker solution by mixing 1 μl Bead Linker (50 mM) to 49 μl low TE buffer. Place the tube with enriched beads in magnetic rack, remove supernatant, resuspend in 100 μl 1× TTR buffer, and transfer reaction to a new 1.5 ml LoBind tube. Place in magnetic rack, remove supernatant, repeat 1× TTR resuspension and magnetic rack step, resuspend beads in 178 μl 1× TTR, and add 20 μl Bead Linker solution (1 mM). Sonicate the mixture with Declump 3, add 2 μl Terminal Transferase enzyme (20 U/μl). Vortex, pulse-spin, rotate tube for 2 h at 37 °C (place rotator in incubator), then place tube in magnetic rack, discard supernatant, resuspend in 400 μl 1× TEX, perform magnetic clearing, resuspend in 200 μl 1× TEX. Quantify beads after utilizing program Declump 1 on NanoDrop and compare with SOLiD bead color chart.

3.4.5 Bead Deposition on SOLiD Sequencing Slide and Instrument Run

1. Declump enriched templated beads with program Declump 1, pulse-spin, transfer appropriate volume of beads (as determined after NanoDrop quantification or in a WFA run) to a new 1.5 ml LoBind tube, place in magnetic rack, remove supernatant, resuspend beads in 400 μl Deposition Buffer, vortex well, spin down, place in magnetic rack, discard supernatant. Repeat resuspension in Deposition Buffer and magnetic rack step twice according to the number of "fields" on the sequencing slide to be filled with beads; the beads need to be resuspended in different volumes, 300 μl per well in a 8-well deposition chamber, 400 μl per well in a 4-well, and 550 μl per well in a single-well chamber.
2. Insert a sequencing slide into the slide carrier and place the assembled slide carrier into a deposition chamber base, and on top place the appropriate deposition chamber lid

(1-well, 4-well, and 8-well). Sonicate beads with program Declump 3, pulse-spin and sonicate again with Declump 3, pipette the bead solution up and down, immediately fill bead solution into corresponding well of deposition chamber to avoid clumped beads on slide, seal portholes and incubate at 37 °C for 1.5 h.

3. In the meantime, prepare the SOLiD sequencing instrument and set up SOLiD sequencing run as per manufacturer's protocol. After the incubation step has finished, remove adhesive seals, drain the top of the deposition chamber with Deposition Buffer, use a 1000 ml pipette, press on portholes, and aspirate the entire Deposition Buffer from deposition wells until freshly layered Deposition Buffer is drawn into the wells, loosen lid of deposition chamber, immediately place the slide carrier assembly on the flow cell of the SOLiD instrument. Avoid drying out the slide, close flow cell, and proceed with the specific steps of the instrument run as described in the manufacturer's protocol.

3.5 Computational analysis of Fosmid Sequences and Haplotype Assembly

The bioinformatics procedure starts from the raw reads obtained from barcoded high-throughput sequencing of each fosmid pool. The main steps include identification of heterozygous sites, fosmid detection, calling of fosmid-specific genotypes at heterozygous sites, and haplotype assembly using the predicted fosmids and their allele calls as virtual reads. Detailed steps to perform this procedure are summarized in Fig. 6 and described as follows:

3.5.1 De-Indexing Barcoded Fosmid Pools

Separate sequence reads per fosmid pool, identifying the barcode sequences in the first 5 bp of each read. Reads sharing the same barcode belong to the same fosmid pool. For SOLiD reads, this is done using Bioscope 1.3. For Illumina reads, this can be done with the "Deconvolute" command of NGSEP.

3.5.2 NGS Read Alignment

Align the reads per pool against the reference genome. For SOLiD reads, use Bioscope 1.3. For reads sequenced using other platforms such as Illumina, align the reads using software tools for alignment of standard WGS reads such as bwa [17] or bowtie2 [14]. Then, sort the alignments by reference sequence and position using the sort command of samtools or the Picard software. In both platforms the final output of this step is a file in SAM or BAM format for each pool, containing the information of the reads aligned to the reference genome (*see* **Note 33**).

3.5.3 Identification of Heterozygous Variants

Merge the BAM files obtained for each pool using the merge command of samtools or Picard. Then, identify variants against the reference genome, again using Bioscope 1.3 for SOLiD reads. Alternatively (*see* **Note 34**), variants can be identified from aligned standard WGS read data by the use of the NGSEP pipeline [15], samtools [17], or the GATK pipeline [16] for other platforms such

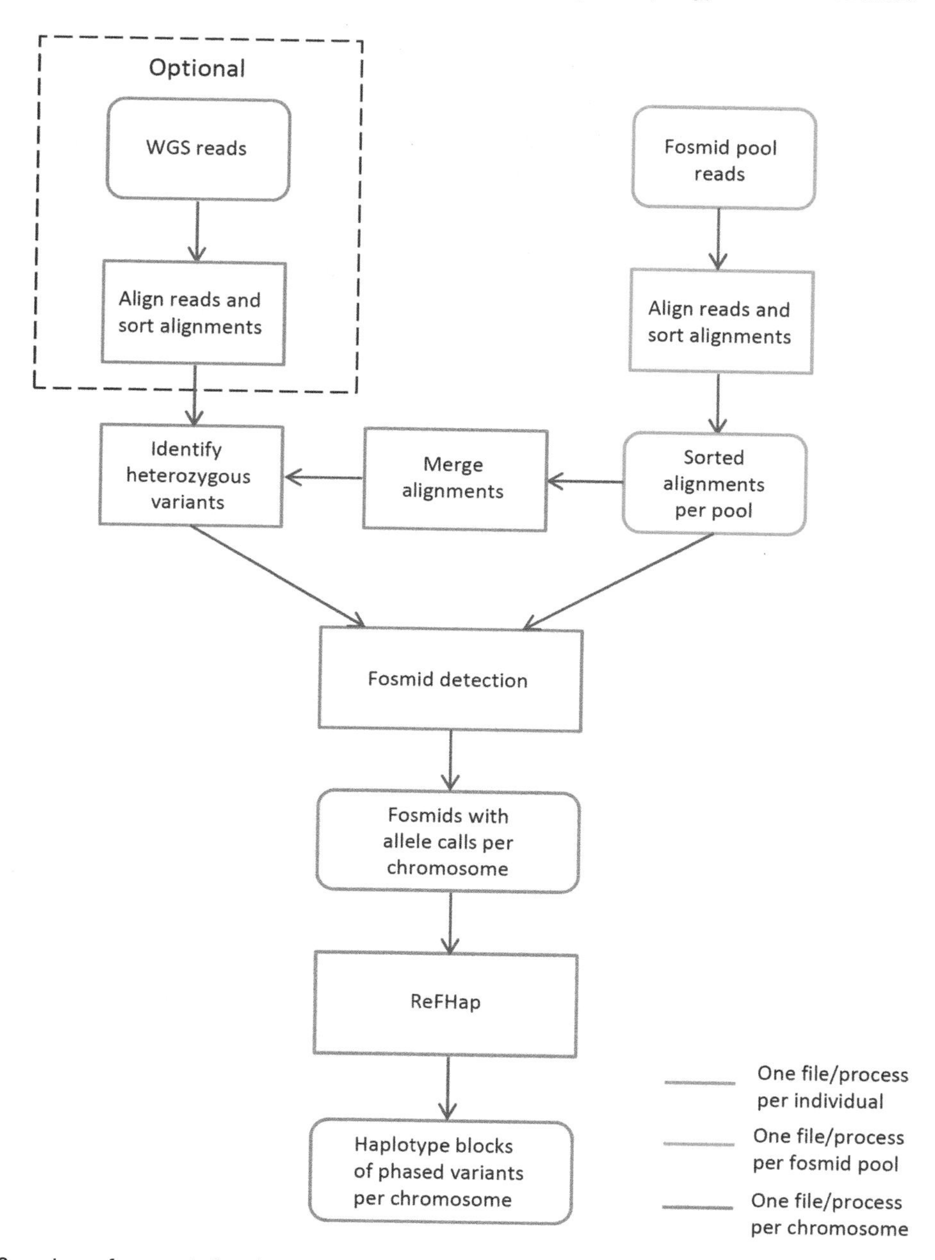

Fig. 6 Overview of computational steps involved in analysis of fosmid pool-based sequences and phasing. Complementary analyses of whole genome sequencing (WGS) data are presented as an option

as Illumina. Detailed instructions, scripts, and recommended parameters for command line usage of NGSEP can be found at (http://sourceforge.net/projects/ngsep/files/training/). Regardless of the sequencing platform and analysis pipeline used at this stage, the final output is a list of variants in VCF format and one sorted BAM file for each fosmid pool (*see* **Note 35**).

3.5.4 Fosmid Detection

Run the fosmid detection program following the specific instructions in the README file. This program receives the list of variants against the reference genome and an XML file describing, for each sequenced fosmid pool, the identification number (id), average coverage, and a path to the BAM file containing the aligned and sorted reads. A GFF file with the format produced by Bioscope 1.3 including predicted heterozygous deletions for the individual can be added as a parameter. If, moreover, files with allele calls in the same format are registered for each pool at the XML file, the fosmid detector can include these large deletions in the final haplotypes. With all this information, the fosmid detector performs the tasks described in detail earlier [3]. The following internal steps are being performed by the fosmid detector and summarized here:

(a) Extract the heterozygous variants from the VCF file obtained in the previous section.

(b) Predict the physical locations of the fosmids sequenced in each pool from the aligned reads of the pool and call one (usually homozygous) fosmid-specific genotype (FSG) for each heterozygous variant covered by one predicted fosmid. To predict fosmids, the software divides the genome in nonoverlapping segments of a fixed length (1 kb by default) and processes the alignments to calculate the read coverage of each segment. Then, it tags bins as candidate members for membership to a fosmid if the read coverage exceeds a minimum threshold (0.4 by default). Fosmids are called collecting nearly adjacent bins (with a maximum gap of 9 bins by default) extending between a minimum and a maximum threshold (by default 3–60 kb). Stretches that are longer than the maximum length are split into two or more predicted fosmids with lengths within the minimum and maximum thresholds. FSGs can be called either with NGSEP or with the allele coverage procedure described [3] (*see* **Note 35**). Although most FSGs should be homozygous, heterozygous FSGs are possible if two fosmids from the two parental homologues of a chromosome overlap within one pool. Consequently, bins showing heterozygous calls within a pool are tagged, and the tagged fosmids split into two fosmids, the first predicted from the covered bins extending upstream of the first tagged bin, and the second from the covered bins following the last tagged bin. This step is important to avoid switch errors produced by overlaps of complementary fosmids. After these steps, fosmids containing only one homozygous FSG are removed to obtain the set that will be used for single individual haplotyping.

(c) Finally, the fosmid detector allows generation of the input matrix for phasing separately for each chromosome, with the fosmids (one per row) sorted by the position of the first genotyped variant, and the allele calls (one per column) designated by physical genome position. This procedure results in the production of the variation matrixes that serve as input files for haplotype assembly by use of ReFHap.

3.5.5 Phasing by Use of ReFHap

For each chromosome, take the input file for ReFHap generated by the fosmid detector as described above (4c) and run ReFHap following the instructions available in the README file. Details of the algorithms implemented in ReFHap are available in Duitama et al., 2010 [11], and benchmarks with other tools/SIH algorithms can be found in Duitama et al., 2011 [5]. For each chromosome, ReFHap outputs the blocks of variants that could be phased by tiling the fosmids (*see* Fig. 7a), and for each block, it outputs the chromosomal position of each phased variant, followed by the allele in the first haplotype, and the allele in the second haplotype (*see* Fig. 7b).

a

chr1	FosSeq	ctg	1	40999	1	+	id=chr1_BC26_6958:posn=1-40999:reads=970:bins=40:cov=0.99:poolId=BC26
chr1	FosSeq	ctg	1	30999	1	+	id=chr1_BC28_10738:posn=1-30999:reads=482:bins=30:cov=0.65:poolId=BC28
chr1	FosSeq	ctg	1	52999	1	+	id=chr1_BC29_12761:posn=1-52999:reads=3026:bins=52:cov=2.41:poolId=BC29
chr1	FosSeq	ctg	1	54999	1	+	id=chr1_B17_37073:posn=1-54999:reads=1260:bins=54:cov=1.09:poolId=B17
chr1	FosSeq	ctg	1	30999	1	+	id=chr1_B22_44150:posn=1-30999:reads=685:bins=30:cov=1.05:poolId=B22
chr1	FosSeq	ctg	1	59999	1	+	id=chr1_B23_45772:posn=1-59999:reads=1376:bins=59:cov=1.1:poolId=B23
chr1	FosSeq	ctg	1	25999	1	+	id=chr1_B27_52229:posn=1-25999:reads=423:bins=25:cov=0.77:poolId=B27
chr1	FosSeq	ctg	1	55999	1	+	id=chr1_B28_53864:posn=1-55999:reads=1915:bins=55:cov=1.64:poolId=B28
chr1	FosSeq	ctg	1	38999	1	+	id=chr1_B30_57409:posn=1-38999:reads=1843:bins=38:cov=2.26:poolId=B30
chr1	FosSeq	ctg	1	45999	1	+	id=chr1_B35_65046:posn=1-45999:reads=1084:bins=45:cov=1.1:poolId=B35
chr1	FosSeq	ctg	1	29999	1	+	id=chr1_B37_68417:posn=1-29999:reads=995:bins=29:cov=1.54:poolId=B37
chr1	FosSeq	ctg	1	47999	1	+	id=chr1_B39_71786:posn=1-47999:reads=1302:bins=47:cov=1.27:poolId=B39
chr1	FosSeq	ctg	1	39999	1	+	id=chr1_B40_73387:posn=1-39999:reads=1677:bins=39:cov=1.96:poolId=B40
chr1	FosSeq	ctg	1000	42999	1	+	id=chr1_BC22_1785:posn=1000-42999:reads=635:bins=42:cov=0.63:poolId=BC22
chr1	FosSeq	ctg	1000	38999	1	+	id=chr1_B13_31958:posn=1000-38999:reads=521:bins=38:cov=0.66:poolId=B13
chr1	FosSeq	ctg	1000	28999	1	+	id=chr1_B19_40285:posn=1000-28999:reads=528:bins=28:cov=0.89:poolId=B19
chr1	FosSeq	ctg	1000	29999	1	+	id=chr1_B29_55605:posn=1000-29999:reads=576:bins=29:cov=0.96:poolId=B29
chr1	FosSeq	ctg	1000	27999	1	+	id=chr1_B33_61712:posn=1000-27999:reads=345:bins=27:cov=0.59:poolId=B33
chr1	FosSeq	ctg	1000	53999	1	+	id=chr1_B36_66810:posn=1000-53999:reads=1522:bins=53:cov=1.34:poolId=B36
chr1	FosSeq	ctg	1000	10999	1	+	id=chr1_B38_70131:posn=1000-10999:reads=139:bins=10:cov=0.64:poolId=B38
chr1	FosSeq	ctg	2000	28999	1	+	id=chr1_BC30_15089:posn=2000-28999:reads=324:bins=27:cov=0.5:poolId=BC30
chr1	FosSeq	ctg	4000	22999	1	+	id=chr1_BC27_8937:posn=4000-22999:reads=132:bins=19:cov=0.29:poolId=BC27
chr1	FosSeq	ctg	4000	29999	1	+	id=chr1_B25_48983:posn=4000-29999:reads=205:bins=26:cov=0.36:poolId=B25
chr1	FosSeq	ctg	6000	56999	1	+	id=chr1_B14_33493:posn=6000-56999:reads=983:bins=51:cov=0.93:poolId=B14

b

BLOCK: offset: 2 len: 2 phased: 2		
536602	0	1
556655	1	0
********		
BLOCK: offset: 4 len: 231 phased:225		
715887	1	0
716595	1	0
724429	1	0
747799	0	1
747911	1	0
748300	1	0
748331	1	0
766409	0	1
779714	1	0
779737	1	0
780559	1	0
786902	1	0
804531	1	0
805284	0	1
805386	1	0
805419	1	0
805558	1	0
805591	0	1
805892	1	0
805897	1	0
805898	1	0
805905	0	1
805897	1	0
...		

Fig. 7 Examples of phasing output files. (**a**) Output file fosmid contigs. A part of a typical output file is shown, listing per chromosome the detected contigs as an output of the fosmid detection program (*see* Subheading 3.5.4). The chromosomal start and end positions per contig are given in columns 4 and 5 by indicating the specific bins (first and last) as intervals of fixed length (1000 bp) that are covered by the fosmid contig. In the following columns, additional properties such as number of reads, bins, read coverage and fosmid super-pool identification number are shown. (**b**) Output file ReFHap with phased variants. A part of a typical output file is shown, listing the haplotype blocks per chromosome with the numbers of phased variants per block. In the first column from the *left*, the chromosomal position numbers of heterozygous variants are given, and in the adjacent, second and third columns the two haplotypes, with the variant alleles (different from the reference sequence) denoted by "1" and the reference alleles denoted by "0." Thus, the second column shows the specific combinations of variants from top to bottom for "Haplotype 1" and the third column the specific combinations of variants constituting "Haplotype 2"

4 Notes

1. Optimally, one kit would be sufficient to generate ten individual fosmid libraries. In practice, the kit will allow generating three to four libraries.
2. Do not vortex the sample to avoid shearing the HMW gDNA.
3. Do not transfer any flocculent material.
4. We find that the gDNA cannot be resuspended in buffer, if the alcohol is not completely evaporated.
5. Alternatively, the Tube Rotisserie Rotator (VWR) can be used, which is also employed during SOLiD sequencing library preparation.
6. Alternatively run a large 1% agarose gel of at least 20 cm length to allow for sufficient resolution, at 30 V/cm overnight.
7. Since under conditions of manual shearing, i.e., variable pressure and speed, the shearing results will turn out to a certain extent variable, check the results on an agarose gel of at least 20 cm length overnight. In order to standardize our manual shearing procedure (which we used at the outset), we constructed an automated prototype, which is not generally available. Therefore, collecting information from experts in the field [18, 19], we have chosen to refer to commercially available HydroShear devices here and the settings that have been experimentally verified.
8. Avoid unnecessary pipetting or vortexing after end-repair, pipet carefully and gently to avoid mechanical damage.
9. If no preparative comb is available, tape 2 wells of a 1.5 mm thick 20-well comb together, keep one well to each side of sample empty, first and last well are used for size ladder. The slots of the preparative comb have a volume of about 200 μl.
10. The gel is very fragile, therefore, do not lift the gel, but transfer it on the gel tray of the electrophoresis chamber.
11. Sheared gDNA appears as a smear; try to cut out gel pieces as exactly as possible to minimize subsequent gelase digestion. Do not stain the gel with ethidium bromide and avoid UVB illumination, which will damage the DNA and significantly reduce ligation efficiency.
12. Scrape glycerol stock of EPI100 bacteria with heat sterilized inoculating loop only briefly before streaking over agar plate to ensure growth of single colonies.
13. Prepare Epi100 bacteria for mass infection 2 days before advancing to the phage packaging step.

14. In order to evaporate condensed water due to storage, place the agar plates in a laminar flow bench one hour before use, remove the cover plates, and let them dry.
15. It is recommendable to produce a total of 1.5×10^6 phages to have a bit of excess later during the fosmid clone partitioning step.
16. The amplification of fosmid clones to high numbers in liquid culture can introduce preferential amplification, that is, overgrowth of few fosmid clones in a fosmid pool, and may lead to significant library complexity reduction. Our experimental results have shown, however, that this was only the case if the second amplification step, required to generate sufficient copies of the ~15,000 fosmid clones in a super-pool before isolation of fosmid DNA for NGS, was also made in liquid cultures. In that case, library complexity was substantially reduced. If the second required amplification step was, however, performed using agar plates, both haploid genomes were found to be nearly completely represented, with an equal physical coverage of both haplotypes. Specifically, we have performed high-throughput SNP typing of the 4 Mb MHC region in one 96-well plate of the fosmid pool-formatted library, demonstrating presence of MHC fosmid clones according to expectations in the fosmid pools and nearly complete and equal coverage of both MHC haplotypes.
17. The wet-lab-based removal of Epi 100 genomic DNA is only partially sufficient; the remaining Epi100 genome, as evident in NGS reads, needs to be removed computationally.
18. Make sure to check whether the tubes are suitable to withstand the centrifugal forces.
19. Use 5 l bottles with cooled, distilled water stored at 4 °C to be able to start the Covaris system more quickly.
20. To prepare a 4% agarose gel, mix buffer and agarose, let ingredients stand overnight, heat the agarose carefully in microwave, avoid over-boil by checking every 2 s.
21. The E-Gel Size Selection Gel or Flash Gel Recovery System can be used instead; in our experience the recovery was not superior to the gel-based size selection.
22. We created 48 different barcode tags within the P1-Adaptor sequence (which we shortened) before SOLiD barcoded adaptors became commercially available. Thus, 48 pools/sequencing libraries could be parallel processed, and up to 16 barcoded NGS libraries could be multiplexed later in a single sequencing run. To avoid color imbalances on the SOLiD NGS system, all 16 barcoded adaptors should also be employed (per run) in the case where the number of samples is smaller.

Barcode ID		bp	Barcode ID		bp
5′	3′				
P1-BC-1-5′Oligo	AAGAGGATCACCGACTGCCCATAGAGAGGTT	31	P1-BC-25-5′Oligo	TGATGAATCACCGACTGCCCATAGAGAGGTT	31
P1-BC-1-3′Oligo	CCTCTCTATGGGCAGTCGGTGATCCTCTT	29	P1-BC-25-3′Oligo	CCTCTCTATGGGCAGTCGGTGATTCATCA	29
P1-BC-2-5′Oligo	AGTGGTATCACCGACTGCCCATAGAGAGGTT	31	P1-BC-26-5′Oligo	AGCCCGATCACCGACTGCCCATAGAGAGGTT	31
P1-BC-2-3′Oligo	CCTCTCTATGGGCAGTCGGTGATACCACT	29	P1-BC-26-3′Oligo	CCTCTCTATGGGCAGTCGGTGATCGGGCT	29
P1-BC-3-5′Oligo	GTTATAATCACCGACTGCCCATAGAGAGGTT	31	P1-BC-27-5′Oligo	AAACTTATCACCGACTGCCCATAGAGAGGTT	31
P1-BC-3-3′Oligo	CCTCTCTATGGGCAGTCGGTGATTATAAC	29	P1-BC-27-3′Oligo	CCTCTCTATGGGCAGTCGGTGATAAGTTT	29
P1-BC-4-5′Oligo	GCGGTCATCACCGACTGCCCATAGAGAGGTT	31	P1-BC-28-5′Oligo	TACTACATCACCGACTGCCCATAGAGAGGTT	31
P1-BC-4-3′Oligo	CCTCTCTATGGGCAGTCGGTGATGACCGC	29	P1-BC-28-3′Oligo	CCTCTCTATGGGCAGTCGGTGATGTAGTA	29
P1-BC-5-5′Oligo	TGTAAGATCACCGACTGCCCATAGAGAGGTT	31	P1-BC-29-5′Oligo	CTAGGGATCACCGACTGCCCATAGAGAGGTT	31
P1-BC-5-3′Oligo	CCTCTCTATGGGCAGTCGGTGATCTTACA	29	P1-BC-29-3′Oligo	CCTCTCTATGGGCAGTCGGTGATCCCTAG	29
P1-BC-6-5′Oligo	GGGACAATCACCGACTGCCCATAGAGAGGTT	31	P1-BC-30-5′Oligo	CGAAGAATCACCGACTGCCCATAGAGAGGTT	31
P1-BC-6-3′Oligo	CCTCTCTATGGGCAGTCGGTGATTGTCCC	29	P1-BC-30-3′Oligo	CCTCTCTATGGGCAGTCGGTGATTCTTCG	29
P1-BC-7-5′Oligo	TATGCCATCACCGACTGCCCATAGAGAGGTT	31	P1-BC-31-5′Oligo	GTGCGTATCACCGACTGCCCATAGAGAGGTT	31
P1-BC-7-3′Oligo	CCTCTCTATGGGCAGTCGGTGATGGCATA	29	P1-BC-31-3′Oligo	CCTCTCTATGGGCAGTCGGTGATACGCAC	29
P1-BC-8-5′Oligo	GAGGATATCACCGACTGCCCATAGAGAGGTT	31	P1-BC-32-5′Oligo	TGGGTGATCACCGACTGCCCATAGAGAGGTT	31
P1-BC-8-3′Oligo	CCTCTCTATGGGCAGTCGGTGATATCCTC	29	P1-BC-32-3′Oligo	CCTCTCTATGGGCAGTCGGTGATCACCCA	29

P1-BC-9-5′Oligo	TGCGACATCACCGACTGCCCATAGAGAGGTT	31	P1-BC-33-5′Oligo	GGATACATCACCGACTGCCCATAGAGAGGTT	31
P1-BC-9-3′Oligo	CCTCTCTATGGGCAGTCGGTGATGTCGCA	29	P1-BC-33-3′Oligo	CCTCTCTATGGGCAGTCGGTGATGTATCC	29
P1-BC-10-5′Oligo	TAAGCTATCACCGACTGCCCATAGAGAGGTT	31	P1-BC-34-5′Oligo	TAAAGGATCACCGACTGCCCATAGAGAGGTT	31
P1-BC-10-3′Oligo	CCTCTCTATGGGCAGTCGGTGATAGCTTA	29	P1-BC-34-3′Oligo	CCTCTCTATGGGCAGTCGGTGATCCTTTA	29
P1-BC-11-5′Oligo	GACACGATCACCGACTGCCCATAGAGAGGTT	31	P1-BC-35-5′Oligo	GTAGGAATCACCGACTGCCCATAGAGAGGTT	31
P1-BC-11-3′Oligo	CCTCTCTATGGGCAGTCGGTGATCGTGTC	29	P1-BC-35-3′Oligo	CCTCTCTATGGGCAGTCGGTGATTCCTAC	29
P1-BC-12-5′Oligo	GGAAAAATCACCGACTGCCCATAGAGAGGTT	31	P1-BC-36-5′Oligo	TCACATATCACCGACTGCCCATAGAGAGGTT	31
P1-BC-12-3′Oligo	CCTCTCTATGGGCAGTCGGTGATTTTTCC	29	P1-BC-36-3′Oligo	CCTCTCTATGGGCAGTCGGTGATATGTGA	29
P1-BC-13-5′Oligo	TAAGGCATCACCGACTGCCCATAGAGAGGTT	31	P1-BC-37-5′Oligo	TATACCATCACCGACTGCCCATAGAGAGGTT	31
P1-BC-13-3′Oligo	CCTCTCTATGGGCAGTCGGTGATGCCTTA	29	P1-BC-37-3′Oligo	CCTCTCTATGGGCAGTCGGTGATGGTATA	29
P1-BC-14-5′Oligo	GGCAGAATCACCGACTGCCCATAGAGAGGTT	31	P1-BC-38-5′Oligo	TCTTAGATCACCGACTGCCCATAGAGAGGTT	31
P1-BC-14-3′Oligo	CCTCTCTATGGGCAGTCGGTGATTCTGCC	29	P1-BC-38-3′Oligo	CCTCTCTATGGGCAGTCGGTGATCTAAGA	29
P1-BC-15-5′Oligo	TGAATGATCACCGACTGCCCATAGAGAGGTT	31	P1-BC-39-5′Oligo	GTGAGTATCACCGACTGCCCATAGAGAGGTT	31
P1-BC-15-3′Oligo	CCTCTCTATGGGCAGTCGGTGATCATTCT	29	P1-BC-39-3′Oligo	CCTCTCTATGGGCAGTCGGTGATACTCAC	29
P1-BC-16-5′Oligo	GACGTTATCACCGACTGCCCATAGAGAGGTT	31	P1-BC-40-5′Oligo	GGGTTAATCACCGACTGCCCATAGAGAGGTT	31
P1-BC-16-3′Oligo	CCTCTCTATGGGCAGTCGGTGATAACGTC	29	P1-BC-40-3′Oligo	CCTCTCTATGGGCAGTCGGTGATTAACCC	29

(continued)

(continued)

P1-BC-17-5′Oligo	CACCGCATCACCGACTGCCCATAGAGAGGTT	31	P1-BC-41-5′Oligo	GGATTGATCACCGACTGCCCATAGAGAGGTT	31
P1-BC-17-3′Oligo	CCTCTCTATGGGCAGTCGGTGATGCGGTG	29	P1-BC-41-3′Oligo	CCTCTCTATGGGCAGTCGGTGATCAATCC	29
P1-BC-18-5′Oligo	GGATGAATCACCGACTGCCCATAGAGAGGTT	31	P1-BC-42-5′Oligo	GTACTAATCACCGACTGCCCATAGAGAGGTT	31
P1-BC-18-3′Oligo	CCTCTCTATGGGCAGTCGGTGATTCATCC	29	P1-BC-42-3′Oligo	CCTCTCTATGGGCAGTCGGTGATTAGTAC	29
P1-BC-19-5′Oligo	TAACTGATCACCGACTGCCCATAGAGAGGTT	31	P1-BC-43-5′Oligo	AGGGTTATCACCGACTGCCCATAGAGAGGTT	31
P1-BC-19-3′Oligo	CCTCTCTATGGGCAGTCGGTGATCAGTTA	29	P1-BC-43-3′Oligo	CCTCTCTATGGGCAGTCGGTGATAACCCT	29
P1-BC-20-5′Oligo	AGCTTTATCACCGACTGCCCATAGAGAGGTT	31	P1-BC-44-5′Oligo	ATGATCATCACCGACTGCCCATAGAGAGGTT	31
P1-BC-20-3′Oligo	CCTCTCTATGGGCAGTCGGTGATAAAGCT	29	P1-BC-44-3′Oligo	CCTCTCTATGGGCAGTCGGTGATGATCAT	29
P1-BC-21-5′Oligo	GGTTCCATCACCGACTGCCCATAGAGAGGTT	31	P1-BC-45-5′Oligo	TGGCTCATCACCGACTGCCCATAGAGAGGTT	31
P1-BC-21-3′Oligo	CCTCTCTATGGGCAGTCGGTGATGGAACC	29	P1-BC-45-3′Oligo	CCTCTCTATGGGCAGTCGGTGATGAGCCA	29
P1-BC-22-5′Oligo	AGATGAATCACCGACTGCCCATAGAGAGGTT	31	P1-BC-46-5′Oligo	GTCGAAATCACCGACTGCCCATAGAGAGGTT	31
P1-BC-22-3′Oligo	CCTCTCTATGGGCAGTCGGTGATTCATCT	29	P1-BC-46-3′Oligo	CCTCTCTATGGGCAGTCGGTGATTTCGAC	29
P1-BC-23-5′Oligo	TGCTTGATCACCGACTGCCCATAGAGAGGTT	31	P1-BC-47-5′Oligo	GGATGGATCACCGACTGCCCATAGAGAGGTT	31
P1-BC-23-3′Oligo	CCTCTCTATGGGCAGTCGGTGATCAAGCA	29	P1-BC-47-3′Oligo	CCTCTCTATGGGCAGTCGGTGATCCATCC	29
P1-BC-24-5′Oligo	CGGTATATCACCGACTGCCCATAGAGAGGTT	31	P1-BC-48-5′Oligo	TAAGTGATCACCGACTGCCCATAGAGAGGTT	31
P1-BC-24-3′Oligo	CCTCTCTATGGGCAGTCGGTGATATACCG	29	P1-BC-48-3′Oligo	CCTCTCTATGGGCAGTCGGTGATCACTTA	29

The unique barcode sequence tags are presented in color; forward and reverse oligos are presented; BC barcode.

23. Alternatively use AMPure Beads to size-select and purify adaptor-ligated DNA fragments as described in the SOLiD User Manual. In our experience, the FlashGel Recovery System (Lonza) and E-Gel SizeSelect system (Thermo Fisher Scientific) did not work as efficiently as the manual gel-based size selection.
24. Do not heat-dissolve the gel as recommended in the manufacturer's protocol to protect the DNA from denaturation and heteroduplex formation.
25. It would be possible to cut out the vector sequence with *NotI* first, and to recover insert DNA with a gel-based size selection.
26. To avoid chimerism, the DNA concentration should be low and the reaction volume high enough to increase the chances that ligation will occur between the two ends of one DNA molecule, rather than two different DNA molecules.
27. DNA polymerase I activity is highly temperature-sensitive, chilled reagents and ice bathing are crucial to create optimal tag length. Thus, a higher temperature, or increased incubation time, may lead to an extension of the size of the fragments.
28. Consistent with the SOLiD sequencing protocol, barcoded adaptors for mate-paired sequencing were not established at the time.
29. Use differently colored caps, if several fosmid pool DNA samples are processed.
30. Run as few cycles as necessary to achieve visible gel bands; over-amplification can lead to redundant library molecules and reduces the library complexity. Start with 5 cycles, if DNA concentration is about 400 ng/μl, start with up to 8 cycles, if DNA concentration is about 50 ng/μl. If necessary, pause PCR reaction and check an aliquot on a gel before running additional PCR cycles.
31. For paired-end libraries, the final concentration needs to be 50 pg/μl and for mate-paired libraries, the final concentration needs to be 96 pg/μl to correspond to 500 pM.
32. Do not magnet the P2-enriched beads before denaturing buffer has been added.
33. The use of the standard SAM or BAM format at this step, as well as the VCF format at the next step, is helpful to achieve independence from the sequencing platform for fosmid detection and haplotyping.
34. The identification of variants against the reference genome can be achieved directly from fosmid pool NGS and analysis by merging the BAM files from each pool as described here, or, alternatively, by analysis of WGS data, in the case where WGS is performed in complementation to fosmid pool NGS, *see* also Fig. 6.

35. In the latter case, if n1 and n2 are the largest and second largest allele coverage, respectively, and t1 and t2 are two corresponding minimum thresholds (with defaults 1 and 2, respectively), a homozygous FSG is called if $n1 \geq t1$ and either $n2 < t2$ or $n1 \geq 2 \times n2$. If the first condition does not hold, the FSG is left uncalled. If the second condition does not hold, a heterozygous FSG is called.

References

1. Hoehe MR (2003) Haplotypes and the systematic analysis of genetic variation in genes and genomes. Pharmacogenomics 4:547–570. doi:10.1517/phgs.4.5.547.23791
2. Tewhey R, Bansal V, Torkamani A, Topol EJ, Schork NJ (2011) The importance of phase information for human genomics. Nat Rev Genet 12:215–223. doi:10.1038/nrg2950
3. Suk EK, McEwen GK, Duitama J, Nowick K, Schulz S, Palczewski S, Schreiber S, Holloway DT, McLaughlin S, Peckham H et al (2011) A comprehensively molecular haplotype-resolved genome of a European individual. Genome Res 21:1672–1685. doi:10.1101/gr.125047.111
4. Kitzman JO, Mackenzie AP, Adey A, Hiatt JB, Patwardhan RP, Sudmant PH, Ng SB, Alkan C, Qiu R, Eichler EE et al (2011) Haplotype-resolved genome sequencing of a Gujarati Indian individual. Nat Biotechnol 29:59–63. doi:10.1038/nbt.1740, nbt.1740 [pii]
5. Duitama J, McEwen GK, Huebsch T, Palczewski S, Schulz S, Verstrepen K, Suk EK, Hoehe MR (2012) Fosmid-based whole genome haplotyping of a HapMap trio child: evaluation of single individual haplotyping techniques. Nucleic Acids Res 40:2041–2053. doi:10.1093/Nar/Gkr1042
6. Peters BA, Kermani BG, Sparks AB, Alferov O, Hong P, Alexeev A, Jiang Y, Dahl F, Tang YT, Haas J et al (2012) Accurate whole-genome sequencing and haplotyping from 10 to 20 human cells. Nature 487:190–195. doi:10.1038/nature11236
7. Kaper F, Swamy S, Klotzle B, Munchel S, Cottrell J, Bibikova M, Chuang HY, Kruglyak S, Ronaghi M, Eberle MA et al (2013) Whole-genome haplotyping by dilution, amplification, and sequencing. Proc Natl Acad Sci U S A 110:5552–5557. doi:10.1073/pnas.1218696110
8. Lo C, Liu R, Lee J, Robasky K, Byrne S, Lucchesi C, Aach J, Church G, Bafna V, Zhang K (2013) On the design of clone-based haplotyping. Genome Biol 14:R100. doi:10.1186/gb-2013-14-9-r100
9. Hoehe MR, Church GM, Lehrach H, Kroslak T, Palczewski S, Nowick K, Schulz S, Suk EK, Huebsch T (2014) Multiple haplotype-resolved genomes reveal population patterns of gene and protein diplotypes. Nat Commun. 2014 Nov 26;5:5569. doi: 10.1038/ncomms6569
10. Burgtorf C, Kepper P, Hoehe M, Schmitt C, Reinhardt R, Lehrach H, Sauer S (2003) Clone-based systematic haplotyping (CSH): a procedure for physical haplotyping of whole genomes. Genome Res 13:2717–2724. doi:10.1101/Gr.1442303
11. Duitama J, Huebsch T, McEwen G, Suk E-K, Hoehe MR (2010) ReFHap: a reliable and fast algorithm for single individual haplotyping. Proceedings of the first ACM international conference on bioinformatics and computational biology ACM, Niagara Falls, New York, pp. 160–169
12. 1000 Genomes Project Consortium, Auton A, Brooks LD, Durbin RM, Garrison EP, Kang HM, Korbel JO, Marchini JL, McCarthy S, McVean GA et al (2015) A global reference for human genetic variation. Nature 526:68–74. doi:10.1038/nature15393
13. Li H, Durbin R (2009) Fast and accurate short read alignment with Burrows-Wheeler transform. Bioinformatics 25:1754–1760. doi:10.1093/bioinformatics/btp324
14. Langmead B, Salzberg SL (2012) Fast gapped-read alignment with bowtie 2. Nat Methods 9:357–359. doi:10.1038/nmeth.1923
15. Duitama J, Quintero JC, Cruz DF, Quintero C, Hubmann G, Foulquie-Moreno MR, Verstrepen KJ, Thevelein JM, Tohme J (2014) An integrated framework for discovery and genotyping of genomic variants from high-throughput sequencing experiments. Nucleic Acids Res 42:e44. doi:10.1093/nar/gkt1381
16. DePristo MA, Banks E, Poplin R, Garimella KV, Maguire JR, Hartl C, Philippakis AA, del Angel G, Rivas MA, Hanna M et al (2011) A framework for variation discovery and genotyping using next-generation DNA sequencing data. Nat Genet 43:491–498. doi:10.1038/ng.806
17. Li H (2011) A statistical framework for SNP calling, mutation discovery, association mapping and population genetical parameter

estimation from sequencing data. Bioinformatics 27:2987–2993. doi:10.1093/bioinformatics/btr509

18. Nedelkova M, Maresca M, Fu J, Rostovskaya M, Chenna R, Thiede C, Anastassiadis K, Sarov M, Stewart AF (2011) Targeted isolation of cloned genomic regions by recombineering for haplotype phasing and isogenic targeting. Nucleic Acids Res 39:e137. doi:10.1093/nar/gkr668
19. Donahue WF, Ebling HM (2007) Fosmid libraries for genomic structural variation detection. Curr Protoc Hum Genet Chapter 5:Unit 5.20. doi:10.1002/0471142905.hg0520s54

Part VI

High throughput Haplotyping of Rare Variants

Chapter 14

Discovery of Rare Haplotypes by Typing Millions of Single-Molecules with Bead Emulsion Haplotyping (BEH)

Elisabeth Palzenberger, Ronja Reinhardt, Leila Muresan, Barbara Palaoro, and Irene Tiemann-Boege

Abstract

Characterizing polymorphisms on single molecules renders the phase of different alleles, and thus, haplotype information. Here, we describe a high-throughput method to genotype hundreds-of thousands single molecules in parallel using bead-emulsion haplotyping (BEH). Haplotyping via BEH is an emulsion-PCR-based method that was adapted to amplify multiple DNA fragments on paramagnetic, microscopic beads within a compartment formed by an aqueous-oil emulsion. This generates beads covered by thousands of clonal copies from several polymorphic regions of an initial DNA molecule that are then genotyped with fluorescently labeled probes. With BEH, up to three different polymorphisms (or more if several polymorphisms are within an amplicon) can be typed within a fragment of several kilobases in a single experiment, rendering haplotype information of a very large number of initial single molecules. The high throughput and digital nature of the method makes it ideal to quantify rare haplotypes or to assess the haplotype diversity in complex samples.

Key words Bead-emulsion haplotyping (BEH), Single-molecule amplification, Emulsion PCR, Digital PCR, High-throughput haplotyping, Allelic phase

1 Introduction

The phase of a series of consecutive polymorphisms on one DNA strand is defined as haplotype. This can be either a full sequence of a chromosome ("chromosome haplotype"), or a sequence of a smaller region containing more than one informative or heterozygous polymorphism. Haplotyping is a valuable tool for mapping disease genes or complex combinations of alleles, especially when polymorphic combinations in *cis* or *trans* influence the phenotype (reviewed in refs. [1, 2]). Moreover, major insights into biological processes such as recombination, which depend on the formation of new combinations of parental chromosomes, have been obtained

Irene Tiemann-Boege and Andrea Betancourt (eds.), *Haplotyping: Methods and Protocols*, Methods in Molecular Biology, vol. 1551, DOI 10.1007/978-1-4939-6750-6_14, © Springer Science+Business Media LLC 2017

from haplotype information of closely spaced genetic markers using, for instance, pooled sperm typing [3] or linkage disequilibrium inferences from population data [4].

However, within the field of haplotyping one of the most challenging aspects has been the characterization of rare haplotypes, which has been extremely difficult since rare haplotypes need to be singled out from a large pool of wild types found in tissues, blood, or other DNA sources. Additional limitations such as long amplicons lengths that need to span several polymorphisms (in humans ~1000 nt for an average SNP pair), jumping PCR [5, 6], and detection and visualization of single molecules, combined with the necessary high throughput, has complicated the task. A few existing methods already address individual aspects of these challenges, e.g., haplotyping of single molecules [7], high-throughput amplification of millions of PCR reactions [8, 9], or bridging the distance between polymorphisms with long-range haplotyping by fusion-PCR [10]. These are brought together into bead-emulsion haplotyping (BEH), described in this chapter. Previously, BEH was implemented to measure mutations or jumping PCR artifacts linked to the fidelity of the polymerase that can be an important source of artifacts when used to detect rare haplotypes [11, 12].

BEH is a high-throughput PCR method that is able to characterize in a single, cost-effective experiment, millions of haplotypes in parallel. BEH can identify rare haplotypes at a frequency of less than 10^{-5} of a handful of heterozygous sites separated up to ~7 kilobases [12]. DNA fragments ranging from less than 100 to ~7000 bp are hybridized to primer-coated beads (at a ratio of less than one DNA fragment to one bead to ensure single molecule amplification) and are amplified in individual compartments using emulsions that isolate individual aqueous micro-PCR reactors in an oil-phase [8, 9, 13–15]. This single molecule amplification is also known as digital PCR. One of the PCR-primers is linked to the beads via a biotin-streptavidin interaction and ensures that the resulting amplification products stay bound to the bead rendering clonal copies of an initial single molecule (*see* Fig. 1).

The principle of BEH is based on the multiplex amplification of several polymorphic sites on a single DNA template in emulsion-PCR. This is achieved using primer pairs, in which primers have an additional universal tail on the 5′ end (e.g., 26 nucleotides [12]) complementary to the primer on the magnetic bead). Thus, during emulsion PCR several different amplicons containing different polymorphic sites can be amplified on the bead. Up to three sites can be analyzed with this approach (or more if several polymorphisms are within an amplicon). The allele of the polymorphism is determined by genotyping the amplified products on the bead using allele-specific probes coupled to different fluorophores [12, 15]. The labeled beads, which fluoresce in different channels, can

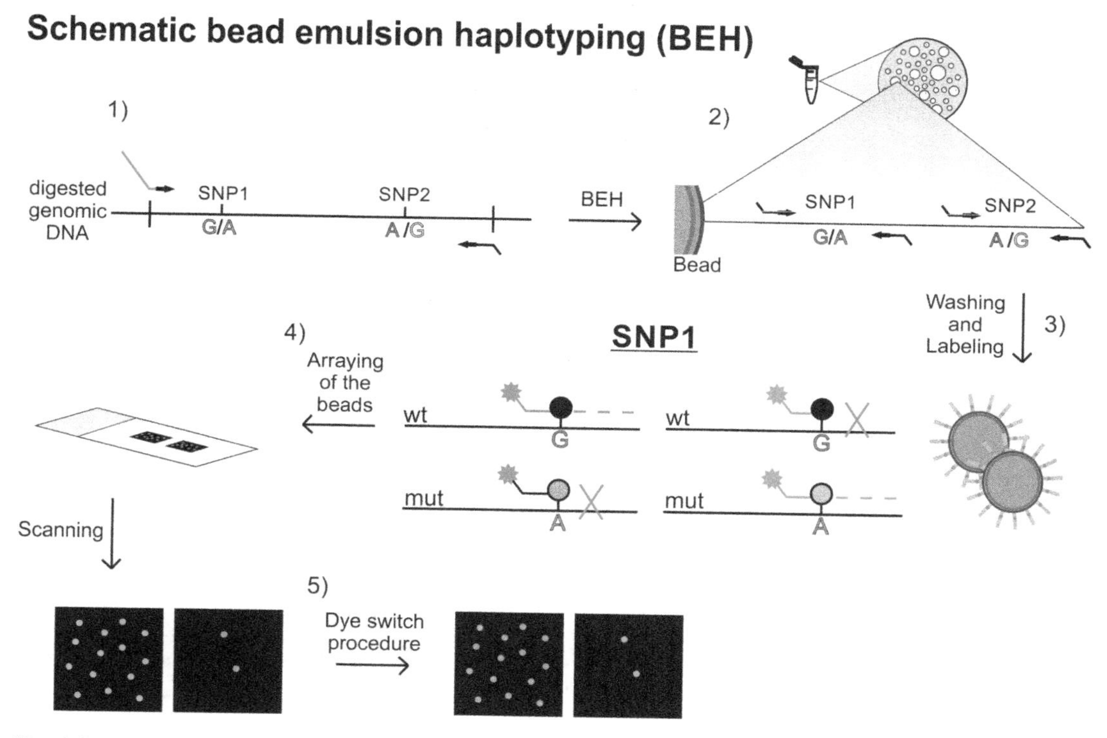

Fig. 1 Overview of bead-emulsion haplotyping (BEH). (*1*) In the first step, the digested genomic DNA is amplified for a few cycles to increase the number of starting molecules to 10^6 or 10^7 templates. (*2*) The templates are then hybridized to the paramagnetic beads and two to three different loci are amplified within the aqueous droplets of a water-in-oil emulsion. During the amplification, the beads get covered with thousands of clonal copies of the initial template. (*3*) The beads are washed and labeled by the extension products of allele-specific oligonucleotides coupled with specific fluorophores. (*4*) The beads are arrayed with a polyacrylamide matrix in a monolayer on a slide and scanned by an automated microscope. (*5*) The results of the first scan are verified by a dye switch procedure (*5*)

then be counted by scanning the beads arrayed on a monolayer with an inverse, automated fluorescence microscope. The resulting images are then analyzed by a mathematical software that extracts the color information of each bead and sorts the beads into different population and phased haplotypes [12].

2 Materials

2.1 Equipment

1. PCR workstation with filtered airflow and UV-light (*see* **Note 1**).
2. PCR thermocycler (*see* **Note 2**).
3. Thermocycler with an in situ block (*see* **Note 3**).
4. Magnetic Particle Concentrator (MPC) (*see* **Note 4**).
5. Vortexer/Mixer (*see* **Note 5**).

6. Bio Rad CFX384™Real-Time System type C1000 thermocycler (*see* **Note 6**).
7. Mixing/heating block (*see* **Note 7**).
8. Microcentrifuge (*see* **Note 8**).
9. TissueLyser II (*see* **Note 9**).
10. Epifluorescent, automated microscope (*see* **Note 10**).
11. 3–4 Coplin jars.

2.2 Plastic Consumables

1. 96-well or 384-well plates (*see* **Note 11**).
2. Adhesive optical clear rt-PCR plate seals (*see* **Note 12**).
3. 0.5–10 μL/2–20 μL/20–200 μL/100–1000 μL pipettors.
4. Filter tips and non-filter tips 0.5–20 μL/20–200 μL/100–1000 μL.
5. Siliconized 0.2 mL PCR reaction tubes (*see* **Note 13**).
6. Siliconized 2 mL reaction tubes with round bottom.
7. Siliconized 0.6 mL/1.6 mL/2 mL reaction tubes.
8. 15 and 50 mL Greiner tubes (for solvent aliquots during breakage of emulsion).

2.3 Software

1. NCBI PrimerQuest (*see* **Note 14**).
2. IDT OligoAnalyzer (*see* **Note 15**).
3. NEB-cutter (*see* **Note 16**).
4. CFX-rt-PCR manager (*see* **Note 17**).
5. Metamorph (*see* **Note 18**).
6. MatLab analysis program.

2.4 Materials

2.4.1 Pre-BEH Amplification

1. Nuclease-free water (Molecular grade).
2. Deoxynucleotide (dNTP) Solution Mix.
3. Phusion Hot Start II High-Fidelity DNA polymerase and 5× Phusion GC buffer.
4. Genomic DNA, plasmid DNA.
5. Restriction enzymes and buffers.
6. Primers (*see* Table 1).
7. 50× EvaGreen, only if PCR is performed in rt-PCR thermocyler.

2.4.2 Bead Preparation

1. Dynabeads® M-270 or M-280 paramagnetic streptavidin-coated beads (*see* **Note 19**).
2. 1 mM dual-biotinylated primer; in this case Bead-SNP1-SNP2-SNP_R2 (*see* **Note 20**).

Table 1
Primers used for the different amplification steps during BEH. The primers for the pre-BEH step amplify the genomic region to be haplotyped and flank the polymorphisms of interest. Letters in bold denote the universal tail sequence that links the PCR product to the bead

Primers for pre-BEH	
Name	**Sequence**
3509 bp fragment:	
FGFR3_SNP_F3	5′- AGC AGG TAA CGA CTC TGT CCC ATG C -3′
FGFR3_SNP_R2	**5′-AGA GCA GGA CCC CAA AGG ACC AGC-3′**
2264 bp fragment:	
F-SNP1-OUT	5′-GGCCTCAACGCCCATGTCTTTGCA-3′
FGFR3_SNP_R2	**5′-AGA GCA GGA CCC CAA AGG ACC AGC-3′**
Bead primer	
Bead-SNP1-SNP2-SNP_R2	5′ /52-Bio//iSp9/TA TGT CTT TCT CTC ACA TAA **AGA GCA GGA CCC CAA AGG ACC AGC** -3′
Primers for BEH	
F-SNP1-88 bp	5′-GAG CTG GTG GAG GCT GAC GA-3′
R-SNP1-R93-SNP_R2	5′-**AGA GCA GGA CCC CAA AGG ACC AGC** CCA CCA CCA GGA TGA ACA GGA AG -3′
F-SNP2-3	5′-CGG GAC GTG CAC AAC CTC GAC TAC-3′
R-SNP2_BA	5′-**AGA GCA GGA CCC CAA AGG ACC AGC** CAG GCG TCC TAC TGG CAT GA-3′

Table 2
Composition of the bind and wash buffer

Bind and wash buffer	
Tris–HCl (pH 7.4)	10 mM
M EDTA	1 mM
NaCl	2 M

3. Bind and Wash buffer prepared with double-deionized water (*see* **Note 21** and Table 2).
4. TE buffer dissolved in double-deionized water (*see* Table 3).

2.4.3 Template Hybridization to the Beads

1. Attachment solution (*see* Table 4).

Table 3
Composition of the TE buffer

TE (Tris-EDTA) buffer	
Tris–HCl (pH 7.4)	10 mM
M EDTA	1 mM

Table 4
Reaction mix for the attachment of the DNA to the beads

Attachment solution			
Stock concentration	Components	Final concentration	1× [μL]
	Nuclease free water		71.5
6 × 10⁵/μL	Beads	6 × 10⁶ beads	10
10×	Titanium Taq buffer	1×	9.5
50 mM	$MgCl_2$	1.58 mM	3
	Pre-amplified DNA (dilution)	~10⁶ molecules/μL	1
	Total volume		95

2.4.4 Emulsification

1. Aqueous phase (*see* Table 5).
2. MOCK mix (only for the alternative Dow Corning/Sigma Aldrich oil approach) (*see* Table 6).
3. Parafilm.
4. Oil phase of Evonik/Degussa (Tegosoft, Mineral Oil, AbilWE09) (*see* Table 7).
5. Alternative Dow Corning/Sigma Aldrich oil phase Formulation Aid, Fluid, Silicone Oil (*see* Table 8 and **Note 22**).
6. Steel bead 5 mm (*see* **Note 23**).

2.4.5 Breakage of the Emulsion and Clean-Up of Beads

1. 100 % isopropanol.
2. 100 % ethanol.
3. NXS buffer (*see* Table 9).
4. TE buffer (*see* Table 3).
5. 1E buffer (*see* Table 10).
6. 0.1 M NaOH (*see* **Note 24**).
7. Pasteur pipette (*see* **Note 25**).

Table 5
Composition of the aqueous phase

Aqueous phase			
Stock concentration	Components	Final concentration	1× [μL]
	Nuclease-free water		85.6
10×	Titanium Taq buffer	1×	15
50 mM	$MgCl_2$	8 mM	24
10 mM	dNTPs	1 mM	15
500 μM	F-ACH-88 bp	9 μM	2.7
5 μM	R-ACH-R- 93-SNP_R2	50 nM	1.5
500 μM	F-TDII-3	9 μM	2.7
5 μM	R-TDII-BA	50 nM	1.5
2 U/μL	Titanium Taq polymerase	0.03 U/μL	2
	Total volume		150

Table 6
Composition of the MOCK mix

MOCK mix			
Stock concentration	Components	Final concentration	1× [μL]
	Nuclease-free water		151.2
10×	Titanium Taq buffer	1×	24
10 mM	$MgSO_4$	2.5 mM	60
10 %	BSA (Bovine serum albumin)	0.1 %	2.4
1 %	Tween80	0.01 %	2.4
	Total volume		240

Table 7
Composition of the oil phase (Evonik/Degussa/Sigma Aldrich)

Oil phase Evonik/Degussa/Sigma Aldrich			
Stock concentration	Components	Final concentration	1×
100 %	Tegosoft (Evonik/Degussa/Sigma Aldrich)	73 % vol/vol	730 μL
100 %	Mineral oil (Sigma Aldrich)	20 % vol/vol	200 μL
100 %	AbilWE09 (Evonik/Degussa/Sigma Aldrich)	10 % wt/vol	100 mg
	Total		~1000 μL

Table 8
Composition of the oil phase (Dow Corning/Sigma Aldrich)

Oil phase Dow Corning/Sigma Aldrich			
Stock concentration	Components	Final concentration	1× [mg]
100 %	Formulation Aid (Dow Corning)	40 % wt/wt	400
100 %	Fluid (Dow Corning)	30 % wt/wt	300
100 %	Silicone Oil (Sigma Aldrich)	30 % wt/wt	300
	Total		1000 mg

Table 9
Composition of the NXS buffer

NXS buffer	
Tris–HCl (pH 7.4)	10 mM
EDTA	1 mM
Triton-X100	0.1 %
NaCl	100 mM

Table 10
Composition of the 1E buffer

1E buffer	
Tris–HCl (pH 7.4)	10 mM
KCl	50 mM
EDTA	2 mM
Triton-X100	0.01 %

2.4.6 Labeling of Beads with Fluorescent Probes

1. Labeling solution (*see* Table 11) and labeling probes (*see* Table 12).
2. Dye switch labeling solution (*see* Table 13) and labeling probes (*see* Table 14).

2.4.7 Bead-Array

1. Microscope slides 76×26×1 mm (*see* **Note 26**).
2. Lens cleaning tissue 80×100 mm.
3. Cover slips 24×40×1 mm.
4. Gamma-Methalcryloxypropyl-trimethoxysilan (≥99.8 %).
5. Rubber cement glue (*see* **Note 27**).

Table 11
Reaction mix for allele-specific extensions

Labeling solution			
Stock concentration	Components	Final concentration	1× [μL]
	Sigma water		41.5
10×	Titanium Taq buffer	1×	5
10 mM each	dNTPs	0.2 mM	1
100 μM	A488 SNP1wt	1 μM	0.5
100 μM	A592 SNP1mut	1 μM	0.5
100 μM	A647 SNP2wt	1 μM	0.5
100 μM	A532 SNP2mut	1 μM	0.5
2 U/μL	Titanium Taq polymerase	0.02 U/μL	0.5
	Total		50

Table 12
Sequences of all used fluorophores during the first scan. Each sequence is specific for one allele and locus. Asterisks denote phosphorothioate bonds

Target	Fluorophore	Sequence
SNP1wt	AlexaFluor 488	5′-Alex488N/ AG GCA TCC TCA GCT *A*C*G -3′
SNP1mut	AlexaFluor 592	5′-TexRd-XN/ CAG GCA TCC TCA GCT *A*C*A -3′
SNP2wt	TexasRed 647	5′-Alex647N/ CAC AAC CTC GAC TAC TAC A*A*G* A -3′
SNP2mut	AlexaFluor 532	5′-Alex532N/ AC AAC CTC GAC TAC TAC A*A*G*G -3′

6. Polyacrylamide matrix (PAA-gel) (*see* Table 15).
7. Hybridization chamber (Grace Bio-Labs HybriWell-FL™ sealing system, fluoro-"friendly" adhesive chamber).

3 Methods

3.1 Pre-BEH Amplification of Genomic DNA

1. Before starting the BEH, the target region containing the polymorphisms to be haplotyped is first amplified from purified genomic DNA by long-range PCR. The long amplicons are then subjected to BEH. We have tested BEH with amplicons of up to 5 kb [16], but longer fragments such as linearized plasmids (~7–8 kb) can also be directly used in BEH (without prior amplification). For regions with a high

Table 13
Reaction mix for second scan (on slide in situ)

Components for dye switch labeling solution			
Stock concentration	Components	Final concentration	2× [μL]
	Millipore water		49.8
10×	Titanium Taq buffer	1×	6
10 mM each	dNTPs	0.2 mM	1.2
100 μM	A592 SNP1wt	1 μM	0.6
100 μM	A488SNP1mut	1 μM	0.6
100 μM	A532 SNP2wt	1 μM	0.6
100 μM	A647 SNP2mut	1 μM	0.6
2 U/μL	Titanium Taq polymerase	0.02 U/μL	0.6
	Total		60

Table 14
Sequences of all used fluorophores during the dye switch. Each sequence is specific for one allele and locus. Asterisks denote phosphorothioate bonds

Target	Fluorophore	Sequence
SNP1wt	AlexaFluor 592	5′-TexRd-XN/CAG GCA TCC TCA GCT *A*C*A -3′
SNP1mut	AlexaFluor 488	5′-Alex488N/ AG GCA TCC TCA GCT *A*C*G -3′
SNP2wt	AlexaFluor 532	5′-Alex532N/ AC AAC CTC GAC TAC TAC A*A*G*G -3′
SNP2mut	TexasRed 647	5′-Alex647N/ CAC AAC CTC GAC TAC TAC A*A*G* A -3′

Table 15
Composition of the PAA-gel

PAA-gel			
Stock concentration	Components	Final concentration	1× [μL]
1×	TE buffer	0.4×	4
20 %	Rhinohide with Acrylamide (37/1)	7.6 %	4
5 %	TEMED	0.6 %	1.25
0.5 %	APS	0.06 %	1.25
	Total		10.5

GC content, we digest the genomic DNA prior to PCR to improve the amplification efficiency and yield (*see* Fig. 2). The digest of ~200 ng of genomic DNA is performed by adding all necessary components and incubating the reaction on a mixing/heating block at 37 °C for 2–4 h and then 16 °C overnight (*see* Table 16). We carry out the digestion in the PCR buffer so that the digested material can be directly added to the amplification reaction.

The amplification of the genomic DNA (digested or undigested) is carried out with a high-fidelity polymerase with proof-

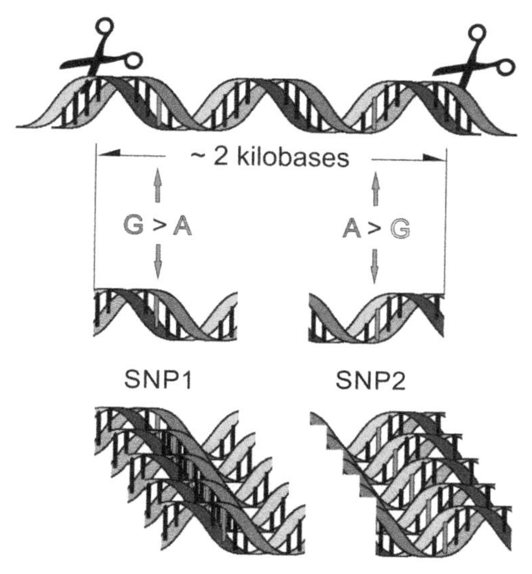

Fig. 2 Pre-BEH amplification. Genomic DNA is digested by restriction enzymes resulting in fragments containing the loci of interest to increase the amplification efficiency for difficult templates. During the amplification step, a universal sequence is introduced that is complementary to the primers attached to the beads

Table 16
Reaction mix for digest of genomic DNA

Genomic DNA Digest			
Stock concentration	Components	Final concentration	1× [μL]
	Nuclease-free water		5.9
5×	Phusion HF buffer	1×	2
100 ng/μL	gDNA	20 ng/μL	2
10 U/μL	Restriction enzyme	1 U/μL	0.1
	Total		10

reading activity. If the amount of amplifiable genomes needs to be monitored, 20 μL of the reaction can be run in an rt-PCR for 35 cycles (do not forget to add EvaGreen). It is important to consider, though, that the amplification products used for BEH only need to be run for 10–15 cycles, since using more cycles can produce chimeric products [12]. Thus, do not use the PCR products of the rt-PCR (run for 35 or more cycles), but perform two reactions instead. Using the same initial master mix (*see* Table 17), one reaction is performed in a regular thermocycler for not more than 15 cycles and the other is performed in a rt-PCR thermocycler (*see* Table 18). Note that 10–15 cycles used to create templates for BEH are not visible in rt-PCR. The final volumes of the reaction can be adjusted, but to save reagents we use a volume of 20 μL for the rt-PCR, and 100 μL for the regular PCR. We prepare a total reaction volume of 120 μL, which is split into 20 μL and 100 μL and amplified in the rt-PCR and regular thermocycler, respectively. Note that highly concentrated genomic DNA (~1 μg per reaction) can inhibit the polymerase, which can be avoided by increasing the ratio of the reaction volume to final DNA concentration.

2. Different dilutions of plasmid DNA are used as standards for the quantification of the number of amplifiable templates. Always include a no-template control, in which water is added instead of DNA. For the rt-PCR, the plate needs to be sealed with optical clear rt-PCR films (*see* **Note 12**).
3. To avoid contamination it is necessary to work in a PCR workstation with filtered airflow and UV-light in this stage for the amplification of genomic DNA. Never bring any PCR product into this PCR workstation.

Table 17
Reaction mix for rt-PCR

Master Mix for regular and rt-PCR			
Stock concentration	Components	Final concentration	1× [μL]
	Nuclease-free water		14.4
5×	GC buffer	×1	12
10 mM each	dNTPs	200 μM	2.4
5 μM each	Primer Mix (ACH + TDII)	1 μM	24
50×	EvaGreen	×0.5	1.2
3 %	DMSO		3.6
2 U/μL	Phusion Hot Start II polymerase	0.02 U/μL	2.4
	Total		120
	DNA (digested in 1xHF buffer)		60

Table 18
2-step PCR-Program used for rt-PCR and regular PCR (only 8 cycles in step 2)

Real time 2-step PCR-program		
Temperature	Time interval	
94 °C	3 min	
94 °C	20 s	×5 (**step 1**)
69 °C	15 s	
72 °C	10 s	
94 °C	15 s	×34 (**step 2**)
70 °C	15 s	
72 °C	10 s	
72 °C	5 min	
65 °C	5 s	
95 °C	∞	

3.2 Bead Preparation

1. First transfer 100 μL of M-270 or M-280 paramagnetic streptavidin-coated beads (*see* **Note 19**) into a siliconized 1.5 mL tube (*see* **Note 13**). Leave it on the MPC for ~1 min. until the beads cluster along the wall touching the magnet (*see* **Note 28**). Then remove the liquid and remove the tube from the MPC, next add 200 μL bind and wash buffer (*see* **Note 29**). Repeat this wash step with 200 μL bind and wash buffer three more times.
2. Resuspend the washed beads in 198 μL bind and wash buffer and add 2 μL of 1 mM dual-biotinylated bead primer (*see* Table 1).
3. Remove the tubes from the MPC and put them into a non-magnetic common rack and incubate the suspension for 20 min at 18–25 °C. It is necessary to mix every 5 min by a short vortexing step or by flicking the tube.
4. Afterwards, place the tubes back in the MPC and wash the beads again two times with 200 μL bind and wash buffer and two times with 200 μL TE buffer.
5. Resuspend in 100 μL TE buffer and use immediately or store at 4 °C (durable for 3 months) (*see* Fig. 3).

3.3 Template Attachment to the Beads

In this step, the amplicons derived from genomic DNA (containing the polymorphic sites of interest) (*see* Subheading 3.1 Pre-BEH) are hybridized via the 5′ universal sequence to the primer

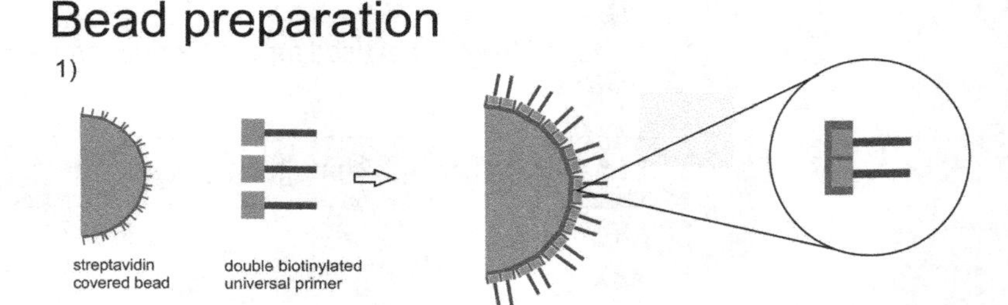

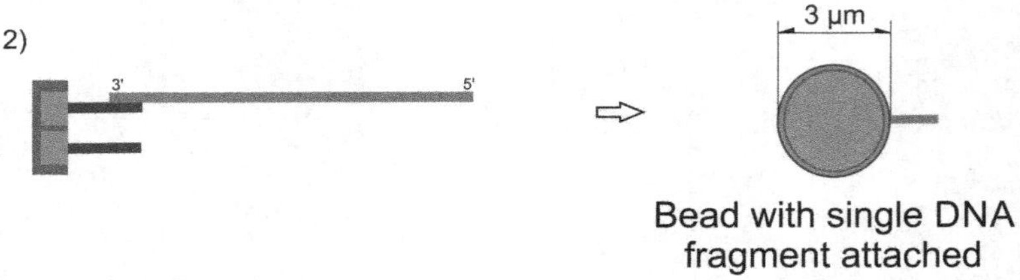

Fig. 3 Preparation of beads. (*1*) Streptavidin coated beads are bound to a double-biotinylated primer (complementary to a universal tail on the BEH primers). (*2*) DNA templates are hybridized to the beads (in a ratio of less than one template per bead) via the universal sequence

immobilized on the beads (*see* Fig. 3) at a ratio of less than one template per bead. Prepare the aqueous phase (*see* Table 5).

1. Then prepare the attachment solution (*see* Table 4) without DNA (*see* **Note 30**).
2. In a separate hood or room, add approximately ~10^6 molecules of DNA template to 6×10^6 beads and the attachment solution (*see* **Note 31**).
3. Incubate the reaction mix with the attachment program (*see* Table 19). After the reaction is completed use the MPC for washing the beads with 150 μL nuclease-free water (*see* **Note 32**). After the beads have been attracted to the MPC, remove the water and add 150 μL of the aqueous phase, vortex and remove the tubes from the MPC. Let the solution sit at least 2–5 min to equilibrate.

3.4 Bead-Emulsion PCR

The emulsion consists of millions of droplets containing ideally one bead per drop on average.

Emulsification Using Evonik/Degussa/Sigma Aldrich components

1. The oil phase is prepared by mixing Tegosoft and Mineral oil (*see* Fig. 4). Then the surfactant AbilWE09 is measured on an analytical scale (*see* Table 7). The oil phase needs to be vortexed for 10 s. followed by a short centrifugation step. This centrifugation step can be replaced by letting the oil phase sit for 5 min or until the bubbles disappeared.

Table 19
PCR-Program for the attachment of the DNA onto the beads

Attachment PCR-program	
Temperature	Time interval
105 °C	Lid
94 °C	2 min
80 °C	5 min
70 °C	1 min
	0.1 °C increments per sec
60 °C	1 min
	0.1 °C increments per sec
50 °C	1 min
	0.1 °C increments per sec
20 °C	1 min
20 °C	∞

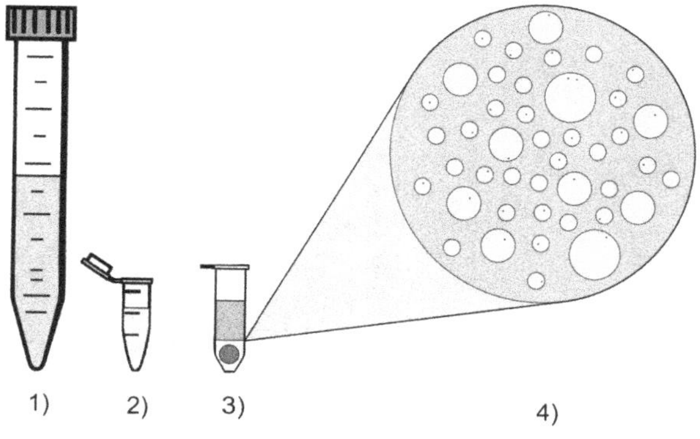

Fig. 4 Preparation of the Emulsion (Evonik/Degussa/Sigma Aldrich approach). 800 μL of oil phase (*1*) is mixed with 150 μL of aqueous phase (*2*) containing the beads with one attached DNA template. The addition of a 5 mm steal bead (*3*) mixed by a 2-step procedure at 15 Hz for 10 s and 17 Hz for 7 s in a TissueLyser II creates a stable emulsion (*4*)

2. Add one 5 mm steel bead (*see* **Note 23**) to a 2 mL round bottom tube (siliconized, *see* **Note 13**). Then add 800 μL oil phase, 150 μL aqueous-phase-beads-mix from the previous attachment step, and seal with Parafilm. Emulsifying the two immiscible liquids, whereby at least one of both needs to contain a surfactant, creates aqueous micro-compartments that

Table 20
PCR-program used for BEH (for Titanium Taq polymerase)

Emulsion PCR-program (for Titanium Taq polymerase)		
Temperature	Time interval	Comment
94 °C	2 min	
94 °C	15 s	×55
65 °C	15 s	
68 °C	70 s	
68 °C	2 min	
8 °C	∞	

contain H_2O, buffer, $MgCl_2$, dNTPs, forward and reverse primers, polymerase, and the bead carrying one single DNA fragment. In case of a stable emulsion, these aqueous droplets are the micro-PCR reactors in which clonal products of an initial template will be produced.

3. Place the tubes into the adaptors and run the TissueLyser II program 15 Hz for 10 s and 17 Hz for 7 s. Consider that the usage of alternative oil-phases requires optimized emulsification procedures (*see* **Notes 33** and **34**). When using higher frequencies, the emulsion usually gets finer; whereas, with lower frequencies the emulsion contains larger droplets. The emulsion appears as a white, smooth, and homogenous phase. The contents of the 2 mL tubes are aliquoted (80–100 μL) into smaller 200 μL PCR tubes that are then placed in the thermocycler and cycled with the PCR-program shown in Table 20 (*see* **Note 35, Fig. 5**).

3.5 Alternative Emulsification with Dow Corning/Sigma Aldrich Oils

1. Due to different densities of the oils, these need to be weighed in. Formulation aid, Fluid, and Silicone oil are mixed in a ratio of 4:3:3 (*see* Table 8). The oil phase is first vortexed for 10 s and then centrifuged at full speed. This centrifugation step can be replaced by letting the oil phase sit for 5 min or until the bubbles disappeared.
2. Add one 5 mm steel bead to a 2 mL round-bottom tube, add 650 μL (62 %) oil phase (*see* Table 8) and 240 μL MOCK mix (*see* Table 6) and seal with Parafilm. Place the tubes into the adaptors of the TissueLyser II (*see* **Note 9**) and apply the program with 25 Hz for 5 min. Remove the steel bead and add 150 μL aqueous phase to the emulsion and apply the program 15 Hz for 5 min (*see* **Notes 33** and **34**).
3. Aliquot (80–100 μL) into smaller 200 μL PCR tubes, place them into the thermocycler, and run the PCR-program in Table 20 (*see* **Note 35**, Fig. 5).

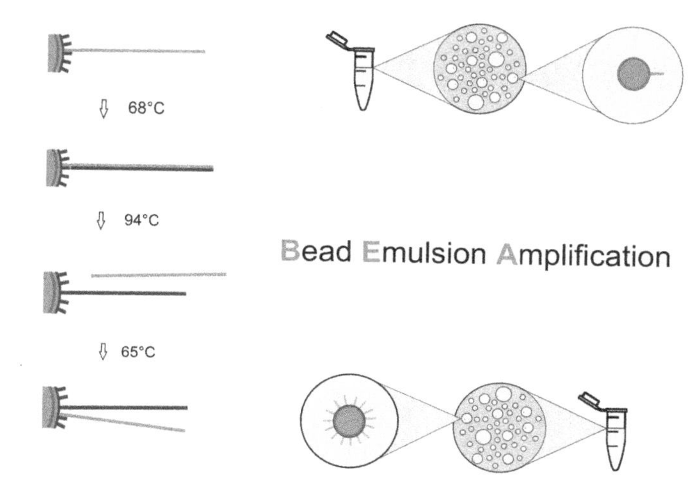

Fig. 5 Bead-Emulsion PCR. Each aqueous droplet of the emulsion ideally contains one single bead with one single DNA template to allow single molecule amplification. The droplets act as separated micro-reactors and prevent different DNA fragments from being amplified on more than one bead

3.6 Breakage of the Emulsion

3.6.1 Washing Steps for Evonik/Degussa/Sigma Aldrich Oils

1. After the emulsion PCR is completed, add a few drops of isopropanol to each PCR reaction tube and mix thoroughly by vortexing so that the emulsion breaks (*see* **Note** 25).
2. Pool the aliquots of the 200 μL tubes into a 1.5 mL tube; fill up with isopropanol and vortex.
3. Centrifuge the tube for 1 min at 9000 × *g* and place in the MPC.
4. Wash and resuspend the beads in ~500 μL isopropanol, then vortex and centrifuge again 1 min at 9000 × *g* (*see* **Note 36**).
5. Place the beads in the MPC again and wash with ~ 500 μL NXS without removing the tube from the MPC. Then exchange the NXS buffer, followed by a thorough vortexing step. In this step, it is important to vortex until no bead clumps are visible anymore. Bead aggregations can occur due to insufficient vortexing and can lead to difficulties in arraying and subsequent scanning of microbeads.
6. Apply a quick spin (17,000 × *g*) and put the tubes back in the MPC.
7. Repeat the NXS wash once more or until the white slurry is gone, respectively.
8. Wash the beads in ~500 μL TE, remove the TE (using the MPC) and rinse with 250 μL freshly prepared 0.1 M NaOH solution from a 2 M stock, and incubate for 1 min at room temperature (*see* **Note 37**).
9. Remove the NaOH solution, wash the beads again with ~500 μL TE, and label them immediately or store them at 4 °C.

3.6.2 Washing Steps for Dow Corning/Sigma Aldrich Oils

1. After the emulsion PCR is completed, pool the content of the 200 μL tubes equally into two siliconized 2 mL round-bottom tubes and add approximately 1.5 mL of ethanol to each tube.
2. Seal the tubes with parafilm. Break the emulsion by applying 30 Hz for 5 min in the TissueLyser II and centrifuge with 17,000 × *g* for 2 min.
3. Take off the supernatant (SN), and rinse again with ~500 μL ethanol.
4. Fill up the tube with ethanol, and homogenize again for 30 Hz for 2 min and centrifuge for 2 min at 17,000 × *g*.
5. Place the beads back in the MPC, remove the SN, fill up with NXS buffer, and repeat the wash of 30 Hz for 1 min in the TissueLyser II.
6. After centrifuging at 17,000 × *g* for 1 min remove the SN, and rinse the beads with TE buffer.
7. Then add 500 μL freshly prepared 0.1 M NaOH solution from a 2 M stock to the bead pellet on the MPC, and incubate for 1 min at room temperature (*see* **Note 37**).
8. Remove the NaOH solution and rinse the beads again with ~500 μL TE buffer.
9. Beads can then be labeled immediately or stored in TE buffer at 4 °C.

3.7 Labeling

1. Up to two different polymorphisms can be screened simultaneously using four different allele-specific extension probes coupled with Alexa fluorophores (A488nm, A592nm, A647nm, or A532nm) (*see* Fig. 6).
2. Prepare labeling solution (*see* Table 11) by mixing 41.5 μL Millipore water, 5 μL Titanium Taq buffer, 1 μL dNTPs, 0.5 μL Titanium Taq polymerase, and 0.5 μL of each of the four labeling probes (*see* Table 12). Once again, the polymerase is added last (*see* **Note 38**).
3. Place the tubes on the MPC, remove the TE buffer, resuspend the beads in 50 μL labeling solution, transfer the whole content into a 200 μL PCR tube, and run the thermocycler with the following program (*see* Table 21, *see* **Note 39**).
4. Meanwhile, prepare ~500 μL 1E buffer in an empty 1.5 mL reaction tube. After the program has finished, do not stop the PCR program, and pipette the hot reaction mix directly from the still hot 75 °C PCR block into the prepared reaction tubes containing 1E buffer to avoid unspecific binding of labeling probes which do not match to the present haplotype (*see* **Note 40**).
5. Centrifuge at 9000 × *g* for 1 min and place the tube in the MPC.

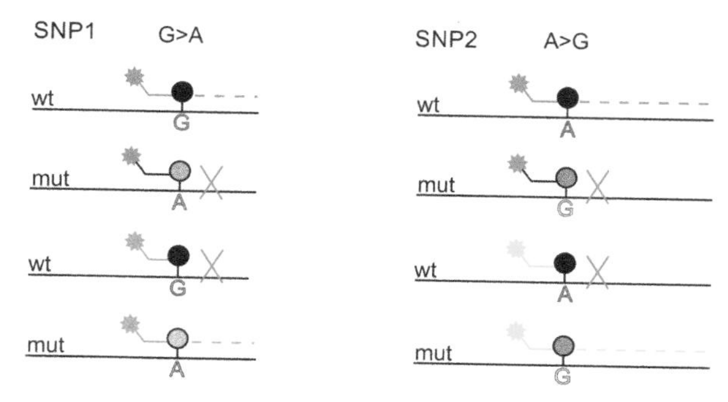

Fig. 6 Principle of labeling by allele-specific extensions. In order to identify the allele amplified on the bead, up to four different fluorescently labeled oligonucleotides (probes) are used for two different loci (2 per allele). The hybridized probe is only elongated by the polymerase if the allele matches the nucleotide at the 3′ end of the probe. A final heating step denatures non-extended probes

Table 21
PCR-Program used for allele-specific extension

Labeling program		
Temperature	Time interval	Comment
94 °C	2 min	
60 °C	5 min	
72 °C	5 min	
75 °C	∞	

6. Take off the SN and wash with ~500 μL TE buffer.
7. Array immediately or store the beads until the arraying step is carried out.

3.8 Array and Scan of the Beads

1. In order to analyze the beads, they are arrayed with a polyacrylamide matrix (PAA gel) that immobilizes the beads in a monolayer onto a slide (*see* **Note 41**), such that the positional information over consecutive washing and scanning steps is maintained (*see* Fig. 7).
2. To array the beads onto a slide, the first step is the preparation of the polyacrylamide gel. In advance, 4 μL of 20 % Rhinohide in Acrylamide (37/1) and 4 μL TE buffer are pipetted together in 200 μL PCR tubes, while TEMED and APS are added shortly before arraying the beads (*see* Table 15 and **Note 42**).
3. Microscope slides are first treated with a layer of Gamma-Methalcryloxypropyl-trimethoxysilan (*see* **Notes 43–45**).

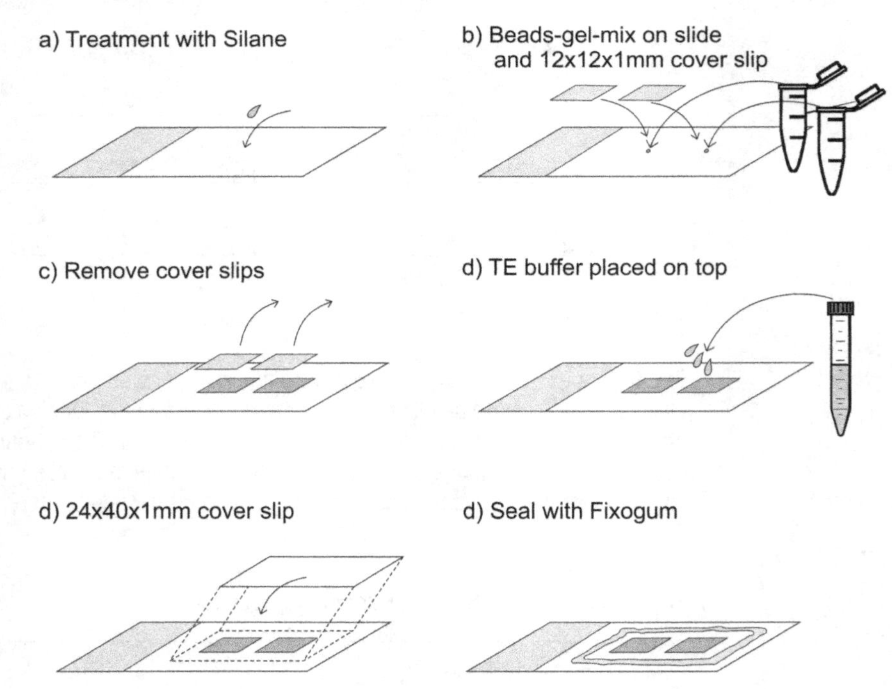

Fig. 7 Arraying of the beads. In order to enable the scanning and the subsequent analysis of the beads, it is necessary to immobilize the beads in a monolayer on a microscope slide. For this purpose, the beads are mixed with a small volume of polyacrylamide gel, arrayed directly on the slide, and covered by a 12 × 12 mm cover slip. The cover slips are then removed, TE buffer is pipetted onto the gels, and the array is covered with TE and a 26 × 40 × 1 mm cover slip sealed with rubber cement glue

4. On the MPC, remove the TE buffer from the beads, remove from the MPC, and let the beads air dry for max. 5 min (*see* **Note 46**).
5. Add 1.25 μL 5 % TEMED and 1.25 μL 0.5 % APS to the already prepared Rhinohide-in-Acrylamide/TE buffer-mix, flick the tube, and take 1.9 μL of the mixture. Add the PAA-gel to the beads, mix by pipetting up and down, and transfer beads to the slide (*see* **Note 47**).
6. Cover the bead drop with a cover slip (12 × 12 mm) and apply soft pressure to spread the gel over the whole area of the cover slip creating the desired monolayer (*see* **Note 48**). Take care to avoid the formation of air bubbles.
7. Wait ~5 min until the gel is polymerized (*see* **Note 49**), and then carefully remove the 12 × 12 mm cover slip. Place a few drops of TE buffer on the sample and cover the array with another cover slip (24 × 40 × 1 mm). Once again, it is important to avoid the formation of air bubbles. Remove the excess of liquid with a piece of tissue and seal the slide with rubber cement glue (*see* **Note 27**).

8. The array is then scanned with an epifluorescent microscope using ~300 raster positions (*see* **Note 50**). Four images (respectively 5) are taken at each raster position in the different fluorescent channels and the bright field mode with the 12-bit 4 K CCD camera. The bright field image is necessary to infer the bead area during the image analysis.
9. If scanning more than two polymorphisms, the array can be washed, the fluorophores need to be stripped off, and the array can be labeled with a new set of probes, as explained in the next step (additional probing and imaging of the array).

3.9 Additional Probing and Imaging of the Array

The probes on the arrayed beads can be stripped off and the beads can be labeled again with another set of probes. This is quite useful when scanning more than two polymorphisms, or for verifying rare haplotypes by a dye-switch approach.

1. Prepare the dye switch solution by mixing the components in Table 13 using the four dye switch labeling probes listed in Table 14.
2. The rubber cement glue is removed from the cover slip sealed on the slide.
3. Carefully remove the cover slip (24 × 40 × 1 mm) by adding a few drops of TE buffer around the edges. The cover slip quickly floats away by the excess TE buffer and can be lifted without loss of beads.
4. Cover the array with TE buffer and use an in situ PCR block or adaptor to strip off the probes by exposing it to 94 °C for ~1 min. Take care that the gel is covered with TE buffer at any time.
5. Remove the slide with the array from the in situ block and rinse with water.
6. Now, let the excess of water air-dry on the array. Prepare the labeling solution (Tables 13 and 14). Place now the hybridization chamber centered on the array. Apply gently pressure with the back of a pen around the edges to make sure that the seal is tight and no labeling solution may leak.
7. Pipette 100 μL dye switch solution into the hybridization chamber through the holes at the two corners of the hybridization chamber. Before sealing the holes with the adhesive films remove all air bubbles by gentle tapping on the surface of the hybridization chamber.
8. Place the slide on the in situ block and use the conditions listed in Table 22.
9. When removing the slide from the hot in situ block (at ~75 °C), immediately strip off the hybridization chamber, and place the slide in a Coplin jar filled with 1E buffer to wash off all non-extended allele-specific probes.

Table 22
PCR-Program used for dye switch

Dye switch program		
Temperature	Time interval	Comment
95 °C	2 min	Deactivation lid heating
63 °C	5 min	
72 °C	5 min	
75 °C	∞	

10. Rinse the array with TE buffer, cover it again with a cover slip (24 × 40 × 1 mm), and seal it with rubber cement glue.
11. When scanning the array for a second or other consecutive time, place the slide at the same position on the slide holder of the microscope and check with Metamorph how well the first image of the first scan aligns with the bead positions of the second scan. If the offset is too large you can adjust the x- and y-positions of the scan area defined in Metamorph for a perfect image overlay.

3.10 Bead Analysis

1. Images taken at each raster position are analyzed using custom-made algorithms written in Matlab that can be obtained per request.
2. In brief, the image analysis consists of the correction of the random shifts between images of the same raster position from consecutive washes. This correction is based on registration of these images, identification of the beads in the bright field via segmentation. Then, the extraction of the fluorescent signal information making use of the segmentation in the previous step is performed.
3. All the images are aligned/registered between raster positions using the bright field image of the first wash as reference. Registration is based on normalized correlation and a summary of all the inter-slide shifts is created at the end of the process. Images that need more than 500 pixel alignment correction are discarded from the analysis.
4. In order to compare the fluorescent signals across channels and washes, first a mask is created via segmentation of the reference bright field image (of the first wash) (*see* Fig. 8). The bright field bead segmentation is based on the selection of significant wavelet coefficients of an isotropic undecimated wavelet transform constructed upon third order B-splines [17]. The selection is made based on the control of the false discovery rate [18].

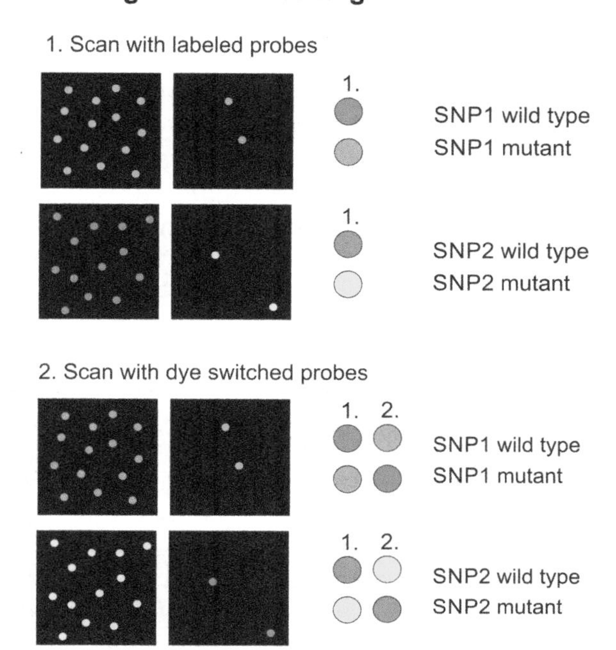

Fig. 8 Analysis of the labeled beads. (*1*) After the first labeling step, the beads carrying different DNA fragments fluoresce in different colors. (SNP1wt = *green*, SNP1mut = *orange*, SNP2wt = *red*, SNP2mut = *yellow*). These signals are imaged by the microscope for each raster position of the array and evaluated by MatLab. (*2*) In order to verify rare alleles, a second wash, labeling, and imaging step is used, in which the colors of the wild type and mutant probes are switched

A post-processing step is applied comprising morphological operations to correct the segmentation and selection of beads fulfilling quality criteria such as eccentricity/elongation and area. Each bead is assigned a unique integer identifier.

5. The same segmentation is applied to bright field images of subsequent washes. The beads in the initial mask are identified as beads in the newly segmented image, if the two bead masks overlap. In case two beads in one image overlap with one bead in the initial mask, the bead with a greater area-overlap is chosen. If a bead does not overlap with a bead in the reference bright field image, the bead is discarded from further analysis.
6. For each bead, the mean intensity of the fluorescent signal is extracted from the corresponding aligned fluorescent image making use of the bright field masks superimposed on the fluorescent image and is saved in a text file in the result directory together with the bead identifier and the computed bead area.

7. The text file contains the average pixel intensity recorded in every fluorescent channel for each bead. These intensities are then used to classify beads with a specific genotype, depending on the classification of the bead into a cluster.
8. Four different bead clusters were used to identify the bead populations based on their signal intensities: (00 = both channels are empty; 10 = first channel is positive; 01 = second channel is positive; 11 = both channels are positive). As a rule of thumb, ~90 % of the beads remain empty. All the details about the clustering statistics used can be found in Boulanger *et al.* [12].

4 Notes

1. It is recommended to use two different workstations located in two different rooms: one for the preparation of stock solutions, master mixes, DNA extractions, and digests, and a second workstation for handling the amplified DNA. Short PCR products easily form aerosols and can quickly spread in the laboratory producing false-positive reactions. These two workspaces are necessary to avoid contamination from PCR products. Laminar flow in the hood reduces aerosol formation and PCR products can be easily destroyed with UV light. For experiments designed to measure rare variants, these contaminating PCR products can influence the results. Routine cleaning of the workstations and the equipment (including decontamination with UV) is essential for accurate and reliable single-molecule measurements.
2. When choosing a thermocycler, make sure that it can run up to 10–100 μL volumes, since reaction volumes of the emulsion PCR range from 80 to 100 μL per 200 μL PCR tube.
3. We use the PeqStar in situ PCR thermocycler. The in situ block is necessary for performing reactions on the microscope slides with the bead array. Alternatively, in situ adaptors for regular thermocyclers are available (e.g. Techne in situ adapter from Tecnprime/ISHA).
4. We recommend the Dynabeads® MPC®-S (Magnetic Particle Concentrator) from Thermo Scientific. MPC contains a strong magnet with two adjustable levels of magnetic strength and enables optimal handling of the paramagnetic beads. It provides space for up to six 1.5 mL/2 mL reaction tubes in parallel. The working volume ranges between 20 μL and 2 mL. Not all magnetic particle concentrators have the same magnet strength and careful selection is recommended to reduce difficulties with handling.

5. We use a vortexer for the mixing steps and a table-centrifuge for spinning down the liquids to avoid spreading tube contents all over the working area. Vortexing is an important step used to properly mix thawed reagents and properly homogenize master mixes.
6. We use the Bio Rad CFX384™Real-Time System type C1000 thermocycler. This equipment is only necessary if the initial amplification of the genomic DNA needs to be monitored and the number of amplifiable templates is quantified.
7. The mixing/heating block is used for the genomic DNA (gDNA) extractions performed at 37 °C.
8. Any micro centrifuge with a g-force of 17,000 can be used. Take care that the tubes are balanced equally since imbalances in the centrifuge will reduce the lifetime of your equipment.
9. A homogenizer is used to create the emulsion since mechanical energy input is essential for the formation of the emulsion compartments. The homogenizer can be an automated homogenizer (e.g., TissueLyser II from Qiagen) that reduces handling differences from different users. Alternatively, a vortexer with an adjustable speed setting of up to 3000 rpm can be used, in which the emulsion is created by tilting the Eppendorf tube with the emulsion and operating at 3000 rpm for 70 s. Compartment sizes in an emulsion vary considerably based on the homogenization speed and time, thus it is important to optimize this step with the equipment of your choice. When using the TissueLyser II, place the tubes at the outermost positions of the adaptors since higher centrifugal forces can be applied resulting in more stable emulsions.
10. We use the automated epifluorescent inverted microscope at a 20× magnification for the image acquisition of the bead array. If an automated microscope is not available, alternatively a flow cytometry with four colors could be used, as done in other bead-emulsion applications [8, 9], with the disadvantage that the positional information is lost and only two polymorphisms can be screened at a time. The components of our microscope are the following: Zeiss Axio Observer.Z1, CoolSNAP-4 K Digital Camera System (TE/CCD-KAI-2040 Interline; 2048×2048 Pixel, 7.4 × 7.4 µm; 12 Bit, 20 MHz Pixelrate; three fullframes/s; thermoelectrical cooling down to −25 °C; about 55 % QE; C-mount connection; high speed PCI Interface), electromechanical shutter with 25 mm aperture inclusive control unit (maximal 40 Hz Frequency, manual control possible, serial interface, TTL Control possible, adapter set for Zeiss microscope, material CuBe for Zeiss microscope, high speed bright field shutter), 120 W mercury short arc lamp (HXP 120× Illumination system with integrated shutter, software controlled

mechanical dimmer, 1.5 m light guide, adapter for liquid fiber) or alternatively a SOLA SE II light engine® that provides much longer and constant illumination times, individual filter cubes optimized to minimize crosstalk between the four different Alexa flours (Chroma HQ480/20; dich Q495LP; HQ510/20 m, HQ530; dich T550lpxr; ET565/25 m, D560/40x; dich 595DCLP; D630/60 m, HQ630/20; dich Q660LP; HW700/75 m), motorized XY stage (100 × 120 mm travel range, speed = 25 mm/s, resolution 0.025 μm, reproducibility +/− 0.75 μm, accuracy = <6 μm), basic MetaMorph Imaging Software for controlling the microscope.

11. 96-well or 384-well plates with white coated wells are recommended to obtain an optimal fluorescence signal during rt-PCR when using DNA intercalating fluorophores.
12. Not every type of plate seal is ideal for the fluorescent signal of an rt-PCR. Microseal "B" adhesive, optical clear rt-PCR films (BioRad) were used to seal the well plates previous to the centrifugation step and the subsequent rt-PCR.
13. When handling DNA or paramagnetic microbeads, siliconized tubes were preferred. Their inner walls are very hydrophobic, which helps to prevent loss of DNA molecules/beads in high-sensitivity assays.
14. The NCBI PrimerQuest program was used to verify that designed primers anneal primarily on the selected target, to avoid unspecific binding and subsequent amplification of other regions. This is particularly important for primers designed in regions with high GC contents, since these primers often have a high 3′ end stability in other regions.
15. The IDT OligoAnalyzer software tool provides information on the melting temperature of the oligo/primer adjusted to the salt concentrations of the reaction, self-dimerization, and hetero-dimer formation, among others.
16. NEB cutter provides a catalogue of appropriate restriction enzymes (RE) that digest the flanking regions of the target to reduce the size of the initial genomic DNA. In addition, a combination of RE with similar activities in the same buffer can be selected.
17. This very intuitive tool analyzes the output files of the CFX-rt-PCR.
18. The Metamorph software controls the microscope to automate the scanning of the bead array. The consecutive scan of more than one bead array can also be programmed with this software.
19. The paramagnetic beads are covered with streptavidin necessary to capture the dual-biotinylated bead primer. Other bead-

emulsion amplification protocols use larger or smaller beads, but in our case, the M-270 or the M-280 beads with a size of ~3 μm were appropriate to capture sufficient amplification products rendering beads that have a 3–5x higher intensity than empty beads without an immobilized product. Both, M-270 and M-280 beads work well in BEH. Moreover, with this size ~1.5–2 million beads fit in an array area of ~1 cm^2 without being too crowded. The magnetic properties of the beads are important to properly wash and clean the beads from the emulsion.

20. We use the following primer (notation used by Integrated DNA technologies): (5′/52 Bio//iSp9/**TATGTCTTTCTC TCACATAAA**GAGCAGGACCCCAAAGGACCAGC-3′).

    This primer has four components: (a) a double-biotin group at the 5′ end that attaches the primer to the streptavidin coating on the beads. Note that if only a single biotin is used, the primer does not properly stay attached, rendering beads with little or no amplification product; (b) a triethylene glycol spacer that reduces the steric hindrance at the surface of the beads, followed by (c) a ~21 bp random sequence (in bold) serving as an additional spacer; (d) a universal sequence found also as a 5′ universal tail on each of the forward or reverse primers of the individual polymorphic loci. This universal sequence is also present in the initial DNA template (genomic DNA, plasmid, or PCR amplicons).

21. We use the Milli-Q® Ultrapure Water system to purify and deionize our tap water.

22. The alternative oil phase of Dow Corning/Sigma Aldrich (Formulation Aid, Fluid, Silicone Oil), also used by the original 454 sequencing method [19], forms a stable, more homogenous emulsion with compartment sizes ranging from 10 to 60 μm in diameter. Keeping the emulsion droplets evenly distributed and small ensures that the majority of the emulsion droplets contain just one bead per microreactor, necessary for the monoclonal amplification of DNA on the bead. The compartment sizes hardly change during the PCR, and the emulsion does not break during the PCR. Emulsion stability is also highly dependent on the surfactant, which has amphiphilic chemical properties and helps to maintain the interface between the aqueous phase in the surrounding continuous oil phase. BSA (Bovine serum albumin) and Tween80 are used in the Dow Corning/Sigma Aldrich protocol as surfactants and help stabilizing the emulsion droplets. Moreover, aggregation processes like coalescence or sedimentation do not occur as frequently as when using Evonik/Degussa/Sigma Aldrich components (less effective surfactant). Furthermore, the Dow Corning/Sigma Aldrich components also enhance the efficiency

of the PCR resulting also in a more densely covered bead. More amplified DNA on the beads leads to stronger illumination signals that facilitate the scanning and the subsequent analysis with MatLab for the distinction of the bead populations.

In comparison, the oil phase of Evonik/Degussa/Sigma Aldrich (Tegosoft, Mineral Oil, AbilWE09) used by the BEAMing method [8, 9, 14]) results in less stable emulsions with a reduced robustness and heterogeneous compartment sizes ranging between 10 and 200 μm. Due to this heterogeneity in droplet sizes, it is likely that more than one bead is placed in one droplet, and at the same time, many functional emulsion droplets remain empty. Emulsions performed with the Evonik/Degussa/Sigma Aldrich components break easier during PCR and show sedimentation visible as bead aggregates at the bottom of the PCR tubes or as a phase separation after PCR. The only advantage of using these components is the easier washing procedure, which results in a cleaner array of the microbeads into a monolayer on the microscope slides.

23. A steel bead of 5 mm can help to create well size-distributed emulsions in a defined range. Other types of beads might also work, but the droplet sizes of the emulsion might change and the homogenization program needs to be adapted. It is also useful to work with a two-step program: an initial step with a 5 mm steel bead (e.g., to prepare the MOCK mix) followed by an emulsification step without a steel bead (e.g. final emulsion in the alternative Dow Corning/Sigma Aldrich approach). The conditions for a suitable emulsion stability need optimization and many trials until reproducible and stable emulsions can be performed.
24. It is necessary to prepare the 0.1 M NaOH freshly every time the BEH protocol is performed. The 0.1 M NaOH is prepared from a 2 M stock, which can be used for a week.
25. During the washing and breakage of the emulsion, the danger of cross-contamination is low, because the amplification step for the DNA during PCR is already completed. Therefore, we use Pasteur pipettes to add isopropanol, NXS buffer, TE buffer, or NaOH. Those pipettes can be used for several weeks and different experiments, until they have to be exchanged. In this way, a large amount of plastic pipette tips can be saved. For discarding the supernatant, a 1 mL pipette is used and the tip can be reused for the whole washing procedure without changing.
26. Choose slides with a labeling area and a pen with ink that does not dissolve with the used solvents (isopropanol or ethanol). Dissolved ink can stain the beads creating false positives.

27. Rubber cement glue is an aqueous-based glue that can be easily removed from the microscope slide and is used to seal the 24 × 40 × 1mm cover slips on the array during the long scans. If the cover slip on the array is not sealed, the TE buffer on the array dries out, which leads to shrinkage of the PAA gel, a change in focus and unusable images. Morever, drying of the TE buffer causes the PAA gel to distort. We apply this glue with a pipette tip placed on the glue opening to facilitate dispensing the glue. Drying time is about 10 min and should occur in a dark room to protect the fluorescent probes from the light.
28. Let the magnetic beads cluster in a small area, and then carefully take off the supernatant without touching the beads. Check the color of the fluid in the pipette tip to minimize the loss of beads. If you see a reddish color, put the supernatant back into the tube, wait another minute to let the beads cluster again.
29. Do not leave the beads on the MPC without any liquid because dry beads tend to form clumps. This clumping is more pronounced, the more DNA is available on the beads.
30. If several experiments should be carried out simultaneously, prepare the amount of needed solution as a single master mix. Calculate the necessary reaction volumes by multiplying each volume by a factor of 0.2 to account for waste volumes.
31. Do not dilute in steps larger than 1:100 since this might compromise the pipetting accuracy of very large or very small volumes. The smaller the volume of the dilution step, the more accurate the dilution. Furthermore, it is important to change the pipette tip when going from higher dilutions to lower ones.
32. Prepare 1.5 mL tubes with 150 μL nuclease-free water placed in the MPC. Then, transfer the completed attachment reaction to the prepared 1.5 ml tube with water and let the beads aggregate on the magnet. Rinse the empty tube with the same water taken from the washed beads in case some beads remained stuck on the wall. Scratching the wall with the pipette tip can help to recover all the beads.
33. The TissueLyser program for the Evonik/Degussa/Sigma Aldrich oil phase is the following: 15 Hz for 10 s and 17 Hz for 7 s (both with 5 mm steel bead). Alternatively, vortexing at full vortexing speed (e.g. 3000 rpm) tilting the tube for 70 s on the Vortexer can also be used for creating an emulsion. The TissueLyser program for the Dow Corning/Sigma Aldrich oil phase is 25 Hz for 5 min with MOCK mix and a 5 mm steel bead, then remove the steel bead and homogenize for an additional 15 Hz for 5 min.

34. When using Formulation Aid, Silicon Oil, and Mineral Oil components, it is necessary to make a MOCK mix (does not contain primers or polymerase). The MOCK mix is used to create an extremely stable microfine emulsion by mixing with the oil phase at 25 Hz for 5 min with a 5 mm steel bead. This serves also to further stabilize the aqueous phase within the continuous oil phase. The aqueous phase is added after the formation of the microfine emulsion and followed by a second TissueLyser step with 15 Hz for 5 min without steel bead. The second step helps to distribute larger aqueous emulsion droplets which work as microreactors in the PCR reaction. The surfactant Tween80 and the BSA in the MOCK mix enable the stabilization of the emulsion.
35. *BEH program*: We use the Titanium Taq polymerase which is a 5′-exonuclease deficient hot start Taq polymerase with an TagStart™ antibody. The program starts with an initial 2 min 94 °C step to activate the polymerase. Next, a second 94 °C step for 15 s denatures the double-stranded DNA. The primer anneals at 65 °C for 15 s followed by an extension time of 68 °C for 70 s. This program is cycled for 55 times and finalized by a 68 °C step for 2 min to complete the elongation of unfinished DNA strands. The completed reaction remains at 8 °C and can be stored overnight without any problems.
36. If the beads do not resuspend completely, repeat the washing step in NXS buffer, as well as the quick spin, once or twice. Ideally, the beads form a sharply bordered and discrete cluster, which moves smoothly when the tube is rotated in the MPC. Do not resuspend by pipetting the beads, as they tend to stick irreversibly inside the tip. When you use another oil phase like Dow Corning/Sigma Aldrich, be aware that different washing steps, as well as different bead handling steps, are necessary.
37. Low concentration of NaOH is used to denature the DNA on the beads from double-strands into single-strands. In order to prepare the 0.1 M NaOH, we freshly dilute it every time from a 2 M stock (stock can be used for a week). Pipetting steps during this washing procedure require careful handling. Particularly, adding or removing of NaOH too fast can lead to the loss of beads since they become very sticky and easily attach to the wall of the tips. The same is true for the subsequent TE buffer washing step.
38. Stocks of the labeling probes should be aliquoted in suitable amounts of 20–30 μL to avoid stock contamination, and should be stored at −20 °C. Aliquots for labeling can be stored at 4 °C up to a month.
39. The initial 94 °C step is necessary for the activation of the polymerase. Polymerases with proof reading activity can be used as

well, because the bases at the 3′-end of the allele-specific extension probe are protected with PTOs (phosphorothioate bonds) that cannot be cleaved by the proof-reading activity of the polymerase. At 60 °C, the labeling probes hybridize to the immobilized DNA amplicons on the beads and only probes perfectly matching the template at the 3' end will get extended. A rise in the temperature to 75 °C leads to the denaturation of non-elongated probes, which fall off. The labeling can also be performed by a simple hybridization step, in which the allele is placed in the middle of the probe and not at the 3' end, as is done for the allele-specific extensions.

40. For higher efficiency, rinse the PCR tube with 1E buffer and apply sheer forces to break up the labeled beads and detach the beads from the wall of the tube. This handling increases the recovery of the beads.
41. The slide can carry one or two gel matrices (maximum two arrays per slide). Each of these gel arrays has an area of ~1 cm^2. Make sure that the two arrays are spaced close enough to fit under the cover slip 26 × 40 × 1 mm and the hybridization chamber.
42. Do not use the gel components TE buffer, Rhinohide in Acrylamide, TEMED, and APS longer than one week. For 0.5 % APS, it is advisable to dilute a 10 % stock that lasts up to a month. Furthermore, all components need to be on ice during preparation in order to slow down the polymerization of the gel.
43. Avoid using colored foil pens (permanent markers) because the color of these pens can dissolve during the isopropanol washing step and color the beads, even the ones without a PCR product. This can lead to a high false-positive rate distorting the experimental results. Alternatively, a pencil can be used with the disadvantage that the pencil rubs easily from the tubes and the labeling needs to be repeated several times. Using Parafilm during different protocol steps can also remove the pencil writings.
44. Use pre-cleaned slides, otherwise dirt specks can interfere with the proper attachment of the polyacrylamide gel of the bead-array and also cause disruptions in the bead monolayer. We use slides made of soda-lime glass, pre-cleaned, ready for use, size: 76 × 26 (3 × 1 in.), thickness of approx. 1.0 mm, and a dim rim for labeling the samples.
45. The microscope slide needs to be treated with silane (Gamma-Methalcryloxypropyl-trimethoxysilan) to ensure the binding between the bead-gel matrix and the glass surface. Use nitrile gloves for handling because silane is a highly toxic substance. Pipette 2 μL of silane on the slide and spread it with a lens

cleaning tissue. Eventual striations can be removed with the back of the glove. Silane treatment is carried out shortly before the beads array because it is very sensitive to oxygen.

46. Remove remaining water droplets on the wall of the tube with a 10 μL pipette. Take care that no water is close to the bead pellet, as this would dilute the gel matrix and prevent the proper polymerization of the gel. If a droplet covers the bead cluster itself, wait a few minutes to let the droplet evaporate (Attention: waiting too long can cause bead clumping).
47. Mix a small aliquot of 1.9 μL of the PAA gel with the beads, stir carefully with the pipette tip, and move the mixture with rotating movements to the border or the bottom of the tube. Take up as many beads as possible by setting the pipette volume slightly higher. Try to keep the beads as close as possible to the tip of the pipette and transfer them onto the slide.
48. When you place the cover slip onto the bead-gel droplet, it is very important to hold the cover slip with the tweezers at an angle. Start touching the bead droplet with the lower edge and then slowly lower the cover slip towards the slide. The beads will distribute evenly by the capillary forces of the cover slip. Small bubbles can be removed by gently tapping onto the cover slip.
49. The state of the gel polymerization can be observed by the formation of a rim under the cover slip or by the polymerized PAA residues remaining in the PCR tubes.
50. Scanning of the array with an inverted epifluorescent microscope with a mercury lamp needs a preheating time for about 5 min to reach a maximum light intensity. Do not switch on and off the light continuously, since the lamp life could decrease dramatically. Alternatively, the use of a LED lamp (light-emitting diode) has the advantage that it always provides the same light intensity throughout the lifetime of the lamp. Moreover, LED lamps do not need calibration measurements after a certain number of hours, as needed for the halogen lamps. The LED lamp delivers more accurate results, needs less scanning time, and saves energy.

Acknowledgments

This work was supported by the "Austrian Science Fund" (FWF) P25525-B13 and P 23811-B12 to I.T-B. We also want to thank our team and colleagues, especially Angelika Heissl, for her input in the manuscript.

References

1. Snyder MW, Adey A, Kitzman JO, Shendure J (2015) Haplotype-resolved genome sequencing: experimental methods and applications. Nat Rev Genet 16(6):344–358. doi:10.1038/nrg3903
2. Tewhey R, Bansal V, Torkamani A, Topol EJ, Schork NJ (2011) The importance of phase information for human genomics. Nat Rev Genet 12(3):215–223. doi:10.1038/nrg2950, nrg2950 [pii]
3. Tiemann-Boege I, Calabrese P, Cochran DM, Sokol R, Arnheim N (2006) High-resolution recombination patterns in a region of human chromosome 21 measured by sperm typing. PLoS Genet 2(5):e70. doi:10.1371/journal.pgen.0020070
4. Myers S, Bottolo L, Freeman C, McVean G, Donnelly P (2005) A fine-scale map of recombination rates and hotspots across the human genome. Science 310(5746):321–324
5. Yu W, Rusterholtz KJ, Krummel AT, Lehman N (2006) Detection of high levels of recombination generated during PCR amplification of RNA templates. Biotechniques 40(4):499–507, doi:000112124 [pii]
6. Meyerhans A, Vartanian JP, Wain-Hobson S (1990) DNA recombination during PCR. Nucleic Acids Res 18(7):1687–1691
7. Xiao M, Wan E, Chu C, Hsueh WC, Cao Y, Kwok PY (2009) Direct determination of haplotypes from single DNA molecules. Nat Methods 6(3):199–201. doi:10.1038/nmeth.1301, nmeth.1301 [pii]
8. Diehl F, Li M, He Y, Kinzler KW, Vogelstein B, Dressman D (2006) BEAMing: single-molecule PCR on microparticles in water-in-oil emulsions. Nat Methods 3(7):551–559. doi:10.1038/nmeth898
9. Li M, Diehl F, Dressman D, Vogelstein B, Kinzler KW (2006) BEAMing up for detection and quantification of rare sequence variants. Nat Methods 3(2):95–97
10. Turner DJ, Tyler-Smith C, Hurles ME (2008) Long-range, high-throughput haplotype determination via haplotype-fusion PCR and ligation haplotyping. Nucleic Acids Res 36(13):e82. doi:10.1093/nar/gkn373, gkn373 [pii]
11. Lahr DJ, Katz LA (2009) Reducing the impact of PCR-mediated recombination in molecular evolution and environmental studies using a new-generation high-fidelity DNA polymerase. Biotechniques 47(4):857–866. doi:10.2144/000113219, 000113219 [pii]
12. Boulanger J, Muresan L, Tiemann-Boege I (2012) Massively parallel haplotyping on microscopic beads for the high-throughput phase analysis of single molecules. PLoS One 7(4):e36064. doi:10.1371/journal.pone.0036064
13. Dressman D, Yan H, Traverso G, Kinzler KW, Vogelstein B (2003) Transforming single DNA molecules into fluorescent magnetic particles for detection and enumeration of genetic variations. Proc Natl Acad Sci U S A 100(15):8817–8822
14. Shendure J, Porreca GJ, Reppas NB, Lin X, McCutcheon JP, Rosenbaum AM, Wang MD, Zhang K, Mitra RD, Church GM (2005) Accurate multiplex polony sequencing of an evolved bacterial genome. Science 309(5741):1728–1732
15. Tiemann-Boege I, Curtis C, Shinde DN, Goodman DB, Tavare S, Arnheim N (2009) Product length, dye choice, and detection chemistry in the bead-emulsion amplification of millions of single DNA molecules in parallel. Anal Chem 81(14):5770–5776. doi:10.1021/ac900633y
16. Kojima T, Takei Y, Ohtsuka M, Kawarasaki Y, Yamane T, Nakano H (2005) PCR amplification from single DNA molecules on magnetic beads in emulsion: application for high-throughput screening of transcription factor targets. Nucleic Acids Res 33(17):e150
17. Starck JL, Fadili J, Murtagh F (2007) The undecimated wavelet decomposition and its reconstruction. IEEE Trans Image Process 16(2):297–309
18. Benjamini Y, Hochberg Y (1995) Controlling the false discovery rate: a new and powerful approach to multiple comparisons. J R Stat Soc Ser B 57:289–300
19. Margulies M, Egholm M, Altman WE, Attiya S, Bader JS, Bemben LA, Berka J, Braverman MS, Chen YJ, Chen Z, Dewell SB, Du L, Fierro JM, Gomes XV, Godwin BC, He W, Helgesen S, Ho CH, Irzyk GP, Jando SC, Alenquer ML, Jarvie TP, Jirage KB, Kim JB, Knight JR, Lanza JR, Leamon JH, Lefkowitz SM, Lei M, Li J, Lohman KL, Lu H, Makhijani VB, McDade KE, McKenna MP, Myers EW, Nickerson E, Nobile JR, Plant R, Puc BP, Ronan MT, Roth GT, Sarkis GJ, Simons JF, Simpson JW, Srinivasan M, Tartaro KR, Tomasz A, Vogt KA, Volkmer GA, Wang SH, Wang Y, Weiner MP, Yu P, Begley RF, Rothberg JM (2005) Genome sequencing in microfabricated high-density picolitre reactors. Nature 437(7057):376–380. doi:10.1038/nature03959

Part VII

Computational Algorithms for Haplotyping

Chapter 15

Computational Haplotype Inference from Pooled Samples

Quan Long

Abstract

Computationally inferring the identities and their relative frequencies from pooled samples that are whole-genome or segmentally genotyped or sequenced (e.g., using next-generation sequencing) in a pool is useful for population genetics analysis. To carry out such analysis, one needs to understand basics of how to use high-performance computing (HPC) facilities and the specifics of corresponding computational tools. Here, we describe the basic knowledge and step-by-step usage of a number of tools for haplotype inference on genotyping or next-generation sequencing data.

Key words Computing cluster, Haplotype, Bioinformatics, Next-generation sequencing, Genotyping

1 Introduction

Pooling multiple individuals together for sequencing or genotyping is sometimes employed in population genetics studies and pathogen genomics. This is either motivated by cost considerations, or by technical difficulties, as in many projects the individuals under investigation are naturally pooled (e.g., pathogen populations) and it is difficult to isolate them experimentally.

Starting from sequencing or genotyping observations that are a collection of signals from multiple individuals, it is possible to infer the haplotypes and/or their relative frequencies computationally. Many bioinformaticians and biostatisticians in the field of computational genomics have made efforts to advance this direction to facilitate downstream genetic research. Dr. Schlotterer et al [1] have published an excellent review of the field. In this chapter, we will focus more on how to make use of these tools by detailed describing a few examples. For each tool, we will briefly introduce its targeting applications, design philosophy, and then elaborate the use of it.

Irene Tiemann-Boege and Andrea Betancourt (eds.), *Haplotyping: Methods and Protocols*, Methods in Molecular Biology, vol. 1551,
DOI 10.1007/978-1-4939-6750-6_15, © Springer Science+Business Media LLC 2017

2 Materials

No "wet-bench" is needed for haplotype inference in silico. The input data for the analysis usually is the outcome of genotyping or sequencing for a segment of genomic region of interest (or the whole genome). A computer cluster and necessary software to build (binary) executable version of the software are required. For experimentalists who have no previous experience with high-performance computing (HPC) facilities and research group with no extensive IT support, this subsection provides an initial description on the basic philosophy and user interface of computer clusters and how to build tools on clusters.

2.1 What Is an HPC Cluster?

Although often intimidating for a biologist, HPC cluster is not as difficult to use as one might think. In particular, for most bioinformatics tools, especially the tools used in the field of genomics, there is no synchronization between different jobs. So advanced techniques such as message passing (by MPI) or shared memory (by OpenMP) are not relevant here. This simplifies the situation significantly. Therefore, from the perceptive of a biologist who needs to analyze high-throughput data, it is sufficient to think of an HPC cluster as a bunch of CPU nodes with fast access to (usually centralized) storage that can carry out a large number of computational jobs in parallel.

Since a cluster is shared by many computing jobs submitted by different users, it is crucial to have a scheduling system that ensures that all jobs competing limited resources are fairly allocated and resources are fully utilized. Therefore, workload management platforms have been developed to organize jobs on clusters—just like the operating system to run desktop computers.

2.2 How Can One Access Cluster?

A beginner for HPC facilities usually has to start with commands specified by job schedulers such as Load Sharing Facility (LSF) or Sun Grid Engine (SGE). They help to schedule computational jobs so that the system can be orderly and optimally utilized. Sometimes, the large number of commands and complicated scripts may overwhelm a user. In fact, a beginner only needs to understand a small subset of commands to start with, leaving the use of more advanced features to the future.

1. How to submit my computational job? (bsub, qsub).
2. How to check the status of my submitted jobs? (bjobs, qstat).
3. How to delete jobs (Oops, I mistakenly submitted 10,000 jobs?) (bkill, qdel).
4. I have many jobs with similar parameters; can I submit them in one command? (job array).

One may feel fairly comfortable to carry out a large number of computational tasks on an HPC cluster as long as the above basic commands are correctly used.

2.3 How Can We Install Software on HPC Clusters?

Although one can always send a ticket to the IT group, sometimes, it is easy and fun for us to build tools. Most bioinformatics tools do not involve complicated dependencies therefore are straightforward to build. HPC clusters usually are operated by Linux system, so the method to install software and build tools follows the convention of doing so in Linux system. A standard way that usually works is:

1. Download and decompress the package of the tool.
2. See if there is a file named "Makefile" in the folder. If yes, simply typing

   % make

   (In this chapter, we use "%" to denote the prompt before commands).

   will work. If not, seek if there is something called "configure" or other scripts that needs to be run before making the tool in the folder. If yes, typing

   % ./configure usually will generate the "Makefile."
3. The system sometimes complains that a piece (or pieces) of software necessary to build the focal tool is missing (e.g., gcc). In that case, do not be intimidated. To continue making the tool, find and download the missing program.
4. Once the "make" is successfully done without any error messages, type the executable, and see whether the help documentation is prompted out. If yes, it indicates that you have successfully built the tool.
5. Set up the path so that the system can find the tool when one types the name of it without specifying the absolute path of where it is. This can be achieved by simply copying the executable to the root folder, e.g., /usr/bin/.

There are two scenarios in which we do not have to go through the above procedure:

1. The authors provide binaries for your platform, and therefore one can just download the right version and set up the path.
2. The executable is a Java-based .jar package. Then, one can just run

```
% java -Xmx4g -jar /path/to/the/package/foo.jar [options] <arguments>
```

Here, –Xmx4g specifies that the memory usage of the tool will not exceed 4 Gb.

2.4 Notes for Beginners

In this subsection, we will briefly summarize some useful tips for beginners based on our experience of collaborating with experimentalists.

1. Try it first! For many bioinformatic tools, running the tool on the command line with no arguments usually will give usage information. Leaning the use of a tool by trying different commands interactively might be a more efficient way than reading a "static" users' manual or paper, as long as the help documents prompted by the program are reasonably good.
2. Do not worry if it does not work. Please do check all the details and ensure that they are all correct. Single typos can cause problems when running commands in terminal. A trick is to use "tab" button to fulfill the file names as often as possible to avoid mistakenly typed long file names.
3. Record the working log. Like experimental steps, computational procedures also need to be recorded in a safe and well-organized place. This is an important style of working for both beginners and experienced users. When there is a problem in the middle of the analysis, it is extremely helpful to have the previous steps, including all commands and arguments specified documented for revisit. This also helps to ensure that the computational part of the scientific results is replicable, which is important for post-publication enquiries.

3 Methods

3.1 PoooL & AEM

PoooL is a software that can efficiently estimate haplotype frequencies from DNA pools with arbitrarily large sizes [2]. It was designed to facilitate the application in which the samples are artificially pooled for genotyping. The target application field of PoooL is Genome-Wide Association Study (GWAS), from which an investigator may be interested in knowing the difference of haplotype frequencies in diseases (i.e., cases) and controls populations. To save genotyping cost when there are large cohorts however the budget for genotyping is limited, one may want to combine some samples into a pool and genotype them jointly. When genetic markers are analyzed one at a time, one can calculate the allele frequency regardless which ones are on the same haplotype. However, if multiple markers in the adjacent region have to be analyzed jointly, reconstructing haplotypes computationally is a necessity. This scenario motivated the development of PoooL. PoooL only outputs the haplotype frequencies in the population (e.g., the aggregated frequencies in all pools), instead of the ones in each individual pool. This perhaps is due to the original rationale of comparing population, and not individual pools that are artificially grouped by the investigator without meaningful biological interpretations. It was demonstrated that PoooL performs better than existing alternatives [2].

3.1.1 Assumptions

(1) We know the number of individuals in the pools, and it remains the same in all different pools – this is a reasonable assumption when the pools are artificially designed. (2) The allele frequencies follow asymptotic normality. (3) The analysis is carried out using a computer operated by Windows.

3.1.2 Models

PoooL extends the EM (expectation and maximization) based model into a constrained maximum entropy model, and then solves this problem by an IIS algorithm. A quantity called "importance factor" is introduced to measure the contribution of a haplotype to the odds of observing the data (which is called likelihood in statistics). Under certain assumptions, the algorithm iteratively recalculates the haplotype frequencies while it still keeps the importance factor a constant.

3.1.3 How to use PoooL

1. Prepare two input files: (a) genotype data, and (b) a supporting file specifying the parameters to be read by the models. The genotype data is stored in a file that has T lines and Q columns (T = number of pools and Q = number of markers). Each line specifies the aggregated allele frequencies of all the markers in one pool, while each column records the aggregated allele frequencies of all the pools at the location of a particular marker. The supporting file is used to specify parameters such as pool size, number of pools, and the number of markers. Taking the example from the PoooL website, a typical information file has the following three lines

   that informs the tool that that the pool size is N=100, the number of pools is T=30, and number of markers is Q=3.

   ```
   pool size      >>100
   pool number    >>30
   loci           >>3
   ```

2. Run the program. Since it is Windows-based only, there is no need to build the tool from source code. Instead, the users can just double click the downloaded executable to start the program. The tool will then output the estimated haplotypes and their relative frequencies in the population, as well as the entropy of the data.

A potential disadvantage of PoooL is that it does not support platforms other than Windows. As an alternative, one may use AEM developed by the same research group [3], which is an extension of PoooL implemented in R. AEM is more resource-effective than PoooL and it does not require some conditions that are usually required in genome-wide association studies (e.g., Hardy-Weinberg equilibrium).

The input genotype file for AEM is identical to the one if PoooL. To use AEM, the users do not have to prepare the information file. Instead, the program will figure out the number of pools (T) and markers (Q) from the data and the users can just specify the number of individuals in the pool by typing it in the R command line:

% aem(data, N=10)

where the object "data" is the input genotype, N is the number of individuals in each pool

A variant of AEM is AES that can incorporate a parameter, rho, an overdispersion factor representing inflation factors, e.g., inbreeding coefficient. This rho can be estimated from genotype data.

Here is the R command when rho is known:

% aes(data, N=10, rho=estimated.rho)

3.2 PoolHap

PoolHap is a tool inferring haplotype frequencies based on sequence data, in particular, the data generated by high-throughput next-generation sequencing instruments [4]. (In contrast, PoooL was designed for genotyping data). The target applications include both artificially pooled and naturally pooled samples. It could be useful if the investigators are interested in the change of haplotype frequencies of an evolving population of an organism (e.g., a pathogen) in an eco-system (e.g., a human tissue) that can be (and have to be) sequenced in a pool. Other scenarios include that to contrast the different frequency distributions of haplotypes in difference pools. It was demonstrated that PoolHap could infer frequencies fairly good even when the overall sequencing coverage is pretty low [4].

3.2.1 Assumptions

(1) An important assumption is that the haplotypes in the population have to be known in advance and PoolHap only infers the frequency of each haplotype in each pool. (2) The samples are sequenced by next-generation sequencing (e.g., Illumina platform) and proceed by SAMtools [5], the standard data manipulating tools facilitating next-generation sequencing data analysis. (3) Although the sequencing coverage per base pair could be very low, it is preferable to have whole genome sequencing data, instead of sequences for just one region (e.g., a single gene).

3.2.2 Models

PoolHap leverages a simple regression model to infer the frequencies of known haplotypes. PoolHap deems the total allele frequency of a marker as the dependent variable that is modeled as the linear combination of explanatory variables representing the frequencies of the markers on each haplotype. Therefore, the haplotype frequencies become the regression coefficients that can be solved using the standard least-square method.

3.2.3 How to Use PoolHap

1. Preparing the two input files: known haplotypes file and pileup files out of SAMtools [5]. The known haplotype files contain M+1 columns and N+1 lines. The first line specifies the marker information and the first column denotes haplotype IDs. The marker information takes the format as "Chr_name:location," which has to use the same coordinate system of the reference genome used for mapping reads. For the rest, each line denotes one haplotype and each column denotes the alleles of one marker on all the haplotypes. The alleles are encoded as 0 or 1, denoting reference allele and alternative allele respectively. The pileup file out of SAMtools is standard, and PoolHap will read through column of the bases of all supporting bases on each marker to figure out the allele frequency in the pool.
2. Run "select." If there are many markers available, PoolHap provides an optional function, "select," that can choose a subset of markers that may be most informative in terms of distinguishing different haplotypes. To run "select," one can just type the following command:

```
% java -Xmx2g -jar poolhap.jar select -in <known_hap_
  file> -out
<output_file> -n <#snps> [-diff <smallest_diff_thresh-
  old>] [-round <max_round>]
[-try <number_try>]
```

   The first three mandatory arguments specify the input file, output file, and number of SNPs to be selected. The other three optional parameters specify the starting configuration of the calculation, which is composed of the preferable smallest difference between any pair of haplotypes, the number of iterations from the same initial guess, and the number of initial guesses respectively.
3. Run "infer." To infer haplotype frequencies, one can type

   The input file could be a known haplotype file that is prepared by the user or generated by the "select" function (which will follow the same format). After running "infer," one can see a new file named XXX.poolhap, reporting the haplotype frequencies in the pool.
4. RNA-Seq data analysis. PoolHap aims to analyze many types of pooled analysis as long as the subtypes are known. In RNA-Seq analysis, multiple isoforms of the same gene could be indirectly observed as a mixture of reads mapped to the same regions of the reference genome. If the gene models of all the isoforms are known, PoolHap can deem known gene models as "known haplotypes" and the relative abundance of the expression of the isoforms as "haplotype frequencies," and analyze them as though they were DNA pooled sequencing data. Therefore, one can also infer relative abundance of multiple isoforms in gene expression using PoolHap. The command is:

The -genemodel argument specifies the known gene models in the format of standard GFF file, and the other parameters are self-explanatory by their names.

```
% java -Xmx2g -jar poolhap.jar rnaseq
-genemodel <gene_model_file>
-pileup <pileup_file>
-ref <ref_genome_file>
-rdlen <read_length>
-output <output_file>
[-max_iso <max_iso_num>]
```

3.3 HARP

HARP, Haplotype Analysis of Reads in Pools [6], is another tool that can calculate the frequencies of known haplotypes in either experimentally pooled or naturally mixed samples, targeting similar applications as PoolHap does. From the perspective of running the tools, a difference between PoolHap and HARP is that HARP is based on BAM files [7] that are on sequence level; whereas PoolHap starts its analysis from variants (e.g., SNP data). Therefore, when using HARP, the users can specify many parameters tuning the filtering at the read level. PoolHap does not offer this flexibility; instead assumes that the variants have been called using other tools and filtering criteria.

3.3.1 Assumptions

Similar to PoolHap, HARP assumes that (1) the haplotypes are known in advance and (2) the pools are sequenced by next-generation sequencing platforms and processed by SAMtools.

3.3.2 Models

Although the application scenario and assumptions of HARP are quite similar to the ones of PoolHap, the advantage is that HARP employed an EM algorithm that iteratively updates the parameters (including haplotype frequencies) that may lead to better precession with the cast of computational time.

3.3.3 How to Use HARP

HARP analyzes data in two steps. Firstly, haplotype likelihoods will be calculated; based on the likelihoods, HARP next estimates haplotype frequencies.

1. There are two alternative ways to calculate haplotype likelihoods, facilitating two different ways of data processing.
   (a) In its first data processing mode, HARP assumes that there is only one large haplotype group represented by a single reference genome, and the difference between each haplotype is presented in the form of a set of SNPs. In this mode, the users have to prepare three data sets: a reference genome (in FASTA format), a BAM file containing reads that have been mapped to that reference genome, and a SNP file in the DGRP format which is a comma-separated file that contains genetic variants.

Specifically, from left to right, the columns represent the position, reference bases, and other haplotype bases. (An example DGRP formatted file is available on the HARP website). In addition, the users need to specify the genomic regions to be analyzed.

(b) In the other mode, HARP assumes that there are multiple haplotype groups represented by multiple reference genomes and the samples have been mapped to multiple reference genomes. Under this assumption, users need to prepare multiple reference genomes files (FASTA files) and corresponding BAM files. It is notable that users do not need prepare SNPs file as the genetic polymorphisms have been incorporated in the multiple reference genomes.

Commands in their simple versions of the above two modes could be typed as below:

Mode (a) (single reference genome):

```
% harp like --bam <bam_file> --region <region_of_inter-
   est> --refseq
<ref_genome> --snps <SNP_file>
```

Mode (b) (multiple reference genomes):

```
% harp like_multi --refseqlist <list_of_ref_
   genomes_file> -- bamlist
<list_of_BAM files> --region <region_of_interest>
```

2. Once the execution of functions calculating likelihood ("like" or "like_multi") is done, HARP will generate a binary file named XXX.hlk that contains the results of likelihood calculations. Based on the XXX.hlk file, another HARP function, "freq," will estimate the haplotype frequencies.

The command in its simple version is

```
% harp freq --hlk <hlk_file> --region <region_of_interest>
```

It will generate a .freq file that contains the final results.

HARP offers another convenient alternative way of execution in which all the arguments specified by the users may be passed to HARP by a configuration file (instead of by the command line). Users are allowed to write all the arguments in the file and just type

```
% harp function -c < config_file>
```

This is more convenient, especially to assist users to record the arguments for different runs.

As mentioned above, HARP analyzes data at the sequencing reads level. It provides many optional arguments available for the users to do quality control. More details and examples may be found in the file named "harp_docs.pdf" at the HARP website.

3.4 Others

If one deems the two haplotypes in a diploid individual, i.e., maternal and paternal haplotypes, as the components forming a pool, and multiple diploid individuals as a population of pools, the problem becomes *phasing*, a long-standing field with many developments that aim to reconstruct haplotypes in diploid genomes based on population genetics models. In the early 2000s, researchers developed quite a few tools using various models including Markov Chain Monte Carlo (MCMC) and Hidden Markov Models (HMM), with PHASE [8] and FastPHASE [9] the perhaps most famous milestones. Later, there are also efforts to extend PHASE to estimate population haplotype frequencies from pooled samples [10]. Recently, there are also efforts that integrate linkage information provided by paired-end reads, such as HaplotypeImprover [11], linkSNPs [12], and GATK [13].

4 Notes

1. All these tools may rely on moderate recombination and mutation rates that lead to relatively stable linkage disequilibrium (LD) blocks in the population. In the event that the recombination rate or mutation rate of the species is too high (e.g., HIV), the underlie assumptions of tools that can infer haplotypes become questionable. Tools that only infer haplotype frequencies (assuming that the haplotypes are known) suffer less; however, they loss power when the number of haplotypes is significantly greater than assumed. In general, in addition to knowing how to run the programs, it is also useful to understand a bit of the underlying models, in terms of both population genetics and statistical inference.
2. PoolHap and HARP estimate haplotype frequencies with very great precision; however, they do not infer haplotypes directly, whereas PoooL/AEM can infer the haplotype in the population although not the frequencies in each pool. As PoooL/AEM are designed for the genotyping data targeting GWAS, for the moment, as far as we know, it is not ready to be integrated with downstream frequency estimators offered by other tools. However, it appears to be a good idea to integrate their algorithms into a single tool for the users. We are making efforts toward this direction.

Acknowledgment

We are grateful to the communications with Dr. Yaning Yang on PoooL and the communications with Dr. Darren Kessner on HARP. This work was partially supported by the start-up grant of University of Calgary and NIH grants (HG008451 and AG046170)

References

1. Schlotterer C, Tobler R, Kofler R, Nolte V (2014) Sequencing pools of individuals - mining genome-wide polymorphism data without big funding. Nat Rev Genet 15:749–763
2. Zhang H, Yang HC, Yang Y (2008) PoooL: an efficient method for estimating haplotype frequencies from large DNA pools. Bioinformatics 24:1942–1948
3. Kuk AY, Zhang H, Yang Y (2009) Computationally feasible estimation of haplotype frequencies from pooled DNA with and without Hardy-Weinberg equilibrium. Bioinformatics 25:379–386
4. Long Q, Jeffares DC, Zhang Q, Ye K, Nizhynska V et al (2011) PoolHap: inferring haplotype frequencies from pooled samples by next generation sequencing. PLoS One 6:e15292
5. Li H, Handsaker B, Wysoker A, Fennell T, Ruan J et al (2009) The sequence alignment/map format and SAMtools. Bioinformatics 25:2078–2079
6. Kessner D, Turner TL, Novembre J (2013) Maximum likelihood estimation of frequencies of known haplotypes from pooled sequence data. Mol Biol Evol 30:1145–1158
7. Li H, Durbin R (2009) Fast and accurate short read alignment with Burrows-Wheeler transform. Bioinformatics 25:1754–1760
8. Stephens M, Smith NJ, Donnelly P (2001) A new statistical method for haplotype reconstruction from population data. Am J Hum Genet 68:978–989
9. Scheet P, Stephens M (2006) A fast and flexible statistical model for large-scale population genotype data: applications to inferring missing genotypes and haplotypic phase. Am J Hum Genet 78:629–644
10. Pirinen M, Kulathinal S, Gasbarra D, Sillanpaa MJ (2008) Estimating population haplotype frequencies from pooled DNA samples using PHASE algorithm. Genet Res (Camb) 90: 509–524
11. Long Q, MacArthur D, Ning Z, Tyler-Smith C (2009) HI: haplotype improver using paired-end short reads. Bioinformatics 25: 2436–2437
12. Sasaki E, Sugino RP, Innan H (2013) The linkage method: a novel approach for SNP detection and haplotype reconstruction from a single diploid individual using next-generation sequence data. Mol Biol Evol 30:2187–2196
13. McKenna A, Hanna M, Banks E, Sivachenko A, Cibulskis K et al (2010) The genome analysis toolkit: a MapReduce framework for analyzing next-generation DNA sequencing data. Genome Res 20:1297–1303

INDEX

Irene Tiemann-Boege and Andrea Betancourt (eds.), *Haplotyping: Methods and Protocols*, Methods in Molecular Biology, vol. 1551, DOI 10.1007/978-1-4939-6750-6, © Springer Science+Business Media LLC 2017

D

E

F

G

H

I

K

L

M

CPSIA information can be obtained
at www.ICGtesting.com
Printed in the USA
LVHW061035150919
631110LV00008B/1000/P